THE EARTH'S DENSITY

The Earth's Density

K.E. Bullen
M.A., Sc.D., Hon.D.Sc., F.R.S.
University of Sydney

LONDON
CHAPMAN AND HALL

First published 1975
by Chapman and Hall Ltd
11 New Fetter Lane London EC4P 4EE

© 1975 K.E. Bullen
Typeset by E.W.C. Wilkins Ltd, London and Northampton
and Printed in Great Britain by
Lowe and Brydone Ltd, Thetford, Norfolk

ISBN 0 412 10860 7

All rights reserved. No part of this book
may be reprinted, or reproduced or utilized in any form or
by any electronic, mechanical or other means, now known
or hereafter invented, including photocopying and recording, or
in any information storage and retrieval system, without
permission in writing from the publisher.

Distributed in the U.S.A.
by Halsted Press, a Division
of John Wiley & Sons, Inc., New York

Library of Congress Catalog Card Number 74–19282

Contents

Preface		*page* xi
1. *Noted ancient investigations*		1
1.1	Size of the Earth	1
1.2	Contributions of Newton and contemporaries	4
1.3	The period after Newton	10
	References	11
2. *Determination of the mean density of the Earth*		13
2.1	Bouguer's experiments in Peru	13
2.2	The Schiehallion experiment	15
2.3	Other pendulum and plumb-line experiments	16
2.4	The Michell-Cavendish experiment	16
2.5	Other small-scale experiments	18
	References	19
3. *Spherical harmonics*		21
3.1	Solutions of Laplace's equation	22
3.2	Legendre polynomials	23
3.3	Integrals of products of spherical harmonics	24
3.4	Tesseral harmonics	25
3.5	Expansions in spherical harmonics	27
3.6	Spherical harmonics and Earth oscillation theory	27
	References	28
4. *Theory of the Earth's gravitational attraction*		29
4.1	General theorems on gravitational attraction	29
4.2	Attractions due to some particular mass distributions	32
4.3	MacCullagh's formula	35
4.4	Representation of the Earth's surface	36
4.5	Attraction due to spheroidal Earth model	39
	References	42
5. *The figure and moment of inertia of the Earth*		43
5.1	The geopotential function	44

CONTENTS

5.2	Forms of surfaces of equal density inside rotating Earth model	45
5.3	Relations involving ϵ_a and h	46
5.4	Clairaut's equation on the internal variation of ϵ	48
5.5	Estimation of the moment of inertia of the Earth	51
5.6	Numerical results on the hydrostatic theory	53
5.7	Use of artificial satellites	56
5.8	International reference systems	57
5.9	Ellipticities of internal surfaces of constant density	58
	References	58

6. Early models of the Earth's density variation — 60

6.1	Earth models	60
6.2	Clairaut's equation and the density problem	64
6.3	The Legendre-Laplace density law	64
6.4	Background theory in density determination	67
6.5	Other early model density laws	70
6.6	Numerical results for early models	72
6.7	Some further nineteenth century results	74
6.8	Early evidence on the Earth's rigidity	81
6.9	Early twentieth century models	83
	References	86

7. Representation of elasticity in the Earth — 87

7.1	Stress	88
7.2	Strain and rotation	89
7.3	Model stress-strain relations	91
7.4	Stress-strain relations for the Earth	94
7.5	Interpretation of coefficients in perfect elasticity	95
7.6	Strength of a material	97
7.7	The terms 'fluid' and 'solid'	99
7.8	Pressure and finite strain	99
7.9	Thermodynamic considerations	101
	References	107

8. Seismic wave transmission — 108

8.1	Earthquakes and other sources of seismic waves	108
8.2	Equations of motion of seismic disturbances	110
8.3	Bodily seismic waves	112
8.4	Scalar and vector potentials	114
8.5	Surface seismic waves	115
8.6	Refraction and reflexion of bodily seismic waves	123
	References	127

9. First approximation to seismic P and S distributions in the Earth — 128
9.1 Seismic rays — 129
9.2 Effect of the Earth's ellipticity on seismic travel times — 135
9.3 Normal and abnormal seismic velocity variation — 136
9.4 Bodily seismic phases — 140
9.5 Evolution of travel-time tables — 143
9.6 Derivation of P and S velocity distributions in the Earth — 145
References — 150

10. Earth models of type A — 152
10.1 Introductory theory of density variation in the Earth — 153
10.2 Historical background on the Earth's internal layering — 156
10.3 The regions A, B, C, D, E, F and G — 158
10.4 Density near the Earth's surface — 165
10.5 Early evidence on inhomogeneity inside the upper mantle — 167
10.6 Minimum central density — 168
10.7 Earth models of type A — 169
10.8 Corrections for temperature and inhomogeneity — 172
10.9 Critique of A-type models — 179
References — 180

11. Evidence on compressibility in the Earth — 184
11.1 Compression in the Earth — 185
11.2 Variation of incompressibility in homogeneous regions — 187
11.3 Some further implications of finite-strain theory — 193
11.4 Compressibility-pressure hypothesis — 197
11.5 Theory for inhomogeneous regions — 205
11.6 Degrees of inhomogeneity in particular regions of the Earth — 212
11.7 Solidity of the inner core — 217
11.8 Earth models of type B — 219
References — 223

12. Some second approximations — 227
12.1 P and S velocities in the mantle — 227
12.2 Structure of the outer core — 235
12.3 Structure of the inner core — 239
12.4 Radius of the Earth's core — 244
12.5 The Earth's central density — 246
12.6 Further evidence bearing on rigidity in lower core — 248
12.7 Improved B-type models — 251
References — 255

CONTENTS

13. *Evidence from seismic surface waves* — 261
13.1 Underlying principles in applying surface-wave data — 263
13.2 More complex model structures — 268
13.3 Direct observation of phase velocities — 270
13.4 Allowance for Earth's curvature and gravity — 271
13.5 Evidence on crustal structure — 273
13.6 Evidence on mantle structure — 276
13.7 Surface waves and density variation — 281
13.8 Further remarks — 282
References — 283

14. *Evidence from free Earth oscillations* — 287
14.1 Free oscillations of a dynamical system — 288
14.2 Approach to the theory of Earch oscillations — 290
14.3 Equations of motion of an oscillating Earth model — 294
14.4 Solving the equations of motion — 296
14.5 Observational data — 301
14.6 Early inferences from free Earth oscillation data — 310
14.7 The model HB_1 — 314
14.8 Other models using free Earth oscillation data — 317
14.9 Oscillation evidence on solidity of inner core — 317
References — 319

15. *Miscellaneous developments* — 322
15.1 Equations of state, and related equations, for the Earth's interior — 322
15.2 Some miscellaneous Earth models — 328
15.3 Monte Carlo techniques — 331
15.4 The general problem of 'inverting' observation data — 334
15.5 Density and seismic wave amplitudes — 337
15.6 Implications of wave-scattering investigations — 339
15.7 Deviations from spherical symmetry — 343
15.8 Changes in gravitational constant — 345
References — 345

16. *Optimum and standard Earth models* — 352
16.1 General requirements of Earth models — 352
16.2 Consequence of non-uniqueness — 354
16.3 Approaches to the optimum model problem — 355
16.4 Progress towards an optimum Earth model — 357
16.5 The problem of a standard Earth model — 363
References — 365

17. *Application to other planets and the Moon*	*367*
17.1 Planetary observational data	369
17.2 Assumptions on the Earth's internal composition	373
17.3 Earth, Venus, Mars	380
17.4 Mercury	385
17.5 Moon	388
17.6 Jupiter and Saturn	390
17.7 Uranus and Neptune	393
17.8 Pluto	394
17.9 Further remarks	394
References	398
Index	*405*

Preface

The book attempts to draw together the various strands of evidence that have led to present knowledge of the distribution of density throughout the interior of the Earth. Details are also given of other properties with which the density is closely linked, including pressure, compressibility and compression, rigidity, seismic velocities, Poisson's ratio and gravitational intensity. Questions of thermodynamics and chemical composition and phase enter discussions where they bear more or less sharply on the density determination; but the book does not purport to be a comprehensive text on the Earth's internal temperature distribution and composition. The density distributions of other planets are discussed.

The quest for clues on the Earth's internal densities has long been an exciting one, and an aim of the book is to present a developing story which has fascinated the author over much of his working life. The early chapters refer to key developments from ancient times to around 1930. The later chapters recount, in greater detail, developments since 1930.

The stage has now been reached where numerous published papers bearing on the Earth's density seem to do little more than fidget around the resolving power of long accumulated observational data. So the present seems an appropriate time to try to put some perspective into the story. Of course, the story will never be quite finished: in describing the interior of the Earth, there will always be extra decimal places to add as further significant evidence arises.

I have sought to describe what have appeared to me to be the more important contributions among those that have caught my active interest during the past forty years. Thus the book inevitably reflects my tastes and interests: the subject now touches on so many fields that a wholly satisfying coverage could hardly be expected from a single author. So I apologize in advance to those whose important work may possibly have been overlooked. I mention also that my

PREFACE

approach has been mainly 'macroscopic': only brief reference is made to 'microscopic' approaches through, for example, lattice theory.

Perhaps I should apologize also for making considerable reference to my own work. I have done this because: first, in a subject rather difficult to expound in all its intricacies, I felt I could contrive the best coherence by basing many of the developments on my own approaches, at least in the first approximation; secondly, I would like this book to help correct numerous recent distortions of detail in my past writings (a phenomenon which of course by no means afflicts only myself in this era of scientist population explosion); thirdly, since the book may be my last major effort on the subject, I have sought to make the account of my work as unambiguous as possible. I hope these reasons will help to counter any suggestion that I regard my contributions as more significant than they really are.

The book will probably be found rather more cautious in its attitude to uncertainties than are many current writings. There is a strong tendency for modern writers (including some notable contributors) in the Earth sciences to be unduly black and white in their pronouncements — rather over-steady to 'prove' and 'disprove' and to declare the 'beliefs' and 'disbeliefs' of themselves and others in contexts where cautious assessments in terms of probability would be wiser. (This tendency is not confined to the Earth sciences.) Here, I have striven to avoid words such as 'proof', 'true', 'false', 'right', 'wrong', 'valid', 'invalid', except in formal deductive arguments. In inductive arguments, I have sought to 'infer', not 'deduce'; I have been at pains to distinguish between 'mathematical models' and 'facts', not only with density distributions and the like, but also with (so-called) physical 'laws'; and so on. Perhaps vainly, I cherish the hope that my pattern of writing may make a modest contribution towards improving the appreciation of some points of scientific inference that need to be specially heeded in geophysics.

Because of complex interplay between different lines of evidence on the Earth's density, some degree of overlapping among several of the chapters has been unavoidable; I have tried to arrange the book so that this overlapping is minimal. The numerous cross-references are principally to help readers who may desire ready access to the subject. Many cross-references may be ignored in the first reading: their purpose is to help the readers who want them, not to distract those who don't.

The book necessarily contains a fair quantity of seismological detail since important density findings have come through the help of seismology. But the book is not intended as a text on seismology. Intermittent reference is made to sections in the third edition of my 'Introduction to the Theory of Seismology' (Cambridge University

Press) — referred to as **B** — where additional seismological detail can be found.

The second half of the book includes a considerable variety of numerical detail which, I hope, the reader will not find too oppressive. I felt that only by giving this detail could I fulfil my purpose of exhibiting in suitable perspective the consequences of varying approaches to a complete subject. By way of compensation, I hope that research workers, lecturers and students of solid-Earth geophysics will find helpful the compilation of information given in the book. With such needs in mind, I have, further, taken special pains to provide an extended Index which should give rapid access to particular strands of the information.

Most of the symbolism used is orthodox, but there are occasional deviations. For example, s is sometimes used instead of S for entropy in contexts where S is already used for other purposes. Most of the units are S.I.; but I cannot bring myself to the point of expressing densities in kg/m^3 with the consequent squandering of cyphers in a book where densities appear on nearly every page.

I owe deep gratitude to Miss K. Yamamoto for her superb typing of an often difficult manuscript and her great patience in numerous re-typings needed to bring parts of the book up-to-date over the several years during which it was written. The book might possibly never have emerged had it not been for her great help. I also wish to thank my former colleagues Dr. A.P. Treweek for the information in Chapter 1 on ancient stadia, and Dr. R.A. Haddon for helpful comments on several seismological points.

Sydney K.E. Bullen
29 June 1974

The Earth's Density

Corrigenda

Page	Line	
vi	31	*for* Thermodynamic considerations *read* Thermodynamical conditions
viii	13	*for* Earch *read* Earth
viii	27	*for* observation *read* observational
xii	19	*for* over-steady *read* over-ready
xii	37	*insert* cross threads of the *before* subject
xiii	7	*for* complete *read* complex
58	25	*for* 342×10^{-5} *read* 242×10^{-5}
63	16	*for* interferences *read* inferences
95	29	*for* ordinaty *read* ordinary
137	29	*for* (1960) *read* (1960a)
146	19	*for* 1960; *read* 1960b
147	19	*replace* dash (*after* β) *by* colon
150		*Insert new reference:* Bullen, K.E. (1960a). Note on cusps in seismic travel times. *Geophys. J., R. Astr. Soc.*, **3**, 354–359.
150	12	*insert* b *after* 1960
191	25	*for* (11.8) *read* (11.18)
192	13	*for* 11.2.4 *read* 11.2.5
199	1	*for* 11.4. *read* 11.4.1
205	3	*for* Steward *read* Stewart
212	13	*for* $(\rho/k)\partial k/\partial p$ *read* $(\rho/k)\partial k/\partial p$
280	5	*for* inference *read* interference
318	27	*for* resulst *read* resulted
323	32	*for* suggestion *read* suggestions
346	27	*for* 408–414 *read* 372–378 (*Bardeen reference*)
347	42	*for* D.G. *read* D.J. (*Doornbos second reference*)
379	5	*for* But *read* but
406		*for* Ben-menahem *read* Ben-Menahem
406		Bullen: *for* 181–82 *read* 181–83
409		for *Everndon* read *Evernden*
406		Bullen: *for* 399 *read* 399–400
407		*for* Clairault *read* Clairaut
407		Coefficient(s): *for* J_m *read* J_n
411		*for* Ion-core *read* Iron-core
414		*insert* Nuttli, O., 232; *258, 260*
417		Seismic phase(s): *for* P_m *read* P_n
418		*for* Spheriodal *read* Spheroidal

CHAPTER 1

Noted ancient investigations

The first steps toward determining values of the density and related physical properties inside the Earth are investigations of the Earth's size, shape, and mass. The present chapter gives a broad outline of developments from early days to the nineteenth century. Some of the key investigations of that period are elaborated in later chapters.

1.1 Size of the Earth

1.1.1 *Early results in Greece*

According to Homer (B.C. c. 900–800), the Earth was a convex dish surrounded by the Oceanus stream.

The notion of a spherically shaped Earth appears to have been contemplated in Greece from the time of Anaximander (B.C. 610–547).

Aristotle (B.C. 384–322) wrote in his *De Caelo* (Book II, Ch. 14): "Moreover, those mathematicians who try to compute the circumference of the Earth say that it is 400 000 stadia, which indicates not only that the Earth's mass is spherical in shape but also that it is of no great size compared with the heavenly bodies." It is not known how this estimate of the length c of the Earth's circumference was arrived at. It may have resulted from crude measurements of the depressed height of objects seen across the sea.

There are also uncertainties in interpreting measurements in stadia. A stadion was defined as equal to 600 Greek feet (a later version gave 125 Roman paces). But different feet were used in the ancient world. For example, the most probable length of the Attic (and the Roman) foot has been estimated as 0·2977 m and corresponds to the Olympic stadion of 178·6 m; the Philetaerian (Babylonian-Persian) foot as 0·3308 m, with corresponding stadion 198·4 m; the smaller Ptolemaic foot as 0·31 m, with (Italic) stadion 186 m; the larger Ptolemaic (Phoenician-Egyptian) foot as 0·3543 m, with stadion 212·6 m. It can be said that the estimate quoted by Aristotle for c is about double the correct value.

Eratosthenes (B.C. c. 276—194) left the earliest known account of a method for estimating c. At the time of a summer solstice, he measured the difference between the Sun's noon altitude at Syene (the modern Aswan) and Alexandria as one-fiftieth of 360°. (The latitude and longitude of Aswan and Alexandria are now known to be 24°6′N 32°51′E and 31°9′N 29°53′E, respectively.) Eratosthenes ignored the slight difference in longitude of the two cities and took the distance between them as 5000 stadia, thus deriving $c = 250\,000$ stadia (see Fig. 1.1). Having regard to the uncertainty of the length of the stadion, it is likely that Eratosthenes' estimate was correct within about 25 per cent.

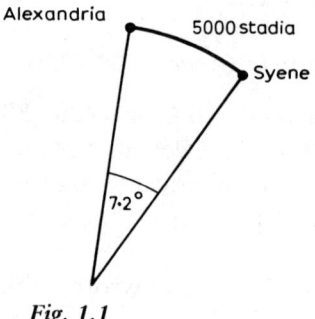

Fig. 1.1

Refinements in estimating c were introduced by later Greeks, notably Hipparchus (B.C. c. 190—125). A revision by Posidonius (B.C. c. 135—51) reducing previous estimates of the length L ($= c/360$) of a degree of latitude to the too-low value of 500 stadia was adopted by Claudius Ptolemy (A.D. c. 100—161) and became part of the Ptolemaic system. A statement that India could be reached by sailing west for 70 000 stadia is said to have influenced the plans of Columbus $1\frac{1}{2}$ millenia later.

1.1.2 Chinese work

In A.D. 723, during the Tang Dynasty, the Chinese astronomer-monk

[Refs. on p. 11]

Yi-Hsing (683–727) led a party to measure the lengths of shadows cast by the Sun, and altitudes of the pole star. The measurements were made on solstice and equinox days in thirteen places in China. Yi-Hsing thence calculated the length of one degree of meridian arc as 351·27 li (of Tang Dynasty). Dr. C.Y. Fu of Peking informs me that 1 Tang-li = 1500 Tang-feet, that 1 Tang-foot = 0·2475 m, and that one revolution was divided into $365\frac{1}{4}$ degrees in the Tang dynasty. Thus Yi-Hsing's result gives $L = 132·3$ km, which is about 20 per cent too high.

1.1.3 *Arabian work*

In A.D. 814, the Arabs under Caliph Abdallah al Mamun, son of the famous Harun al-Rashid, arrived at an estimate of about 90 km for L, about 20 per cent too low. Two groups of astronomers were despatched, one north and one south, from a fixed base on the plains of Mesopotamia, each equipped with measuring rods and astronomical instruments, and charged with ascertaining the distance to where the altitude of the pole star had changed by one degree.

Around A.D. 1000, the Arabian astronomer Ebn-Junis measured time with a pendulum, seven centuries in advance of the application to gravity determinations and the figure of the Earth. His astronomical observations were used eight centuries later as evidence on changes in the eccentricity of the Earth's orbit.

1.1.4 *European work to the time of Newton*

The next known work on the subject was carried out after the circumnavigation of the Earth had led to wide acceptance of the spherical shape. Attention was now given to improving measurements of distance over the Earth's surface, and later to improved astronomical measuring devices.

In A.D. 1527, Fernel in Paris counted revolutions of a carriage wheel to measure L, and obtained a result equivalent to $c = 36\,500$ km. With a view to checking Fernel's work, Snell in 1617 applied the idea of triangulation, starting from a measured base line on a frozen surface near Leiden. In 1637, Norwood measured the distance from London to York with a chain and by pacing. Observing also the Sun's meridian altitude, he arrived at 367 196 feet for the length L. In 1669, Picard brought the telescope to bear in measuring angles

[*Refs. on p. 11*]

and, from observations of a star in Cassiopeia, inferred a value of 57 060 toises or 111·2 km for L near Paris. (One toise = 6 French feet ≈ 6·395 English feet.)

Picard's result for L was within nearly 0·1 per cent of the modern value, so that the dimensions of the Earth were now sufficiently reliably determined to permit an estimate of the mean density, should evidence on the mass be forthcoming. The stage had also been reached where some effects of the Earth's oblateness and axial rotation were beginning to show in terrestrial measurements.

The name of Newton now enters prominently, and a separate section will refer to his contributions. Had Newton been working today in his fields of endeavour, he would have been described as a theoretical geophysicist, above all else.

1.2 Contributions of Newton and contemporaries

1.2.1 *Laws of motion and the inverse-square law*

When Newton (*Principia*, 1687) published his account of the laws of motion which bear his name (others, notably Galileo, had previously contributed much on these laws) and also his inverse-square law, he laid the foundations of the dynamical study of the shape and structure of the Earth. Previous endeavours had been, essentially, geometrical. Newton not only used geophysical and planetary observations in arriving at his laws, but also spent much of his scientific life in applying the laws to investigating the Earth's shape and physical properties further.

According to Newton's theory of gravitation, the presence of a particle of mass m at a point O gives rise to a gravitational field around O, such that a particle of mass m' at any other point P is attracted towards O with a force F, where

$$F = Gmm'r^{-2}, \qquad (1.1)$$

G is the constant of gravitation, and $r = OP$. Likewise, the gravitational field due to m' causes m to be attracted towards P with the same force F. For a given gravitational field, the intensity at any point P is defined as the force due to the field on a particle of unit mass at P, or as the acceleration caused in an otherwise free particle at P. The gravitational intensity at P arising from the presence of m at O is thus Gmr^{-2}.

[*Refs. on p. 11*]

A large body of theory has been developed from (1.1) to yield expressions for the gravitational intensities due to various extended distributions of mass, for example, the Earth.

1.2.2 *Newton and the mass of the Earth*

Let g be the magnitude of the acceleration, relative to axes rotating with the Earth, of a small body falling freely near the Earth's surface. The laws of motion (as was recognized before Newton) are historically linked with the determination of values of g. Then, through the inverse-square law, g also became connected with the gravitational intensity near the Earth's surface.

Consider a simplified Earth model, taken to be spherically symmetrical, in which effects of the Earth's axial rotation are neglected. Let O be the centre, a the radius, and M the mass of the model, and let f be the gravitational intensity at a point P distant r from O (Fig. 1.2). It can be deduced from (1.1) — see §4.2.3 — that

$$f = GMr^{-2} \quad (r \geqslant a); \qquad f = Gmr^{-2} \quad (r \leqslant a), \qquad (1.2)$$

where m is the mass inside the sphere of centre O and radius r.

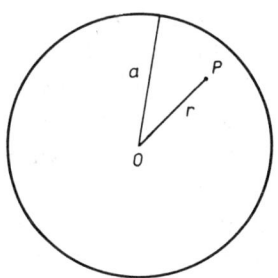

Fig. 1.2

For points on the surface ($r = a$), (1.2) gives $f = g = GMa^{-2}$. Hence, when observational values of a and g became available, a useful estimate could immediately be made of GM, and thus an important step taken towards estimating M and the mean density of the Earth.

Improved estimates of GM are obtained by using Earth models which take account, among other things, of the Earth's axial rotation and consequent equatorial bulge. Details for such models are given in Chapters 4 and 5. The currently best estimate of GM (see §5.3.5) is $3 \cdot 986 \times 10^{14} \, \text{m}^3 \, \text{s}^{-2}$.

When GM is known, separate values of both G and M are yielded by any experiment which determines either G or M alone. Newton suggested two methods for doing this: (i) measuring the attraction between two bodies in the laboratory to determine G directly; (ii) measuring the deflection of the plumb-line alongside a mountain of calculable mass M' to estimate M/M'.

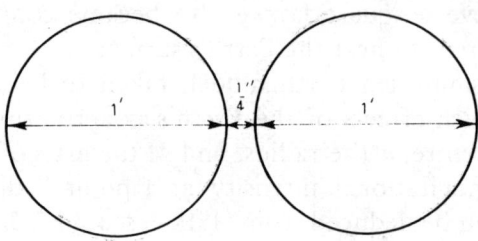

Fig. 1.3

Unfortunately an early miscalculation led to a serious underestimate of the practical possibilities. Newton considered the attraction between two spheres (Fig. 1.3), each of mean density equal to that of the Earth and of diameter 1 foot, and wrote that if they were "distant but by $\frac{1}{4}$ of an inch, they would not, even in spaces void of resistance, come together by the force of mutual attraction in less than a month's time ... . Nay, whole mountains will not be sufficient to produce any sensible effect." Newton had correctly deduced that the gravitational intensity f at the surface of a sphere of diameter D and given mean density σ is proportional to D (actually $f = 2\pi G\sigma D/3$), and hence that, if σ is equal to the mean density of the Earth, f can be determined ($f = gD/2a$) independently of knowledge of G. But he did not complete the calculation correctly. The correct time for his two spheres to come together is about $5\frac{1}{2}$ minutes, one eight-thousandth of a month.

Whatever delay the error may have caused in planning experiments to find M, the fact is that both methods (i) and (ii) were used in the eighteenth century.

Newton also saw that the inverse-square law of attraction would enable the masses and mean densities of the Sun, Earth, and certain other planets to be compared, using observations of the motions of

[Refs. on p. 11]

satellites. In this way, he succeeded in determining the ratio of the masses and mean densities of several of the bodies, but not the absolute values.

Newton nevertheless made a remarkable surmise on the Earth's mean density: "If the whole consisted of water only, whatever was of less density ... would emerge and float above. And upon this account, if a globe of terrestrial matter, covered on all sides with water, was less dense than water, it would emerge somewhere; and the subsiding water falling back, could be gathered to the opposite side. And such is the condition of our Earth, which, in great measure, is covered with seas ... . And however the Planets have been form'd while they were yet in fluid masses, all the heavier matter subsided to the centre. Since, therefore, the common matter of our Earth on the surface thereof, is about twice as heavy as water, and a little lower, in mines is found about three or four, or even five times more heavy; it is probable that the quantity of the whole matter of the Earth may be five or six times greater than if it consisted all of water, especially since I have before shewed that the Earth is about four times more dense than Jupiter."

1.2.3 *Early evidence of the Earth's oblateness*

The determination of the Earth's oblateness is nowadays closely linked with the determination of the Earth's moment of inertia I, the principal subject of Chapter 5. But useful estimates of the oblateness had been made well before either M or I had been determined.

The earliest known suggestion that the Earth's shape is oblate appears in a work by the historian-geographer Strabo about B.C. 5. This work refers to a supposition noted by an earlier historian Polybius (B.C. *c.* 203—120) that the Earth's equatorial regions are elevated. By Newton's time, visual observations by Dominique Cassini and Flamstead had indicated that the figure of Jupiter deviates significantly from a sphere, and the same was presumed to apply to other planets.

The hypothesis of the diurnal rotation of the Earth, after early controversies from the time of Aristarchus (B.C. *c.* 310—230) and its development by Copernicus in A.D. 1530—43, was well accepted by Newton's time, even though celebrated demonstrations such as Foucault's pendulum provided in 1851 were still far into the future.

[*Refs.* on *p. 11*]

In 1673, Huygens gave a satisfactory account of centrifugal force, and Newton applied this to the question of the Earth's oblateness.

Let a_1 and a_3 be the major and minor semi-axes of the surface of the Earth, assumed spheroidal (Fig. 1.4), and a the mean radius, and

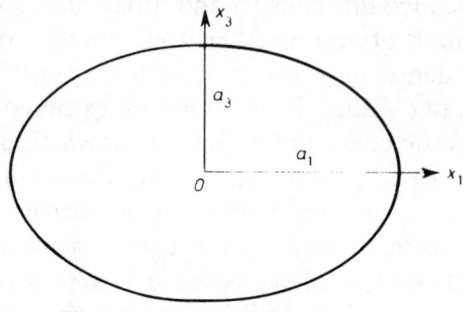

Fig. 1.4

let Ω be the angular velocity of the Earth about the polar axis. The ellipticity or 'flattening' ϵ is defined by

$$\epsilon = (a_1 - a_3)/a_1. \tag{1.3}$$

Let h be the ratio of the centrifugal and gravitational forces at the equator, i.e. of $a\Omega^2$ and GMa^{-2}, approximately. Newton computed h as 1/289. Then, treating the Earth as a rotating fluid mass of constant density, and assuming the spheroidal shape for the surface, he arrived by a formally valid argument at the result $\epsilon = 5h/4$ (see §5.5.3), and hence $\epsilon^{-1} \approx 230$. This was the first numerical estimate of the Earth's surface ellipticity. (Newton also worked on the figures of other planets.) Newton recognized that his estimate of ϵ^{-1} would need to be corrected to take account of the variation of density inside the Earth. But he incorrectly thought that a decrease would be required, whereas in fact the assumption of constant density yields a lower, not upper, limit to ϵ^{-1}.

By this time, pendulum clocks were being used by astronomers, and in 1672 Richer found that a clock beating seconds in Paris (latitude 49°N) lost about $2\frac{1}{2}$ minutes per day at Cayenne (5°N), where he had to shorten the pendulum by more than a line (1/12 French inch). Similar clock deviations were later noted by Varin and Des Hayes at Gorée (15°N) and elsewhere. A Paris Academy member had meanwhile conjectured that a body would weigh less

[Refs. on p. 11]

at the equator than at the poles. Newton, who is reported to have heard of Richer's discovery "accidentally at a Royal Society meeting" in 1682, then calculated the relative weights, according to his theory, of a body at Paris, Gorée, Cayenne, and the Equator, getting general concordance with the clock observations. This work supplied new evidence both on ϵ and on the theory of gravitation.

In 1690, Huygens noted that the plumb-line would point normally at the surface of a rotating self-gravitating fluid. He also made an estimate of the ellipticity ϵ. He treated the acceleration of a freely falling body as centrally directed and constant in magnitude over the Earth's surface, and obtained for the surface an equation equivalent to

$$g\sqrt{(x_1^2 + x_3^2)} - \tfrac{1}{2}\Omega^2 x_1^2 = \text{constant},$$

where the axes Ox_1 and Ox_3 are taken as in Fig. 1.4. This was a first step towards justifying Newton's assumption of the spheroidal shape. (The final justification was supplied in work of Stirling, Clairaut, and Maclaurin around 1740, and put in good form when Euler in 1755 gave the first clear exposition of hydrostatic equilibrium). By treating g as constant and centrally directed over the Earth's surface, Huygens in effect had assumed the mass of the Earth to be concentrated at the centre. His deduced value for ϵ^{-1}, namely 577 (see §5.5.3), is actually an upper limit.

Subsequently, much effort, theoretical and observational, was put into trying to resolve the discrepancy between the Newton and Huygens estimates for ϵ for the Earth. It was not appreciated for some time that ϵ^{-1} must lie between 230 and 577. In due course the intimate connection with the Earth's internal density distribution was generally realized.

Newton sought to improve his estimate for ϵ from observational work of others on the length L of a degree in different latitudes. Early series of measurements, while leading to improved estimates of a, did not, however, yield ϵ reliably. One series, gathered by Dominique and Jacques Cassini (father and son) of France, of which much notice was taken at the time, in fact implied that the Earth is a prolate spheroid, and satisfactory results were not obtained until much later. In 1735–6, Clairaut, Bouguer, de la Condamine, de Maupertius, and other noted French investigators embarked on expeditions to the Arctic Circle in Lapland and the Equator in Peru, and obtained clear evidence that the surface of the Earth is oblate.

[Refs. on p. 11]

1.3 The period after Newton

Geophysical investigators from the time of Newton to at least a century after his death in 1727 were predominantly British and French. The outstanding exceptions were Huygens (1629–95) of Holland, Euler (1707–83) of Switzerland, Boscovich (1711–87) in Italy, and, much later, Gauss (1777–1855) of Brunswick and Hanover, and Bessel (1784–1846) of Prussia.

Referring to French work around 1730, Todhunter (1873) wrote: "The Academy of Sciences at Paris seems to have selected the problem of the Figure of the Earth as peculiarly its own. But the success hitherto attained scarcely corresponded to the labour which had been expended."

Building on the foundations which Newton had principally supplied, French as well as British work achieved, however, a rare brilliance during the following century. Todhunter's remark suggests that questions of national prestige were becoming influential, and one is tempted to draw analogies with the present time. It would seem that the great leap forward in the eighteenth century in knowledge of the solid Earth owes as much to British and French rivalry as does the current leap outward into space to U.S.A. and U.S.S.R. rivalry.

On the observational side, the French successes began with the expeditions to Lapland and Peru, and culminated in the ambitious attempt over the period 1790–1820 to establish, in the metre, an enduring international standard of length based on measurements of the Earth. (The recent world-wide adoption of S.I. units is the victorious culmination of a battle by France begun more than 200 years ago!) In Peru, Bouguer and his collaborators founded the science of gravity observations and made the first attempt to measure the mass of the Earth. But possibly the greatest practical achievement of the century was the British success when Cavendish actually determined the mass to fair precision. This work will be described in some detail in Chapter 2.

On the theoretical side, there was notable British work on the attraction of rotating spheroids by Stirling (1735), Maclaurin (1742), and Simpson (1743). But the French theoretical work outstripped all other, starting from the publication in 1743 of Clairaut's brilliant work on the theory of the figure of the Earth. There was hardly a noted French mathematician who did not contribute significantly

[Refs. on p. 11]

to theoretical geophysics in the following century. Some of the names are d'Alembert (1717–83), Coulomb (1736–1806), Lagrange (1736–1813), Laplace (1749–1827), Legendre (1752–1833), Biot (1777–1862), Poisson (1781–1840), and Cauchy 1789–1857). The importance of this work will be seen in the following chapters.

By the early nineteenth century, the mean density of the Earth was known within about 1 per cent, the theory of gravitational attraction had been put into elegant mathematical form and applied to a great variety of problems on the rotating Earth, the science of geodesy had been placed on a firm footing, the Earth's surface ellipticity was known within moderately close bounds, and some theory on the propagation of disturbances in a deformable medium, of possible application to the Earth, had been developed.

A good part of later nineteenth century work was concerned with improving numerical detail in the light of astronomical observations which were now becoming increasingly refined. The dynamics of the solar system was worked out in fuller detail and applied, notably by Lord Kelvin (1824–1907) and Sir George Darwin (1845–1912), to studies of the Earth's tides (bodily as well as oceanic), and to such matters as observed deviations of the Earth from simple axial rotation. Using tidal observations, Kelvin made his famous calculation showing that the mean rigidity of the Earth exceeds that of ordinary steel.

There was a limit, however, to what could be learned about the detailed structure of the Earth's interior until seismology, by providing direct information on certain mechanical properties at specific depths below the surface, led to a further great leap forward in knowledge of solid-Earth geophysics in the twentieth century.

Later chapters will trace the historical development of knowledge of the internal distribution of the density and related properties in the Earth up to the present time.

REFERENCES

Clairaut, A.C. (1743). *Théorie de la Figure de la Terre*. Paris.
Darwin, Sir George (1910). *Figures of Equilibrium of Rotating Liquid and Geophysical Investigations* (Scientific Papers, Vol. III). Cambridge University Press.
Newton, I. (1687). *Philosophiae Naturalis Principia Mathematica.* Royal Society, London.

NOTED ANCIENT INVESTIGATIONS

Pratt, J.H. (1865). *A Treatise on Attractions, Laplace's Functions, and the Figure of the Earth* (3rd Ed.). Macmillan, London.

Todhunter, I. (1873). *History of the Theories of Attraction and the Figure of the Earth*, Vols. I and II. Macmillan, London.

CHAPTER 2

Determination of the mean density of the Earth

A description will now be given of various experimental efforts to measure the mass M and mean density of the Earth, including the early work of Bouguer in Peru, the Cavendish success sixty years later, and more recent results.

2.1 Bouguer's experiments in Peru

2.1.1 *The Quito experiments*

Let a be the radius, σ the mean density, and f the gravitational intensity at the surface (sea-level) of the Earth, treated as spherically symmetrical apart from a superposed extensive table-land $ABCD$

Fig. 2.1

(Fig. 2.1) of height h and density σ'. By (1.2), the gravitational intensity f' at points of AB is given by

$$f' = fa^2(a+h)^{-2} + f'' \approx f(1 - 2h/a) + f'', \qquad (2.1)$$

where f'', the contribution from the table-land itself, is (see §4.2.4) given by

$$f'' \approx 2\pi G \sigma' h \approx (3h/2a)\sigma' f/\sigma, \qquad (2.2)$$

since $f = GMa^{-2} = 4\pi Ga\sigma/3$. By (2.1) and (2.2),

$$f'/f \approx 1 - 2ha^{-1} + 3h\sigma'(2a\sigma)^{-1}. \qquad (2.3)$$

[Refs. on p. 19]

The result (2.3) was obtained by Bouguer, and the second and third terms on the right are associated with what are now, in the reduction of gravity observations, called the free-air and Bouguer corrections, respectively.

Bouguer and his collaborators applied (2.3) when making, in Peru, the first attempt to determine σ. Over the period 1737–40, they measured the length l of the seconds pendulum at a station of altitude 1466 toises at Quito (latitude 0°.25 S), treated as on a table-land, on the nearby summit of Pichincha (altitude 2434 toises), and on the island of Inca in the Esmeralda River, 30 to 40 toises above sea-level, some 60 km north-west of Quito. At these three places, they obtained $l = 438·88$, $438·69$, and $439·21$ French lines, respectively. Application of (2.3) to the Quito and Inca values, l being proportional to the local gravity value, gave $\sigma/\sigma' \approx 4·5$.

The Quito experiments demonstrated that a large mountain mass does exert a measurable attraction on a small mass. In indicating for the Earth a greater mean density than that of the Cordilleras, the experiments also enabled Bouguer to refute lingering ideas that the Earth is hollow or full of water. But, as Bouguer realized, the experiment had too many uncertainties to yield a closely reliable estimate of σ. Among other things, there was no evidence at the time on the degree of uncertainty in the assumptions underlying (2.3). More than a century was to elapse before high mountain ranges and table-lands were found to be often 'compensated' by significant deficiencies in density in the underlying region. But the Quito experiments were a bold first and historic step in long-sustained efforts to measure the Earth's mean density.

2.1.2 *The Chimborazo experiment*

Bouguer also sought in 1738 to estimate σ by measuring the deflection δ of the plumb-line on the side of an attracting mountain, and selected Chimborazo (altitude 6250 m, latitude 1° 25' S) for the purpose. A first station was set up on the southern slope at altitude 2400 toises (4678 m) in about the same longitude as the estimated centre of mass of the mountain. The apparent meridian (astronomical) altitudes, which would exceed the real altitudes by $-\delta$ and $+\delta$, respectively, were there observed for a group of northern and a group of southern stars. At a second station, about 3500 toises to the west and 174 toises lower, similar observations were taken of the same

[*Refs. on p. 19*]

stars with a view to providing equations from which the unknown true meridian altitudes could be eliminated.

The volume of mountain above the stations was estimated. Taking the density of the matter in this volume as σ', Bouguer then calculated that the deflection δ should be $(\sigma'/\sigma)\,1'43''$. However, the mean of very discordant measurements gave $\delta \approx 8''$, which would entail a quite unsatisfactory value of σ/σ' in excess of 12 units.

The experiments had been carried out under conditions of extreme privation: toilsome journeys; intense cold affecting instruments as well as men; wind and sand storms. Bouguer accepted that the experiments had value only as trials, and he expressed the hope that a suitable mountain might be found in France or Britain for a repetition.

2.2 The Schiehallion experiment

In 1772, Nevil Maskelyne proposed a repetition of the Chimborazo experiment in Scotland, and a Royal Society Committee selected Mt. Schiehallion (1010 m) in Perthshire for the purpose. The mountain has a short east-west ridge, with steep slopes to the north and south. (The name was spelt Schehallien in Maskelyne's paper. Referring to differences in spelling, Dr. P.L. Willmore tells me that he has the impression that "the early English explorers had some difficulty in transliterating the sounds made by the local inhabitants, particularly as the climate in the area is very conducive to colds".)

Careful surveys were first made during the period 1774–6. Two observing stations were then set up on the same meridian, one on the northern and the other on the southern slope. At each station, Maskelyne made some 170 observations of the apparent zenith distances of more than 30 stars, and arrived at a mean difference of $54''\cdot 6$ for the two stations. The excess of this difference over the measured latitude difference $42''\cdot 9$ was attributed to the plumb-line deflection caused by the mountain.

Hutton and Cavendish took part in the calculations, which gave $\sigma/\sigma' = 1\cdot 79$. The density σ' of the mountain was estimated as $2\cdot 5$ g/cm^3, thus yielding $4\cdot 5$ g/cm^3 for the mean density σ of the Earth.

In 1811, Playfair measured the densities of rock strata in Schiehallion in finer detail, and raised the estimate of σ to between

[Refs. on p. 19]

4·56 and 4·87 g/cm³. Hutton in 1821 gave his final value as 4·95.

2.3 Other pendulum and plumb-line experiments

Using Bouguer's Quito (pendulum) method in the vicinity of Milan, Carlini in 1821 obtained $\sigma = 4\cdot 39$ g/cm³, raised by Sabine to 4·77 in 1827 and by Giulio to 4·95 in 1841.

Airy sought to estimate σ by measuring the difference in the periods of a pendulum at the top and bottom of a mine. After early attempts in Cornwall in 1826 and 1828 that were frustrated by fire and flood, he obtained $\sigma = 6\cdot 6$ g/cm³ in 1854 at the Harton coal-pit in Sunderland. His method assumed the Earth to be spherically layered and required the assumption of specific density values at depth in the mine. Much later (1883), von Sterneck, experimenting at different depths in silver mines in Saxony and Bohemia, obtained values of σ ranging from 5·0 to 6·3 g/cm³ and demonstrated the degree of unreliability in Airy's assumptions. In the meantime, Pratt (1855) and Airy himself (1855) had put forward ideas on density compensations below the Earth's surface, thus also pointing to the limitations of pendulum experiments for the purpose of measuring the Earth's mean density.

In 1855, James and Clarke repeated the Schiehallion type of experiment on Arthur's Seat in Edinburgh and obtained $\sigma = 5\cdot 3$ g/cm³.

In 1880, Mendenhall measured pendulum periods in Tokyo and at the summit of Fujiyama, and gave $\sigma = 5\cdot 77$ g/cm³.

2.4 The Michell-Cavendish experiment

2.4.1 *Michell's apparatus*

Towards the close of the eighteenth century, the most noted seismologist of the period, John Michell, followed Newton's other suggested approach to the problem (§1.2.2(i)), an approach which avoids the uncertainties attaching to estimates of large terrestrial masses as in the mountain and mine experiments.

Michell constructed a torsion balance to measure directly the gravitational attraction F between spherical masses m_1 and m_2, small enough to be housed in a laboratory. If d is the distance between the centres of the spheres, the gravitation law (§1.2.2)

[*Refs. on p. 19*]

stars with a view to providing equations from which the unknown true meridian altitudes could be eliminated.

The volume of mountain above the stations was estimated. Taking the density of the matter in this volume as σ', Bouguer then calculated that the deflection δ should be $(\sigma'/\sigma)\,1'43''$. However, the mean of very discordant measurements gave $\delta \approx 8''$, which would entail a quite unsatisfactory value of σ/σ' in excess of 12 units.

The experiments had been carried out under conditions of extreme privation: toilsome journeys; intense cold affecting instruments as well as men; wind and sand storms. Bouguer accepted that the experiments had value only as trials, and he expressed the hope that a suitable mountain might be found in France or Britain for a repetition.

2.2 The Schiehallion experiment

In 1772, Nevil Maskelyne proposed a repetition of the Chimborazo experiment in Scotland, and a Royal Society Committee selected Mt. Schiehallion (1010 m) in Perthshire for the purpose. The mountain has a short east-west ridge, with steep slopes to the north and south. (The name was spelt Schehallien in Maskelyne's paper. Referring to differences in spelling, Dr. P.L. Willmore tells me that he has the impression that "the early English explorers had some difficulty in transliterating the sounds made by the local inhabitants, particularly as the climate in the area is very conducive to colds".)

Careful surveys were first made during the period 1774–6. Two observing stations were then set up on the same meridian, one on the northern and the other on the southern slope. At each station, Maskelyne made some 170 observations of the apparent zenith distances of more than 30 stars, and arrived at a mean difference of $54''{\cdot}6$ for the two stations. The excess of this difference over the measured latitude difference $42''{\cdot}9$ was attributed to the plumb-line deflection caused by the mountain.

Hutton and Cavendish took part in the calculations, which gave $\sigma/\sigma' = 1{\cdot}79$. The density σ' of the mountain was estimated as $2{\cdot}5$ g/cm^3, thus yielding $4{\cdot}5$ g/cm^3 for the mean density σ of the Earth.

In 1811, Playfair measured the densities of rock strata in Schiehallion in finer detail, and raised the estimate of σ to between

[Refs. on p. 19]

4·56 and 4·87 g/cm³. Hutton in 1821 gave his final value as 4·95.

2.3 Other pendulum and plumb-line experiments

Using Bouguer's Quito (pendulum) method in the vicinity of Milan, Carlini in 1821 obtained $\sigma = 4\cdot39$ g/cm³, raised by Sabine to 4·77 in 1827 and by Giulio to 4·95 in 1841.

Airy sought to estimate σ by measuring the difference in the periods of a pendulum at the top and bottom of a mine. After early attempts in Cornwall in 1826 and 1828 that were frustrated by fire and flood, he obtained $\sigma = 6\cdot6$ g/cm³ in 1854 at the Harton coal-pit in Sunderland. His method assumed the Earth to be spherically layered and required the assumption of specific density values at depth in the mine. Much later (1883), von Sterneck, experimenting at different depths in silver mines in Saxony and Bohemia, obtained values of σ ranging from 5·0 to 6·3 g/cm³ and demonstrated the degree of unreliability in Airy's assumptions. In the meantime, Pratt (1855) and Airy himself (1855) had put forward ideas on density compensations below the Earth's surface, thus also pointing to the limitations of pendulum experiments for the purpose of measuring the Earth's mean density.

In 1855, James and Clarke repeated the Schiehallion type of experiment on Arthur's Seat in Edinburgh and obtained $\sigma = 5\cdot3$ g/cm³.

In 1880, Mendenhall measured pendulum periods in Tokyo and at the summit of Fujiyama, and gave $\sigma = 5\cdot77$ g/cm³.

2.4 The Michell-Cavendish experiment

2.4.1 *Michell's apparatus*

Towards the close of the eighteenth century, the most noted seismologist of the period, John Michell, followed Newton's other suggested approach to the problem (§1.2.2(i)), an approach which avoids the uncertainties attaching to estimates of large terrestrial masses as in the mountain and mine experiments.

Michell constructed a torsion balance to measure directly the gravitational attraction F between spherical masses m_1 and m_2, small enough to be housed in a laboratory. If d is the distance between the centres of the spheres, the gravitation law (§1.2.2)

[*Refs. on p. 19*]

gives $F = Gm_1 m_2 / d^2$. Thus the constant of gravitation G could be found from measurements of F, m_1, m_2, and d.

It has already been pointed out (§1.2.2) that GM is known to good accuracy in terms of g and a, where M and a are the mass and mean radius of the Earth, and g is the gravitational intensity at the surface. Hence the determination of G also gives M and σ.

Michell's apparatus (Fig. 2.2) included a horizontal rod AB of centre C and length 6 feet ($2l$ say), suspended from a fixed point O by a 40-inch vertical wire OC. From each of A and B, a 2-inch sphere of lead (mass m_1) was suspended by a short vertical wire, and the whole system (S say) was housed in a narrow wooden case. Arrangements were made for bringing up, outside the case, on opposite sides of the plane OAB, two 8-inch lead spheres (each of mass m_2) into positions such that each would attract, by the same amount $Gm_1 m_2 / d^2$, a particular one of the two masses m_1 in a horizontal direction approximately at right-angles to the plane OAB. The wire OC would thus be twisted by a horizontal torque.

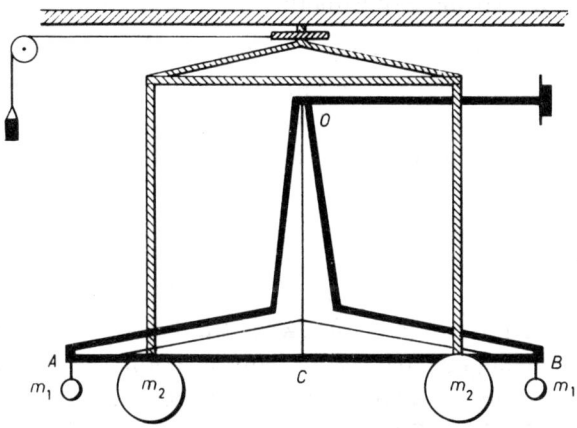

Fig. 2.2. *The Michell-Cavendish apparatus*

Let τ be the torsional rigidity of the wire OC, and θ the angle between the equilibrium positions of AB before and after the larger spheres are brought up. Then

$$\tau\theta = 2Gm_1 m_2 l / d^2. \tag{2.4}$$

Let I be the moment of inertia and T the period of vibration of the system S about OC when the masses m_2 are absent. Then

[*Refs. on p. 19*]

$$T = 2\pi\sqrt{(I/\tau)}. \tag{2.5}$$

The period T could be measured in an auxiliary experiment. Then, through (2.4) and (2.5), G could be determined in terms of the measured quantities $m_1, m_2, l, d, \theta, I$, and T.

Michell died before using the apparatus, which passed to Wollaston and then Cavendish.

2.4.2 Cavendish's determination

Henry Cavendish held closely to Michell's ideas, but reconstructed or modified parts of the apparatus. He placed the apparatus in a closed inner chamber, and observed deflections through a telescope. His celebrated result was published in 1798. The mean of 29 sets of observations after various corrections had been applied (and after Bailey had later corrected an arithmetical error) gave $\sigma = 5 \cdot 448 \pm 0 \cdot 033$ g/cm^3.

2.5 Other small-scale experiments

Using similar apparatus to that of Michell and Cavendish, Reich in Germany obtained $\sigma = 5 \cdot 49$ g/cm^3 in 1837, and 5·58 in 1852. Baily in England obtained 5·67 in 1842. Cornu and Baille in France added further experimental refinements and in 1873 gave interim results of 5·50 and 5·56.

Von Jolly devised a different type of experiment using a common balance and involving the change in g between the top and bottom of a 21-metre tower, and obtained 5·69 in 1881. Richarz and Krigar-Menzel obtained 5·505 in a somewhat similar way in 1898.

Poynting arranged masses m_1 on a common balance with an attracting mass m_2 placed vertically below each of the m_1 in turn, and obtained 5·49 in 1892.

In 1895, Boys modified the Michell-Cavendish experiment by introducing a fine quartz fibre in place of Michell's torsion wire. He was thus able to use smaller masses (which were of gold) and reduce certain extraneous effects by the overall smaller size of the apparatus. His result was 5·527, in close agreement with results obtained by Braun and Eötvös independently in 1896, also using torsion balances.

More recently, Heyl (1930), Zahradnicek (1933), and Heyl and Chrzanowski (1942) — see Heiskanen and Vening Meinesz (1958) — obtained estimates of G which entail $\sigma = 5 \cdot 517$, 5·528, and

$5 \cdot 514$ g/cm^3 respectively. Jeffreys (1939, 1970), applying statistical criteria to 25 determinations of G by Boys and Heyl, gave $G = (6 \cdot 670 \pm 0 \cdot 004) \times 10^{-11}$ m^3 kg^{-1} s^{-2}. Taking $GM = 3 \cdot 986 \times 10^{14}$ m^3 s^{-2} (§5.3.5) then gives $M = 5 \cdot 977 \times 10^{24}$ kg, and hence $\sigma = 5 \cdot 517$ g/cm^3, the uncertainties in M and σ being proportionately the same as for G.

REFERENCES

Airy, G.B. (1855). On the computations of the effect of the attraction of the mountain masses as disturbing the apparent astronomical latitude of stations in geodetic surveys. *Phil. Trans.*, **145**, 101–104.

Baily, F. (1842). An account of some experiments with the torsion-rod for determining the mean density of the Earth. *Mem. Roy. Astron. Soc.*, **14**, 1–120, i–ccxlviii.

Bouguer, P. (1749). *La Figure de la Terre.* Paris.

Boys, C.V. (1895). On the Newtonian constant of gravitation. *Phil. Trans.*, A, **186**, 1–72.

Cavendish, H. (1798). Experiments to determine the density of the Earth. *Phil. Trans.*, **58**, 469–526.

Cornu, A. and Baille, J. (1878). Sur la mésure de la densité moyenne de la Terre. *C.R. Acad. Sci. (Paris)*, **86**, 699–702.

Eötvös, R. von (1896). *Wied. Ann.*, **59**, 354.

Heiskanen, W.A. and Vening Meinesz, F.A. (1958). *The Earth and its Gravity Field*, McGraw-Hill, New York.

Heyl, P.R. and Chrzanowski, P. (1942). A new determination of the constant of gravitation. *J. Res. Nat. Bur. Standards*, **29**, 1–31.

Hutton, C. (1778). An account of the calculations made from the survey and measures taken at Schehallien, in order to ascertain the mean density of the Earth. *Phil. Trans.*, 689–788.

James, R.E. and Clarke, R.E. (1856). On the deflection of the plumb-line at Arthur's Seat and the mean specific gravity of the Earth. *Phil. Trans.*, 591–606.

Jeffreys, H. (1939). *Theory of Probability* (1st Ed.), Clarendon Press, Oxford.

Jeffreys, Sir H. (1970). *The Earth* (5th Ed.), Cambridge University Press.

Jolly, P. von (1878). Die Anwendung der Waage auf Probleme der Gravitation. *Abhand. der k. Bayer. Acad. der Wiss.*, 2 cl., **13**(1), 157–176; *Wied Ann.*, **5**, 112–134.

Maskelyne, N. (1775). An account of observations made on the Mountain Schehallien for finding its attraction. *Phil. Trans.*, p. 500.

Poynting, J.H. (1894). *The Mean Density of the Earth*, Charles Griffin, London.

Pratt, J.H. (1855). On the attraction of the Himalaya Mountains and of the elevated regions beyond and upon the plumb-line of India. *Phil. Trans.*,

145, 53–100.
Richarz, F. and Krigar-Menzel. (1898). *Abhand. der Königl. Preuss. Akad.*, Berlin.

CHAPTER 3

Spherical harmonics

The shape of the Earth makes it desirable to use spherical polar coordinates in describing a number of its properties, including those connected with the gravitational field (Chapter 4) and with free Earth oscillations (Chapter 14). The analysis in terms of spherical polar coordinates involves consideration of spherical harmonics. The present chapter gives an outline of results needed in applications in this book. More complete analytical detail can be found in standard works on spherical harmonics.

Referred to an origin O, in our context the centre of the Earth, the spherical polar coordinates of a point P (Fig. 3.1) will be taken

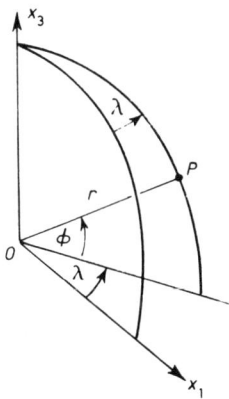

Fig. 3.1

as (r, ϕ, λ), where ϕ and λ correspond to (geocentric) latitude and longitude; thus $\phi = \tfrac{1}{2}\pi - \theta$, where θ is the colatitude, which is taken as the second coordinate in some other contexts. The coordinates (r, ϕ, λ) are connected with the cartesian coordinates x_i by

$$x_1 = r \cos \phi \cos \lambda; \quad x_2 = r \cos \phi \sin \lambda; \quad x_3 = r \sin \phi. \quad (3.1)$$

3.1 Solutions of Laplace's equation

Gravitation theory involves discussion (see §4.1.4) of Laplace's equation

$$\nabla^2 y = 0, \quad (3.2)$$

where $\nabla^2 \equiv \Sigma(\partial^2/\partial x_j^2)$. In terms of (r, μ, λ), where $\mu = \sin \phi$, Laplace's equation becomes

$$\frac{\partial}{\partial r}\left(r^2 \frac{\partial y}{\partial r}\right) + \frac{\partial}{\partial \mu}\left\{(1-\mu^2)\frac{\partial y}{\partial \mu}\right\} + (1-\mu^2)^{-1}\frac{\partial^2 y}{\partial \lambda^2} = 0. \quad (3.3)$$

Towards solving (3.3), substitute $y = RS$ as a trial solution into (3.3), where R is a function of r alone, and S of μ and λ alone. This gives

$$R^{-1}\frac{d}{dr}\left(r^2 \frac{dR}{dr}\right) = n(n+1), \quad (3.4)$$

$$S^{-1}\frac{\partial}{\partial \mu}\left\{(1-\mu^2)\frac{\partial S}{\partial \mu}\right\} + S^{-1}(1-\mu^2)^{-1}\frac{\partial^2 S}{\partial \lambda^2} = -n(n+1), \quad (3.5)$$

where the separation into (3.4) and (3.5) and the forms of the right sides of (3.4) and (3.5) are taken for convenience, with n undetermined. Since the left sides of (3.4) and (3.5) are independent of μ and λ, and of r, respectively, n is constant.

For an assigned n, the general solution of (3.4) is

$$R = A_n r^n + B_n r^{-(n+1)}, \quad (3.6)$$

where A_n and B_n are integration constants. Let $S = S_n(\mu, \lambda)$ be a solution of (3.5). Superposing for different n, we then obtain as a solution of (3.3) of some generality (sufficient for our principal applications)

$$y = \sum_{n=0}^{\infty} \{A_n r^n + B_n r^{-(n+1)}\} S_n(\mu, \lambda). \quad (3.7)$$

Any particular solution of Laplace's equation given by a term of (3.7) is called a *solid spherical harmonic* and the coefficient $S_n(\mu, \lambda)$ is called a *surface (spherical) harmonic of degree n*.

3.2 Legendre polynomials

Legendre's polynomials $P_n(\mu)$ are defined for positive integer n by the terminating series

$$P_n(\mu) = \frac{1.3.5 \ldots (2n-1)}{n!} \left\{ \mu^n - \frac{n(n-1)}{2(2n-1)} \mu^{n-2} \right. $$
$$\left. + \frac{n(n-1)(n-2)(n-3)}{2.4.(2n-1)(2n-3)} \mu^{n-4} - \ldots \right\}. \quad (3.8)$$

The following properties of the P_n can be derived from (3.8) by routine methods [see e.g. Jeffreys and Jeffreys (1950)].

3.2.1 The first five Legendre polynomials are: .

$$P_0 = 1; \quad P_1 = \mu; \quad P_2 = \tfrac{1}{2}(3\mu^2 - 1);$$
$$P_3 = \tfrac{1}{2}(5\mu^3 - 3\mu); \quad P_4 = \tfrac{1}{8}(35\mu^4 - 30\mu^2 + 3). \quad (3.9)$$

3.2.2 For any positive integer n,

$$P_n(1) = 1; \quad P_n(-1) = (-1)^n; \quad P_{2n+1}(0) = 0. \quad (3.10)$$

3.2.3 Rodrigues' formula is

$$P_n(\mu) = \frac{1}{2^n.n!} \frac{d^n(\mu^2-1)^n}{d\mu^n}. \quad (3.11)$$

From (3.11), it can be deduced (using Rolle's theorem) that the n zeroes of $P_n(\mu)$ are all real and distinct and lie in the range $-1 < \mu < 1$.

3.2.4 An asymptotic approximation to $P_n(\sin\phi)$ for n large and ϕ not too near $\pm\tfrac{1}{2}\pi$ is

$$P_n(\sin\phi) \sim \{\tfrac{1}{2}\pi(n+\tfrac{1}{2})\cos\phi\}^{-\tfrac{1}{2}} \cos\{(n+\tfrac{1}{2})\phi - \tfrac{1}{2}n\pi\}. \quad (3.12)$$

3.2.5 In the case of symmetry about an axis (taken to be Ox_3), we have $\partial/\partial\lambda = 0$, and (3.5) becomes

$$(1-\mu^2)d^2S/d\mu^2 - 2\mu dS/d\mu + n(n+1)S = 0, \quad (3.13)$$

[Refs. on p. 28]

which is Legendre's equation. By substitution, or otherwise, it can be checked that $S = P_n(\mu)$ is a solution of (3.13). In this case, the solution (3.7) of Laplace's equation becomes

$$y = \sum_{n=0}^{\infty} \{A_n r^n + B_n r^{-(n+1)}\} P_n(\mu). \tag{3.14}$$

Thus if S_n is independent of λ, it is proportional to $P_n(\mu)$. Also, $P_n(\mu)$ is a surface harmonic. [$P_n(\mu)$ is a zonal surface harmonic -- see §3.4.]

3.2.6 By direct expansion, or otherwise, it can be shown that, for $r' < r$,

$$(r^2 - 2\mu r r' + r'^2)^{-\frac{1}{2}} = r^{-1} \sum_{n=0}^{\infty} \{(r'/r)^n P_n(\mu)\}. \tag{3.15}$$

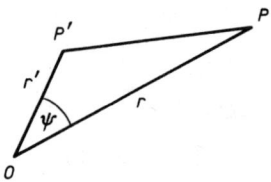

Fig. 3.2

The expansion (3.15) is important in the next chapter because, when $\mu = \cos \psi$, the left side is equal to $(P'P)^{-1}$, where $P'P$ is as in Fig. 3.2; $P'P$ is involved in the expression (4.4) for the potential function U.

3.3 Integrals of products of spherical harmonics

In the following subsections, the double integrals are over the surface of the unit sphere (an element of area of which is equal to $r^2 \cos \phi \, d\phi \, d\lambda$, where $r = 1$, i.e. is equal to $d\mu \, d\lambda$).

3.3.1 If S_m and S_n are two surface harmonics and $m \neq n$, it can be shown that

$$\iint S_m(\mu, \lambda) S_n(\mu, \lambda) \, d\mu \, d\lambda = 0. \tag{3.16}$$

[Refs. on p. 28]

3.3.2 Let ψ be the angle between the radius vectors to the points (r, μ, λ) and (r', μ', λ'), so that

$$\cos\psi = \sin\phi\sin\phi' + \cos\phi\cos\phi'\cos(\lambda - \lambda')$$
$$= \mu\mu' + \sqrt{\{(1-\mu^2)(1-\mu'^2)\}}\cos(\lambda - \lambda'). \quad (3.17)$$

Then it can be shown that for any surface harmonic $S_n(\mu, \lambda)$

$$\iint S_n(\mu, \lambda) P_n(\cos\psi)\, d\mu\, d\lambda = 4\pi(2n+1)^{-1} S_n(\mu', \lambda'). \quad (3.18)$$

3.3.3 Take the particular case of §3.3.2 where there is symmetry about Ox_3. Then, by §3.2.5, $S_n(\mu, \lambda)$ is proportional to $P_n(\mu)$. Take also $\phi' = \tfrac{1}{2}\pi$ in particular, so that $\mu' = 1$ and $\cos\psi = \mu$ by (3.17). The integrand in (3.18) is then independent of λ and, since $\int d\lambda = 2\pi$, we obtain

$$\int_{-1}^{1} P_n(\mu) P_n(\mu)\, d\mu = 2(2n+1)^{-1} P_n(1). \quad (3.19)$$

Since $P_n(1) = 1$, we have, on combining (3.19) with the companion result corresponding to (3.16),

$$\tfrac{1}{2}(2n+1) \int_{-1}^{1} P_m(\mu) P_n(\mu)\, d\mu = \delta_{mn}, \quad (3.20)$$

where δ_{mn} is equal to unity $(m = n)$ or zero $(m \neq n)$.

3.4 Tesseral harmonics

When symmetry with respect to the polar axis Ox_3 cannot be assumed, i.e. in contexts where S may depend on the longitude λ as well as on the latitude ϕ, it is convenient for a number of purposes to consider solutions of (3.5) of the form

$$S = \Phi(\phi)\Lambda(\lambda), \quad (3.21)$$

Theory developed from this procedure is required in Chapter 14.

Substituting (3.21) into (3.5) and separating the variables yields

$$\Lambda^{-1} d^2\Lambda/d\lambda^2 = -m^2, \quad (3.22)$$

$$(d/d\mu)\{(1-\mu^2)\, d\Phi/d\mu\} + \{n(n+1) - m^2(1-\mu^2)^{-1}\}\Phi = 0, \quad (3.23)$$

where m is an undetermined constant. The equation (3.23) is

[Refs. on p. 28]

Legendre's Associated Equation; it reduces to Legendre's equation (3.13) when S is independent of λ, i.e. when $m = 0$.

For an assigned m, the general solution of (3.22) is

$$\Lambda = C_m \cos(m\lambda + \epsilon_m), \tag{3.24}$$

where C_m and ϵ_m are integration constants.

A solution of (3.23) is $\Phi = P_n^m(\mu)$, where m is a positive integer and

$$P_n^m(\mu) = (\mu^2 - 1)^{\frac{1}{2}m} d^m P_n(\mu)/d\mu^m; \tag{3.25}$$

$P_n^m(\mu)$ is called the Associated Legendre Function, of degree n and order m, of the First Kind. [Legendre and Associated Legendre Functions of the Second Kind, which are also solutions of (3.23), are relevant in contexts where $\mu \geqslant 1$.] By Rodrigues' formula (3.11),

$$P_n^m(\mu) = \frac{1}{2^n . n!} (\mu^2 - 1)^{\frac{1}{2}m} \frac{d^{n+m}(\mu^2 - 1)^n}{d\mu^{n+m}}. \tag{3.26}$$

In our contexts, n is zero or a positive integer, and (3.25) then gives $P_n^m = 0$ for $m > n$; thus it is appropriate to take $m \leqslant n$.

By (3.21), (3.24), and (3.26), it follows that the function $S_n^m(\mu, \lambda)$ is a spherical harmonic of degree n and order m, where

$$S_n^m(\mu, \lambda) = K_n^m \cos(m\lambda + \epsilon_m)(1 - \mu^2)^{\frac{1}{2}m} d^{n+m}(1 - \mu^2)^n/d\mu^{n+m}, \tag{3.27}$$

K_n^m is a constant, m and n are positive integers, and $m \leqslant n$.

The function $S_n^m(\mu, \lambda)$ vanishes where any of the three factors (other than K_n^m) on the right side of (3.27) vanishes. The first of these factors vanishes where $m\lambda + \epsilon_m = (k + \frac{1}{2})\pi$, k being an integer, and so vanishes on m planes all intersecting on the polar axis Ox_3, the angle between any pair of consecutive planes being π/m. The second factor vanishes where $\mu = \pm 1$, i.e. on the polar axis. The third factor vanishes on a series of right circular cones coaxial with the polar axis; by Rolle's theorem, since $d^n(\mu^2 - 1)^n/d\mu^n$ has n real distinct zeroes in the range $-1 < \mu < 1$ (§3.2.3), the cones (taken as starting from O) are $n - m$ in number; they are symmetrically situated above and below the equator.

Over the surface of any sphere $r = $ constant, S_n^m thus vanishes on a series of $2m$ equally spaced meridian lines and on a series of $n - m$ circles of latitude symmetrically placed with respect to the equator. The two sets of circles intersect orthogonally and the general pattern is as shown in Fig. 3.3. Because of the form of this pattern, S_n^m is

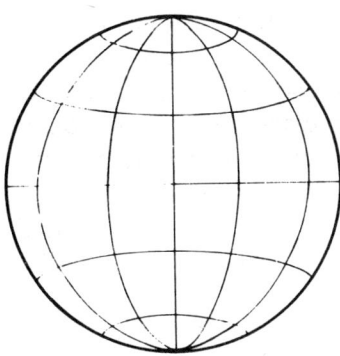

Fig. 3.3

called a *tesseral harmonic*. [The particular case S_n^0, which is proportional to $P_n(\mu)$, has its zeroes confined to circles of latitude and is called a *zonal harmonic*. The case S_n^n has its zeroes confined to meridian lines and is called a *sectorial harmonic*.]

3.5 Expansions in surface harmonics

Under fairly wide conditions, it can be shown that a given function $F(\mu, \lambda)$ can be expressed as an expansion in surface harmonics. Thus

$$F(\mu, \lambda) = \sum_{n=0}^{\infty} S_n(\mu, \lambda)$$
$$= \sum_{n=0}^{\infty} \sum_{m=0}^{n} \{C_n^m \cos(m\lambda + \epsilon_m) P_n^m(\mu)\}, \quad (3.28)$$

where the C_n^m are constants.

In the particular case where F is a function of μ only,

$$F(\mu) = \sum_{n=0}^{\infty} C_n P_n(\mu). \quad (3.29)$$

Using (3.20), the coefficients C_n in (3.29) are found to be given by

$$C_n = \tfrac{1}{2}(2n + 1) \int_{-1}^{1} F P_n \, d\mu. \quad (3.30)$$

3.6 Spherical harmonics and Earth oscillation theory

By (3.7) and §3.4, the function y, where

$$y = r^n P_n^m(\mu) \cos m\lambda, \quad (3.31)$$

[Refs. on p. 28]

is a solid spherical harmonic, satisfying Laplace's equation (3.3). In Chapter 14, we shall be investigating the Earth's free oscillations in terms of spherical polar coordinates. For this purpose it is convenient to assume, for displacement components, trial forms which are based on (3.31), but are a little more general.

A typical trial form will consist of a product of a function $R(r)$ of r, a function of μ and λ closely connected with S_n^m, and a function of the period τ of the particular mode of oscillation. The function R depends on the degree and order numbers m and n in S_n^m and also on a third number l connected with the zeroes of $R(r)$; for given m and n, the values $l = 0, 1, 2, ..$ give the fundamental oscillation and successive overtones; each zero of $R(r)$ corresponds to a nodal surface (a surface at which the displacement permanently vanishes) which is a sphere of centre O, and each unit increase in l increases the number of such spherical nodal surfaces by one.

The trial form is therefore suitably written as

$$_l R_n^m(r) \, F_n^m(\mu) \, \cos m\lambda \, \exp(\iota \gamma t), \qquad (3.32)$$

where $F_n^m(\mu)$ is identical with or closely related to $P_n^m(\mu)$, and $\gamma = 2\pi/\tau$. The character and significance of the trial forms will be more fully appreciated when specific detail is given in §14.2.

The nodal surfaces for (3.32) consist in general of three sets, namely, the planes through Ox_3 and the cones referred to in §3.4, and a set of concentric spheres.

In general, the periods τ of the family of oscillations given by (3.32) depend on l, m, and n. In special cases τ may turn out to be independent of one (or more) of these parameters; the oscillating system is then said to be *degenerate* with respect to that parameter.

REFERENCES

Bath, M. (1968). *Mathematical Aspects of Seismology*, Elsevier, Amsterdam.
Jeffreys, Sir H. and Jeffreys, Bertha Swirles (1950). *Methods of Mathematical Physics* (2nd Ed.), Cambridge University Press.

CHAPTER 4

Theory of the Earth's gravitational attraction

After the determination of the Earth's mass, the next major contribution towards knowledge of the internal density distribution was the evaluation of the Earth's moment of inertia. Essential details of the theory of the Earth's gravitational attraction needed for both this and later purposes will occupy the present chapter.

4.1 General theorems on gravitational attraction

4.1.1 *Attraction due to a general body of matter*

Let B be a body of attracting matter. (B may consist of one or more separate pieces, though in our context a single continuous body is usually involved.)

Referred to an origin O (Fig. 4.1), let $x'_i (i = 1, 2, 3)$ be the coordinates of any point P' of B, ρ the density at P', and $\delta\tau$ a

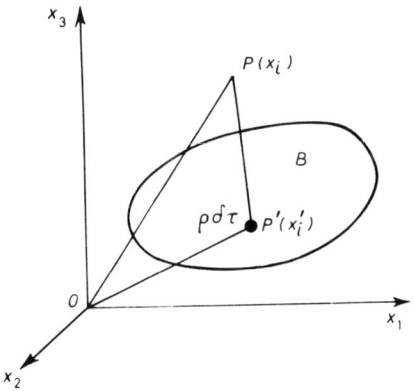

Fig. 4.1

volume-element containing P'. Let x_i be the coordinates of any point P at which the gravitational intensity f_i, or attractive force per unit mass, due to B is sought.

Newton's law gives, as the contribution to f_i from the element of mass $\rho\delta\tau$ at P', a gravitational intensity of magnitude $G\rho\delta\tau(P'P)^{-2}$ and direction PP', where G is the constant of gravitation. The magnitude of $P'P$ is $\sqrt{\{\Sigma(x_i - x_i')^2\}}$, and the direction-cosines are: $(x_i - x_i')/P'P$.

Hence

$$f_i = -\iiint G\rho(x_i - x_i')(P'P)^{-3}\,d\tau$$
$$= -\iiint G\rho(x_i - x_i')\{\Sigma(x_j - x_j')^2\}^{-3/2}\,d\tau, \qquad (4.1)$$

the integrals being through the total volume occupied by B.

4.1.2 Gauss's total normal intensity theorem

Let S be any closed surface (which may enclose none, part, or all of B), and let l_i be the direction-cosines of the outward normal at any point P of S (Fig. 4.2).

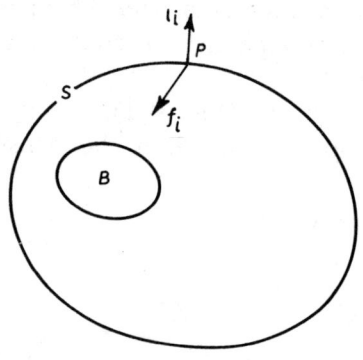

Fig. 4.2

Gauss's theorem states that

$$\iint \sum_{i=1}^{3} (f_i l_i)\,dS = -4\pi G m, \qquad (4.2)$$

where the integral is over the surface of S, and m is the mass of all matter enclosed inside S. The expression $\Sigma(f_i l_i)$ gives the component, in the direction of the outward normal to S, of the gravitational intensity at P.

To establish (4.2), consider the contribution to $\Sigma(f_i l_i)$ from the element of mass $\rho\delta\tau$ at any point P' of B. This is equal to $-G\rho\delta\tau(P'P)^{-2}\cos\chi$, where χ is the angle between $P'P$ and the normal at P. For a given P', $\iint (P'P)^{-2}\cos\chi\, dS$ is the solid angle subtended by S at P', and so is equal to 4π or zero according as P' is inside or outside S. Hence the contribution to the left side of (4.2) from the element at P' is $-4\pi G\rho\delta\tau$ or zero, respectively, in the two cases. Integrating through the whole volume occupied by B then gives (4.2).

4.1.3 *Gravitational potential function*

The potential U at the point P of coordinates x_i in a gravitational field is defined, apart from an arbitrary additive constant, so that

$$\mathbf{f} = \text{grad } U, \quad \text{or} \quad f_i = \partial U/\partial x_i, \qquad (4.3)$$

where $\mathbf{f}$ is the vector of components f_i. Thus the gravitational potential at P due to the body B may be taken as

$$U = \iiint G\rho(P'P)^{-1}\, d\tau = \iiint G\rho\{\Sigma(x_i - x_i')^2\}^{-1/2}\, d\tau. \qquad (4.4)$$

It is readily checked that (4.1) and (4.4) satisfy (4.3) for points external to B. [When P is internal, the integrands in (4.1) and (4.4) diverge at P. The integrals are then interpreted as the limits, as $\delta \to 0$, of integrals which range through the volume of B apart from a volume of small dimensions δ surrounding P. The integrals converge, but care is needed in some operations involving them.]

4.1.4 *Laplace's equation*

By direct differentiation, it is seen that $\nabla^2 \{\Sigma(x_i - x_i')^2\}^{-1/2}$ is zero except at $x_i = x_i'$. Hence when P is outside B, (4.4) gives Laplace's equation

$$\nabla^2 U = 0. \qquad (4.5)$$

4.1.5 *Poisson's equation*

By Green's lemma and (4.3),

$$\iint \Sigma(f_i l_i)\, dS = \iiint \Sigma(\partial f_i/\partial x_i)\, d\tau = \iiint \nabla^2 U\, d\tau, \qquad (4.6)$$

[Refs. on p. 42]

the volume integrals being here taken through the volume enclosed by S. Hence by (4.2)

$$\iiint (\nabla^2 U + 4\pi G\rho)\,d\tau = 0 \qquad (4.7)$$

where ρ is the density at P.

Since (4.7) applies for any surface S, it follows that Poisson's equation, namely

$$\nabla^2 U = -4\pi G\rho, \qquad (4.8)$$

is satisfied at any point P, whether inside or outside the attracting matter. Poisson's equation includes (4.5) as a particular case.

4.2 Attractions due to some particular mass distributions

The particular results in §§ 4.2.1–4.2.4 below are relevant to the Earth's gravitational field. The equations (4.13) give the external and internal gravitational potentials for an Earth treated as spherically symmetrical, and supply a useful preliminary approximation for the actual Earth.

The result (4.14) given in § 4.2.4 for the gravitational intensity due to a simplified form of table-land has been quoted in § 2.1.1 in connection with the Bouguer correction.

4.2.1 *Spherical mass of constant density*

Take the body B of § 4.1 to be spherical, with centre O, radius q, mass M, and constant density ρ. At any point P, where $OP = r$, let f be the magnitude of the gravitational intensity due to B. Let S be the surface of the sphere through P of centre O, and m the mass enclosed by S. Fig. 4.3 corresponds to the particular case where P is internal to B.

By symmetry, the gravitational intensity has the same magnitude f at all points of S, and its direction is towards O. Applying Gauss's theorem (4.2) over S then gives $4\pi r^2 f = 4\pi G m$, so that

$$f = Gm/r^2. \qquad (4.9)$$

In particular,

$$f = GM/r^2 = 4\pi G\rho q^3/3r^2 \qquad (r \geqslant q), \qquad (4.10\text{a})$$

$$f = 4\pi G\rho r/3 \qquad (r \leqslant q). \qquad (4.10\text{b})$$

By (4.3), the corresponding results for the potential U are

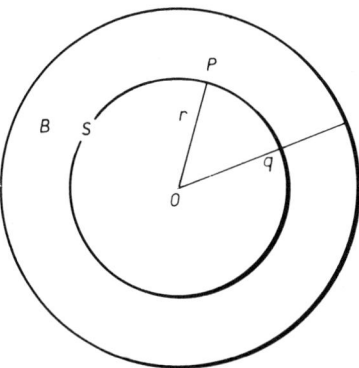

Fig. 4.3

$$U = GM/r = 4\pi G\rho q^3/3r \qquad (r \geq q), \qquad (4.11a)$$
$$U = 2\pi G\rho(3q^2 - r^2)/3 \qquad (r \leq q), \qquad (4.11b)$$

additive constants being selected (for convenience) as the simplest which make U continuous at $r = q$. It is readily checked that U as given by (4.11a) and (4.11b) satisfies Laplace's and Poisson's equations, respectively.

4.2.2 Thin spherical shell of constant density

For a spherical shell of radius q, thickness δq, mass δM, and density ρ, the potential, δU say, is deduced from (4.11) as

$$\delta U = 4\pi G\rho r^{-1} q^2 \delta q \qquad (r \geq q), \qquad (4.12a)$$
$$\delta U = 4\pi G\rho q \delta q \qquad (r \leq q). \qquad (4.12b)$$

Fig. 4.4 corresponds to the case $r > q$.

The corresponding intensities are by (4.3) equal in magnitude to $G\delta M/r^2$ and zero, respectively; these results are also derivable directly using (4.2).

4.2.3 Spherically symmetrical Earth model

Take the attracting body B as in §4.2.1 except that ρ is now a variable function of r and the outside radius is a. This case corresponds to a spherically symmetrical Earth model.

[Refs. on p. 42]

EARTH'S GRAVITATIONAL ATTRACTION [4.2-

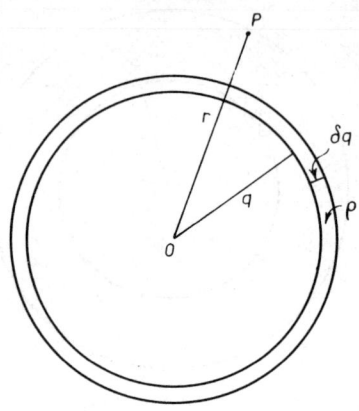

Fig. 4.4

The gravitational potential U at P is obtained by integrating (4.12), giving

$$U = 4\pi G r^{-1} \int_0^a \rho q^2 \, dq \qquad (r \geqslant a), \qquad (4.13\text{a})$$

$$U = 4\pi G r^{-1} \int_0^r \rho q^2 \, dq + 4\pi G \int_r^a \rho q \, dq \qquad (r \leqslant a). \qquad (4.13\text{b})$$

The intensity f at P can be deduced from (4.13), or immediately through (4.2), as Gm/r^2, where m is the mass inside the sphere of radius OP. If P is outside or on the surface of the model, $f = GM/r^2$. If M and a are taken as the mass and mean radius of the Earth, GM/a^2 is thus approximately equal to the acceleration of a freely falling body near the Earth's surface.

4.2.4 Table-land

Let $ABCD$ (Fig. 2.1) represent a section of a table-land AB corresponding to the upper (horizontal) surface, of area A say. Let h be the height DA, and σ' the density, both assumed constant, of the matter inside $ABCD$, and take h/AB as small. Apply Gauss's theorem (4.2) to the whole boundary surface of $ABCD$. To a first approximation, the gravitational intensity f'' due to the table-land at points on the surfaces AB and CD is constant in magnitude and normally directed, and the contributions over AD and BC may be neglected. Gauss's theorem then gives $2f''A \approx (4\pi G) \times (Ah\sigma')$, whence

$$f'' \approx 2\pi G\sigma'h, \qquad (4.14)$$

a result already applied in §2.1.1.

4.3 MacCullagh's formula

The following formula was implicit in work of Poisson and was later put more conveniently by MacCullagh.

4.3.1 Derivation

Let M be the mass and O the centre of mass of a body B of attracting matter (Fig. 4.5). Take O as origin and Ox_i as principal axes of inertia at O; and let (r, μ, λ) be as defined in §3.1. Let P' be any point of B given by (r', μ', λ'), and let P be an external point given by (r, μ, λ). Let the angle POP' be ψ. Let ρ be the density and $\delta\tau$ an element of volume at P'.

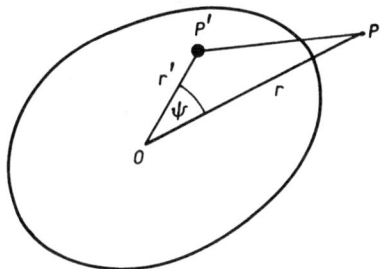

Fig. 4.5

Since $(P'P)^2 = r^2 - 2rr'\cos\psi + r'^2$, we have by (4.4) and (3.15) for the potential U at P due to B

$$U = \iiint Gr^{-1}\{1 + (r'/r)\cos\psi \\ + \tfrac{1}{2}(r'/r)^2(2 - 3\sin^2\psi) + O(r'/r)^3\}\rho d\tau, \qquad (4.15)$$

the volume integration being throughout B.

Since M and O are the mass and centre of mass of B,

$$\iiint \rho d\tau = M; \qquad \iiint \rho r'\cos\psi \, d\tau = 0. \qquad (4.16)$$

Let k_i and k be the radii of gyration of B about the axes Ox_i and OP, respectively. Then, by usual moment of inertia theory,

[Refs. on p. 42]

$$\iiint \rho r'^2 \, d\tau = \tfrac{1}{2} M(k_1^2 + k_2^2 + k_3^2), \tag{4.17}$$

$$\iiint \rho r'^2 \sin^2 \psi \, d\tau = Mk^2. \tag{4.18}$$

Substituting into (4.15), we obtain MacCullagh's formula:

$$U = GMr^{-1}\{1 + \tfrac{1}{2}(k_1^2 + k_2^2 + k_3^2 - 3k^2)r^{-2} + O(b/r)^3\}, \tag{4.19}$$

where b is of the order of the mean radius of B.

4.3.2 *Case of axial symmetry*

The direction-cosines of OP are the coefficients of r in (3.1). Hence, again by moment of inertia theory,

$$k^2 = k_1^2 \cos^2\phi \cos^2\lambda + k_2^2 \cos^2\phi \sin^2\lambda + k_3^2 \sin^2\phi.$$

In the particular case where Ox_3 is an axis of symmetry, $k_1 = k_2$ and

$$k^2 = k_1^2 \cos^2\phi + k_3^2 \sin^2\phi. \tag{4.20}$$

By (3.9), $3\sin^2\phi = 1 + 2P_2(\mu)$. Hence (4.20) gives

$$3k^2 = k_1^2(2 - 2P_2) + k_3^2(1 + 2P_2),$$

$$2k_1^2 + k_3^2 - 3k^2 = 2P_2(k_1^2 - k_3^2);$$

and MacCullagh's formula reduces to

$$U = GMr^{-1}\{1 + (k_1^2 - k_3^2)P_2(\mu)r^{-2} + O(b/r)^3\}. \tag{4.21}$$

4.4 Representation of the Earth's surface

4.4.1 *General assumptions*

We now take account of certain deviations of the Earth from spherical symmetry, but retain symmetry about the polar axis Ox_3. The origin O will again be taken at the centre of mass. The position of a point P will continue to be represented by (r, μ, λ), where $\sin^{-1}\mu$ and λ correspond to the geocentric latitude and longitude, as in Chapter 3.

Let a be the mean radius of the outside surface S. Thus the volume $4\pi a^3/3$ enclosed by S is taken to be the same as the volume enclosed by the sphere S_0 of centre O and radius a.

In accordance with geodetic observation, we take $(r - a)/a$ to be small at all points of S.

[*Refs. on p. 42*]

4.4.2 Details of representation

The assumptions permit us to write the equation of the Earth's surface as

$$r = a\{1 + F(\mu)\}, \qquad (4.22)$$

where F is small.

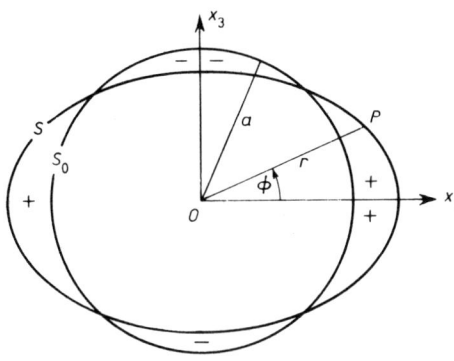

Fig. 4.6

Let R denote the region between S_0 and S, the volume of a portion of R being treated as positive or negative according as S lies above or below S_0. [In Fig. 4.6, the negative and positive portions of R are indicated by (+) and (−).] The total volume of R and its first moment with respect to the equatorial plane are both zero (the latter since O is at the centroid). Hence (see Fig. 4.7), correct to the first order in F,

$$\int_{-\frac{1}{2}\pi}^{\frac{1}{2}\pi} (2\pi a \cos\phi) \times (aF) \times (1, a \sin\phi) a d\phi = 0, \qquad (4.23)$$

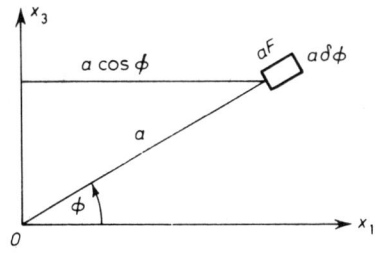

Fig. 4.7

giving
$$\int_{-1}^{1} F\,d\mu = 0; \quad \int_{-1}^{1} F\mu\,d\mu = 0, \qquad (4.24)$$

whence, by (3.9),
$$\int_{-1}^{1} FP_0\,d\mu = 0; \quad \int_{-1}^{1} FP_1\,d\mu = 0. \qquad (4.25)$$

By (3.29), we can write $F(\mu) = \Sigma C_n P_n(\mu)$. Then, by (3.30) and (4.25), $C_0 = C_1 = 0$.

Thus the equation of S can be expressed as
$$r = a\left(1 + \sum_{n=2}^{\infty} \epsilon_n p_n\right), \qquad (4.26)$$

where the p_n are proportional to the P_n, and the ϵ_n are constants.

It is convenient to take $p_2(\sin \phi)$ so that
$$p_2 = -2P_2/3 = \tfrac{1}{3} - \sin^2\phi. \qquad (4.27)$$

It will further be assumed that, for $n > 2$, all the terms $\epsilon_n p_n$ are small compared with ϵ_2, which will now be denoted as ϵ. The equation of S then becomes
$$r = a(1 + \epsilon p_2) = a\{1 + \epsilon(\tfrac{1}{3} - \sin^2\phi)\}. \qquad (4.28)$$

4.4.3 Comparison with spheroid

Consider a spheroid of major semi-axis a_1 and ellipticity ϵ. Its minor semi-axis a_3 is $a_1(1-\epsilon)$, its volume is $4\pi a_1^3(1-\epsilon)/3$, and its mean radius a is given by $a^3 = a_1^3(1-\epsilon)$. The accurate equation of its surface is
$$r^2\{(1-\epsilon)^2 \cos^2\phi + \sin^2\phi\} = a_1^2(1-\epsilon)^2, \qquad (4.29)$$

which agrees with (4.28), correct to $O(\epsilon)$.

Thus neglecting $(\epsilon_3 p_3 + ...)$ is equivalent to taking the Earth as spheroidal to a first approximation, and treating residual deviations (e.g. those associated with mountain ranges, ocean deeps, the so-called 'pear shape', etc.) as of higher order than ϵ. Except where otherwise stated, terms of order ϵ^2 will be neglected.

Correct to $O(\epsilon)$, the Earth's equatorial semi-axis $a_1 = a(1 + \epsilon/3)$, and the polar semi-axis $a_3 = a(1 - 2\epsilon/3)$.

[Refs. on p. 42]

4.5 Attraction due to spheroidal Earth model

We now repeat the calculations of §§4.2.1–4.2.3, except that the surfaces of constant density in the Earth and the outside surface will be assumed to have the form (4.28) instead of being spherical.

4.5.1 Case of nearly spherical mass of constant density

Let B be a body of constant density ρ bounded by the surface S whose equation is

$$r = q\{1 + \epsilon p_2(\mu)\}. \tag{4.30}$$

Let (r', μ', λ') now apply to any point P' on S, and (r, μ, λ) to any other point P (Fig. 4.8). Thus $OP' = q\{1 + \epsilon p_2(\mu')\}$, and q is the mean radius of S.

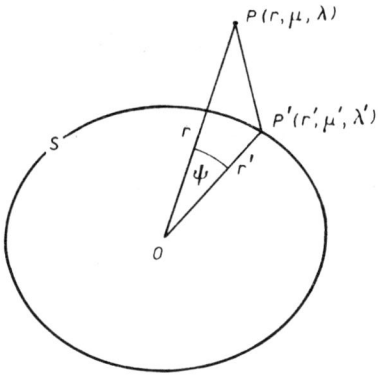

Fig. 4.8

The gravitational potential U at P due to B is the sum of a part U_0 given by the right sides of (4.11), and a part U_1 contributed by the thin distribution of matter between S and the spherical surface of radius q. At P' the radial thickness of this distribution is $\epsilon q p_2(\mu')$, and an element of its volume is $\epsilon q^3 p_2(\mu') d\mu' d\lambda'$.

By (3.15),

$$(P'P)^{-1} = (q^2 - 2qr \cos \psi + r^2)^{-\frac{1}{2}} + O(\epsilon)$$

$$= \sum_{n=0}^{\infty} \{(q^n/r^{n+1}), (r^n/q^{n+1})\} P_n(\cos \psi) + O(\epsilon),$$

[Refs. on p. 42]

the terms before and after the comma inside the curled brackets corresponding to the cases $r > q$ and $r < q$, respectively.

Hence, by (4.4),

$$U_1 = \sum_{n=0}^{\infty} \{(q^n/r^{n+1}), (r^n/q^{n+1})\} G\rho eq^3 \iint p_2(\mu') P_n(\cos\psi)\, d\mu'\, d\lambda'.$$

Since p_2 is a surface harmonic, (3.16) and 3.18) give

$$\iint p_2(\mu') P_n(\cos\psi)\, d\mu'\, d\lambda' = 0\cdot 8\pi p_2(\mu),\ \text{or zero},$$

according as $n = 2$ or $n \neq 2$. Hence

$$U_1 = 0\cdot 8\pi G\rho eq^3 \{(q^2/r^3), (r^2/q^3)\} p_2(\mu) \qquad (4.31)$$

for $r > q$ and $r < q$, respectively.

Since $U = U_0 + U_1$, we then have, from (4.11) and (4.31),

$$U = \tfrac{4}{3}\pi G\rho q^3 \{r^{-1} + 0\cdot 6\epsilon q^2 r^{-3} p_2(\mu)\} \qquad (r>q), \quad (4.32\mathrm{a})$$

$$U = \tfrac{4}{3}\pi G\rho \{\tfrac{1}{2}(3q^2 - r^2) + 0\cdot 6\epsilon r^2 p_2(\mu)\} \qquad (r<q). \quad (4.32\mathrm{b})$$

4.5.2 Thin spheroidal shell of constant density

Take next a thin spheroidal shell of constant density ρ bounded by the surfaces

$$r = q\{1 + \epsilon p_2(\mu)\}; \quad r = (q + \delta q)\{1 + (\epsilon + \delta\epsilon) p_2(\mu)\},$$

and let δU be the potential at P due to this shell, where P is the points P of coordinates (r, μ).

Differentiating (4.32) with respect to q gives

$$\delta U = 4\pi G\rho \{r^{-1} q^2 \delta q + 0\cdot 2 r^{-3} p_2(\mu)\delta(\epsilon q^5)\} \quad (r>q), \quad (4.33\mathrm{a})$$

$$\delta U = 4\pi G\rho \{q\,\delta q + 0\cdot 2 r^2 p_2(\mu)\,\delta\epsilon\} \qquad (r<q). \quad (4.33\mathrm{b})$$

4.5.3 Spheroidal Earth model of variable density

Now consider an Earth model of mean radius a, in which the density ρ varies in such fashion that the internal surfaces of constant density are all of the form (4.30). We again seek the potential U at points P of coordinates (r, μ).

When P is external, U is obtained by integrating (4.33a) with respect to q from 0 to a, ρ being now treated as variable. The result is expressible as

[Refs. on p. 42]

$$U = GMr^{-1}(1 + \tfrac{3}{2}K_2 a^2 r^{-2} p_2) \quad (r > a), \quad (4.34\text{a})$$

where

$$15K_2 = 8\pi(Ma^2)^{-1}\int_0^a \rho\{d(\epsilon q^5)/dq\}dq.$$

[Note that K_2 is independent of the position of P, as expected on comparing (4.34a) with (3.7).]

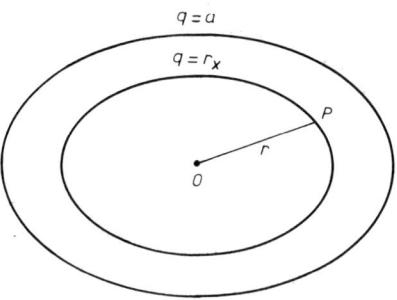

Fig. 4.9

When P is internal (Fig. 4.9), write

$$r = r_x\{1 + \epsilon_x p_2(\mu)\} \quad (4.35)$$

for the equation of the internal surface of constant density through P; r_x and ϵ_x are the mean radius and ellipticity of this surface. We now have to consider contributions to U of the form (4.33a) for $q < r_x$, and (4.33b) for $q > r_x$. The result is

$$U = 4\pi G r^{-1}\left\{\int_0^{r_x} \rho q^2\, dq + 0\cdot 2 p_2 r^{-2}\int_0^{r_x}\rho[d(\epsilon q^5)/dq]\, dq\right\}$$
$$+ 4\pi G\left\{\int_{r_x}^a \rho q\, dq + 0\cdot 2 p_2 r^2 \int_{r_x}^a \rho(d\epsilon/dq)\, dq\right\} \quad (r < a), \quad (4.34\text{b})$$

where r is as in (4.35).

Correct to the first order in ϵ, the gravitational results (4.34) supply an important part of the basis for theory needed in estimating the Earth's moment of inertia. The theory has been extended, notably by Callandreau and G.H. Darwin to take account of higher-order terms in ϵ [see e.g. Cook (1959) and Jeffreys (1970)].

[Refs. on p. 42]

REFERENCES

Callandreau, O. (1889). *Ann. Obs.* (Paris), 1–84.

Callandreau, O. (1897). *Bull. Astron.*, **14**, 214–217.

Cook, A.H. (1959). The external gravity field of a rotating spheroid to the order of e^3. *Geophys. J., R. Astr. Soc.*, **2**, 199–214.

Darwin, Sir George (1900). The theory of the figure of the Earth carried to the second order of small quantities. *Mon. Not. R. Astr. Soc.*, **60**, 82–124.

Jeffreys, Sir H. (1970). *The Earth* (5th Ed.), Cambridge University Press.

Kellogg, O.D. (1953). *Foundations of Potential Theory*, Dover, New York.

CHAPTER 5

The figure and moment of inertia of the Earth

The present chapter is centrally concerned with the estimation of the Earth's moment of inertia and a number of related quantities. As already stated, knowledge of the moment of inertia provides an important constraint on possible distributions of the density ρ inside the Earth.

Theory of the figure of the Earth which bears closely on the moment of inertia determination is outlined in the earlier sections of the chapter. Equations are derived connecting the mean I of the principal moments of inertia with observational quantities such as the Earth's mean radius, mass M, and angular velocity Ω of rotation about the polar axis, the constant G of gravitation, and the precession constant H. The theory also involves the distribution of the ellipticities ϵ of internal surfaces of constant density.

The theory is developed below correct to the first order in ϵ and, where no reference to the contrary is made, it is to be understood that equations are formally correct to at least that order, but not necessarily to higher order. Modern observational precision requires allowance to be made, however, for various second-order terms. The more important second-order terms will be set down without proof, with references to where proofs can be found.

In much of the chapter, and except where otherwise indicated, the stress in the Earth's interior is treated as hydrostatic (see §7.8.1) and therefore represented in terms of the pressure p. A number of equations derived with the help of the hydrostatic theory continue to be important to various aspects of the Earth's density variation. But some numerical results, including the estimate of I, have had to be amended because of significant deviations from the hydrostatic theory. Analyses of the orbits of artificial satellites in 1963 revealed rather dramatically the extent of these deviations and were principally

[Refs. on p. 58]

EARTH'S FIGURE AND MOMENT OF INERTIA

responsible for the revised estimate of I. Details are given towards the end of the chapter.

Included in the chapter are standard numerical values that have been internationally adopted for several constants that are prominent in the present theory.

Throughout the chapter, the Earth is treated as symmetrical about the polar axis. The equatorial, polar and mean radii are denoted as a_1, a_3, and a, and the origin O is taken at the Earth's centre of mass. The subscript a indicates values which variables take at the Earth's surface. Where necessary, the subscript x (see below for details) is introduced in specifying variables at a general internal point.

A reference frame $Ox_1x_2x_3$ is taken fixed to and rotating with the Earth, Ox_3 being the polar axis.

5.1 The geopotential function

Referred to the frame $Ox_1x_2x_3$, let P (Fig. 5.1) be the point (x_1, x_2, x_3) or (r, ϕ, λ), where r, ϕ, and λ are as defined at the start of Chapter 3; the point P may be internal or external to the Earth. The perpendicular PN from P to Ox_3 is equal to $r\cos\phi$, where ϕ is the geocentric latitude. The centrifugal force per unit mass at P, having magnitude $\Omega^2 PN$ and acting in the direction NP, is given by $\text{grad}(\tfrac{1}{2}\Omega^2 r^2 \cos^2\phi)$.

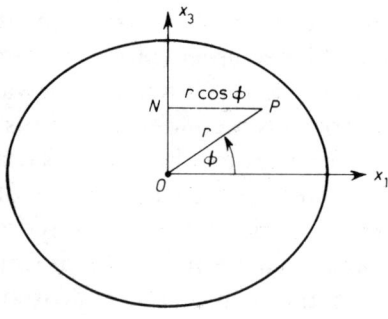

Fig. 5.1

Let U be the gravitational potential at P due to the Earth's attraction. Then the geopotential function Ψ at P is defined as

$$\Psi = U + \tfrac{1}{2}\Omega^2 r^2 \cos^2 \phi = U + \Omega^2 r^2 (2 + 3p_2)/6, \quad (5.1)$$

where U is given by one of the forms (4.34a) and (4.34b), and (as in §4.4.2)

$$p_2 = \tfrac{1}{3} - \mu^2, \quad \mu = \sin \phi. \quad (5.2)$$

Let $g_i (i = 1, 2, 3)$ be the components (referred to $Ox_1 x_2 x_3$) and g the magnitude of the acceleration of a freely falling particle at P. (If P is internal, a small free space surrounding P is here artificially envisaged.) Then $g_i = \text{grad } \Psi$ and surfaces of constant Ψ are normal to the plumb-line. Also

$$g = \{(\partial \Psi / \partial r)^2 + (\partial \Psi / r \partial \phi)^2\}^{\frac{1}{2}}, \quad (5.3)$$

since $\partial / \partial \lambda = 0$ because of axial symmetry.

For internal points, the hydrostatic theory gives

$$\text{grad } p = \rho \text{ grad } \Psi, \quad (5.4)$$

so that internal surfaces of constant Ψ are also surfaces of constant pressure and density. In particular, Ψ has the constant value Ψ_a over the Earth's outside surface.

5.2 Forms of surfaces of equal density inside rotating Earth model

Let r_x be the mean radius and ϵ_x the ellipticity of a surface S_x of constant density inside a rotating Earth model to which the hydrostatic theory applies. The subscript x will denote values which functions of r_x take on the surface S_x. The subscript will be dropped when there is no risk of misinterpretation.

The equation of S_x will be taken [cf. (4.35)] as

$$r = r_x \{1 + \epsilon_x p_2(\mu)\}, \quad (5.5)$$

correct to the first order in ϵ_x. As in §§4.4.2–4.5.3, $\epsilon \equiv \epsilon_2$ and $\epsilon_x \equiv (\epsilon_2)_x$.

It is necessary to show that the representation given by (5.5) is adequate (on the hydrostatic theory). This can be done by replacing (5.5) by the more general form

$$r = r_x \left\{ 1 + \sum_{2}^{\infty} (\epsilon_n)_x p_n(\mu) \right\}, \quad (5.6)$$

and proceeding in other respects as in the following discussion as far as equation (5.26). In place of the single equations (5.24), (5.25), (5.26), which involve only $(\epsilon_2)_x$, there would then be sets of

[*Refs. on p. 58*]

equations involving all the $(\epsilon_n)_x$. From these equations it can be deduced that $(\epsilon_n)_x = 0$ for $n > 2$. The essential underlying reason is that allowance for the Earth's rotation as in (5.1) involves the introduction of p_2, but no higher spherical harmonic. The full discussion is given in Jeffreys (1970, §4.03). We now revert to writing ϵ_2 as ϵ.

As a particular case of (5.5), the equation of the Earth's outside surface is taken as

$$r = a\{1 + \epsilon_a p_2(\mu)\}. \tag{5.7}$$

5.3 Relations involving ϵ_a and h

5.3.1 *The constant h*

It is convenient to introduce the constant h, where (cf. §1.2.3)

$$h = a^3 \Omega^2 / GM. \tag{5.8}$$

Thus h is approximately equal to the ratio of the magnitudes $a_1 \Omega^2$ and GMa^{-2} of the centrifugal and gravitational forces at the equator. To good approximation (see below), $h \approx 0{\cdot}003450$. Thus h is of the order ϵ_a (known from the time of Newton to lie between $0{\cdot}0017$ and $0{\cdot}0044$ — see §1.2.3; see also later).

5.3.2 *Relations involving the external geopotential*

Substituting into (5.1) from (4.34a), namely

$$U = GMr^{-1}(1 + \tfrac{3}{2} K_2 a^2 r^{-2} p_2) \quad (r > a), \tag{5.9}$$

we have, correct to $O(\epsilon_a)$, for points (r, μ) outside the Earth,

$$\Psi = GMr^{-1}(1 + \tfrac{3}{2} K_2 a^2 r^{-2} p_2) + \Omega^2 r^2 (2 + 3p_2)/6. \tag{5.10}$$

Substituting from (5.7) and using (5.8) then gives

$$\Psi_a = GMa^{-1}\{1 - \epsilon_a p_2 + \tfrac{3}{2} K_2 p_2 (1 - 3\epsilon_a p_2) + h(2 + 3p_2)/6\}. \tag{5.11}$$

By §5.1, Ψ_a is constant. Hence the coefficient of p_2 on the right side of (5.11) may be equated to zero, giving

$$3K_2 = 2\epsilon_a - h. \tag{5.12}$$

By (5.8), (5.10), and (5.12), the external geopotential can be expressed as

$$\Psi = GMa^{-2}\{a^2r^{-1} + a^4r^{-3}p_2(\epsilon_a - \tfrac{1}{2}h) + a^{-1}r^2(2 + 3p_2)h/6\}. \tag{5.13}$$

5.3.3 Note on second-order theory

The results (5.9)–(5.13) are all first-order results. Higher-order accuracy is obtained on replacing (5.9) by a series in ascending powers of r^{-1} which corresponds to the series involving B_n on the right side of (3.14). A convenient form for the series is

$$U = GMa_1^{-1}\left\{\frac{a_1}{r} - \sum_{n=2}^{\infty} J_n \left(\frac{a_1}{r}\right)^{n+1} P_n(\mu)\right\}, \tag{5.14}$$

in which the J_n are coefficients. Correct to $O(\epsilon_a)$, it is seen that $K_2 = J_2$. Correct to $O(\epsilon_a^2)$, it can be shown (see Jeffreys, *loc. cit.*) that

$$3J_2 = 2\epsilon_a - h - \epsilon_a^2 + 2h\epsilon_a/7. \tag{5.15}$$

In the estimation of I, the effect of the second-order terms in (5.15) turns out to be significant in our applications, so that (5.15) has to supersede the first-order relation (5.12).

The relations (5.12) and (5.15) show incidentally that J_2 and K_2 are of the same order as ϵ_a and h.

5.3.4 Clairaut's theorem on g

An approximate expression for g_a can be immediately derived using (5.3) and (5.13). The term $(\partial\Psi/r\partial\phi)^2$ in (5.3) is of the second order since ϕ occurs (through p_2) in (5.13) only in terms containing ϵ_a or h as factors. Hence $g = |\partial\Psi/\partial r|$, correct to $O(\epsilon)$. Differentiating (5.13) with respect to r and then using (5.7), we obtain

$$g_a = GMa^{-2}\{1 - 2h/3 - (5h/2 - \epsilon_a)p_2\}. \tag{5.16}$$

In particular, the value g_e at the equator (where $\sin\phi = 0$) is given by

$$g_e = GMa^{-2}\{1 - 2h/3 - (5h/2 - \epsilon_a)/3\}$$
$$= GMa^{-2}\{1 + \epsilon_a/3 - 3h/2\}. \tag{5.17}$$

Hence by (5.2)
$$g_a = g_e\{1 + (5h/2 - \epsilon_a)\sin^2\phi\}, \tag{5.18}$$

as found by Clairaut in 1743.

[Refs. on p. 58]

The second-order theory is found to give

$$g_a = g_e\{1 + (5h/2 - \epsilon_a + 15h^2/4 - 17\epsilon_a h/14)\sin^2\phi$$
$$+ \epsilon_a(15h/8 - 7\epsilon_a/8)\sin^2 2\phi\}. \qquad (5.19)$$

5.3.5 Evaluation of h

Let g_b be the value of g_a in the latitude where $p_2 = 0$, i.e. where $\phi = \sin^{-1}\sqrt{(1/3)}$. Then (5.16) gives

$$g_b = GMa^{-2}(1 - 2h/3). \qquad (5.20)$$

A standard value used in geodesy for Ω is $7\cdot2921151467 \times 10^{-5}$ rad/s, and $g_b = 9\cdot79771$ m/s². The value of a given in equation (5.62) below is $6\cdot371023 \times 10^6$ m. Then (5.8) and (5.20) yield $h = 0\cdot003450$ and $GM = 3\cdot986 \times 10^{14}$ m³s⁻².

[The constant m, where $m = a_1\Omega^2/g_e$, is sometimes used in place of h. It can be shown that $m \approx h(1 + 3h/2) = 0\cdot003468$.]

5.4 Clairaut's equation on the internal variation of ϵ

5.4.1 Expressions involving mean density

Let M_x and σ_x be the mass and mean density of the matter inside the spheroid (5.5). Then (correct to the zeroeth order)

$$\sigma_x = (3M/4\pi r^3)_x = 3r_x^{-3}\int_0^{r_x} \rho q^2 \, dq, \qquad (5.21)$$

$$(d\sigma/dr)_x = -3r_x^{-1}(\sigma_x - \rho_x). \qquad (5.22)$$

References made below to (5.21) and (5.22) may assume (5.21) and (5.22) as they stand, or equivalent equations with the subscript x dropped.

5.4.2 Derivation of Clairaut's equation

For the geopotential Ψ at internal points (r, μ) lying on the surface (5.5), we have by (5.1) and (4.34b), using (5.8) and (5.21) for convenience,

$$(4\pi G)^{-1}\Psi = r^{-1}r_x^3\sigma_x/3 + 0\cdot 2p_2 r^{-3}\int_0^{r_x} \rho[d(\epsilon q^5)/dq]\,dq$$
$$+ \int_{r_x}^a \rho q\, dq + 0\cdot 2p_2 r^2 \int_{r_x}^a \rho(d\epsilon/dq)\,dq + hr^2\sigma_a(2 + 3p_2)/18. \qquad (5.23)$$

[Refs. on p. 58]

5.4] CLAIRAUT'S EQUATION FOR ELLIPTICITIES

In (5.23), we can put $r = r_x(1 + \epsilon_x p_2)$ and therefore $r^{-1} = r_x^{-1}(1 - \epsilon_x p_2)$, etc., to first-order accuracy. Then, since (5.5) is a surface of constant density, and therefore of constant Ψ, we can equate to zero the coefficient of p_2 on the right side. After this is done, we can without ambiguity drop the subscript x. On multiplying through by $3r^3$, we then obtain

$$-\epsilon\sigma r^5 + 0\cdot 6\int_0^r \rho[d(\epsilon q^5)/dq]\,dq + 0\cdot 6 r^5 \int_r^a \rho(d\epsilon/dq)\,dq$$
$$+ \tfrac{1}{2}hr^5\sigma_a = 0. \tag{5.24}$$

Differentiating (5.24) twice with respect to r leads as follows to Clairaut's differential equation (1743) for the variation of ϵ with r in the Earth. The first differentiation gives, after subsequently using (5.22), multiplying through by r^{-4} and reducing,

$$-2\sigma\epsilon - \sigma r(d\epsilon/dr) + 3\int_r^a \rho(d\epsilon/dq)\,dq + 5h\sigma_a/2 = 0. \tag{5.25}$$

The second differentiation gives

$$\sigma r(d^2\epsilon/dr^2) + 6\rho(d\epsilon/dr) - 6\epsilon r^{-1}(\sigma - \rho) = 0.$$

Writing $\tau = \rho/\sigma$, we have Clairaut's equation in the form

$$\frac{d^2\epsilon}{dr^2} + 6\tau r^{-1}\frac{d\epsilon}{dr} - 6(1-\tau)r^{-2}\epsilon = 0. \tag{5.26}$$

An alternative form, obtained using (5.22) again, is

$$\frac{d^2\epsilon}{dr^2} + \left(\frac{6}{r} + \frac{2}{\sigma}\frac{d\sigma}{dr}\right)\frac{d\epsilon}{dr} + \frac{2\epsilon}{r\sigma}\frac{d\sigma}{dr} = 0. \tag{5.27}$$

It is important to note that an acceptable solution of (5.26) or (5.27) has also to satisfy (5.24) and (5.25).

5.4.3 Deduced properties of ϵ

Writing
$$\eta = (r/\epsilon)d\epsilon/dr, \tag{5.28}$$

and putting $r = a$ in (5.25), we obtain

$$2\eta_a = 5h/\epsilon_a - 4. \tag{5.29}$$

[Refs. on p. 58]

It is appropriate to assume $d\rho/dr < 0$ for $r > 0$ in the Earth. This condition could be departed from only if the rate of temperature increase with depth were sufficiently high to offset the rate of density increase due to pressure. (Such a departure is not impossible for a limited range of depth inside the outermost few hundred kilometres of the Earth, but would not be great enough to affect the results of the present theory significantly.) For all r, it can then be shown, using (5.26), that $\epsilon \geqslant 0$ and $d\epsilon/dr \geqslant 0$, and therefore that $\eta \geqslant 0$; also that ϵ/r^3 decreases with r, and hence that $\eta \leqslant 3$. In the vicinity of $r = 0$, ϵ has in general the form $\epsilon_0 + Ar^s$, where $s > 0$; hence, in general, $\eta = 0$ at $r = 0$. [This last property does not apply with the Darwin density law $\rho = Ar^{-n}$, which has a singularity at $r = 0$ – see §6.7.1.3. Derivations of the foregoing general properties of ϵ and η are given in Tisserand (1891) and Jeffreys (1970, §4.03).]

Further properties of ϵ are derived through the help of Radau's transformation.

5.4.4 *Radau's transformation*

With the use of (5.22) and (5.28), and some ad hoc manipulation, Radau (1885) transformed Clairaut's equation into

$$\frac{d}{dr}\{\sigma\sqrt{(1+\eta)}\} + \frac{(5\eta + \eta^2)\sigma}{2r\sqrt{(1+\eta)}} = 0, \tag{5.30}$$

and so was able to rewrite Clairaut's equation as

$$d\{\sigma r^5 \sqrt{(1+\eta)}\}/dr = 5\sigma r^4 \psi(\eta), \tag{5.31}$$

where

$$\psi(\eta) = (1 + \tfrac{1}{2}\eta - \eta^2/10)(1+\eta)^{-\frac{1}{2}}. \tag{5.32}$$

The parameter η is sometimes referred to as *Radau's parameter*.

The function $\psi(\eta)$ has the following properties. At $\eta = 0$, its value is unity. As η increases from zero, ψ increases steadily to a maximum of 1·00074 at $\eta = \tfrac{1}{3}$, and then diminishes steadily to 0·8 at $\eta = 3$. For $0 \leqslant \eta \leqslant 0·6$, $|\psi - 1| < 0·0008$ and the error in putting $\psi(\eta) = 1$ is correspondingly small. It transpires (see §5.9) that $0 \leqslant \eta \leqslant 0·6$ throughout the Earth, so that (for the Earth) a highly accurate approximation to (5.31) and therefore to Clairaut's equation is

$$d\{\sigma r^5 \sqrt{(1+\eta)}\}/dr \approx 5\sigma r^4. \tag{5.33}$$

[Refs. on p. 58]

Radau's approximate equation (5.33) enabled the theory of the figure of the Earth to be greatly advanced. Among other things, the approximation led Darwin to an important approximate relation, (5.45) below, connecting η_a with the Earth's moment of inertia.

Jeffreys (1963) showed that, in the light of modern data, replacing (5.31) by (5.33) introduces an error of only 0·0000006 (on the over side) in estimates of ϵ_a. [Radau's approximation is, however, somewhat less reliable for the larger (outer) planets — see §17.1.2.]

5.5 Estimation of the moment of inertia of the Earth

5.5.1 *Some preliminary relations*

Let A and C be the moments of inertia and k_1 and k_3 the radii of gyration of the Earth about the equatorial and polar axes, Ox_1 and Ox_3, respectively. Thus $A = Mk_1^2$ and $C = Mk_3^2$, and

$$I = \tfrac{1}{3}(2A + C). \tag{5.34}$$

Write $I = y_a Ma^2$, and let y ($\equiv y_x$) denote the coefficient corresponding to y_a for the mean moment of inertia of the matter inside an internal surface (5.5) of constant density.

Thus

$$yMr^2 = (8\pi/3)\int_0^r \rho q^4 \, dq, \tag{5.35}$$

and by (5.21)

$$yr^5 \sigma = 2\int_0^r \rho q^4 \, dq, \tag{5.36}$$

Putting $r = a$ in (5.21) and (5.36) gives, in particular,

$$a^3 \sigma_a = 3\int_0^a \rho q^2 \, dq, \tag{5.37}$$

$$y_a a^5 \sigma_a = 2\int_0^a \rho q^4 \, dq. \tag{5.38}$$

5.5.2 *Relation derived from MacCullagh's formula*

By (4.21), (4.27) and (5.9), we have [correct to $O(\epsilon_a)$],

$$(C - A)/Ma_1^2 = (k_3^2 - k_1^2)/a_1^2 = K_2 = J_2, \tag{5.39}$$

which incidentally shows, as expected, that $(C-A)/Ma^2$ is of the order of ϵ_a. It can be shown that the relation $(C-A)/Ma_1^2 = J_2$ holds to $O(\epsilon_a^2)$.

[*Refs. on p. 58*]

5.5.3 Interlude: the special cases of Newton and Huygens

Correct to $O(\epsilon_a)$, (5.12) and (5.39) give

$$(C-A)/Ma^2 = (2\epsilon_a - h)/3, \qquad (5.40)$$

through which the results of Newton and Huygens on ϵ_a (§1.2.3) can be readily derived.

Newton estimated ϵ_a for an Earth model of constant density. For this model, $A = 0.2 M(a_1^2 + a_3^2)$ and $C = 0.4 Ma_1^2$. Since $a_1 \approx a(1 + \epsilon_a/3)$ and $a_3 \approx a(1 - 2\epsilon_a/3)$, this gives $(C-A)/Ma^2 \approx 0.4\,\epsilon_a$. Then (5.40) gives $\epsilon_a \approx 5h/4$, and hence Newton's result $\epsilon_a^{-1} \approx 230$ on substituting $h = 0.00345$. This model incidentally gives $(C-A)/C \approx \epsilon_a$.

Huygens unwittingly assumed the Earth's mass to be all concentrated at the centre, in which case $A = C = 0$. Then (5.40) gives $\epsilon_a = \tfrac{1}{2}h$, and hence Huygens' result $\epsilon_a^{-1} \approx 580$.

The actual density distribution inside the Earth being intermediate between the distributions assumed by Newton and Huygens, it follows immediately that $230 < \epsilon_a^{-1} < 580$.

5.5.4 Relation between y and η

Integrating (5.31) gives, for all r in the range $0 \leqslant r \leqslant a$,

$$\sigma r^5 \sqrt{(1+\eta)} = 5\psi(\eta') \int_0^r \sigma q^4\, dq, \qquad (5.41)$$

where η' is a number lying between 0 and 3. By (5.22),

$$d(\sigma r^5)/dr = 2\sigma r^4 + 3\rho r^4,$$

so that, by (5.36),

$$\sigma r^5 = 2\int_0^r \sigma q^4\, dq + \tfrac{3}{2} y \sigma r^5. \qquad (5.42)$$

Then, by (5.41) and (5.42),

$$(1 - \tfrac{3}{2}y)\psi(\eta') = 0.4\sqrt{(1+\eta)}. \qquad (5.43)$$

When Radau's approximation holds, (5.43) reduces to the Radau-Darwin approximation

$$1 - \tfrac{3}{2}y = 0.4\sqrt{(1+\eta)}. \qquad (5.44)$$

In particular, putting $r = a$, we have

$$1 - \tfrac{3}{2}y_a = 0.4\sqrt{(1+\eta_a)}. \qquad (5.45)$$

[Refs. on p. 58]

5.5.5 *Relation derived from precession of equinoxes*

A consequence of the Earth's ellipticity of figure and the gravitational fields of the Sun and Moon is that, relative to distant stars, the Earth's axis slowly describes a cone, with semi-apex-angle $\approx 23\frac{1}{2}°$, the period being about 25 800 years. An observed effect, called the precession of the equinoxes, is that the equinoxial points move at a mean rate of $50''\cdot26$ per year. [The equinoxial points are intersections of the ecliptic (the plane of the Sun's orbit in the heavens) and the celestial equator (the plane of which contains the Earth's equator).]

The observations and related dynamical theory yield a well-determined value of the *precession constant* (or 'dynamical ellipticity') H, defined by

$$H = (C-A)/C. \tag{5.46}$$

(In some geodynamical problems, k_1 and k_2 are not taken equal.. Then A is replaced by the mean of Mk_1^2 and Mk_2^2.) The dynamical theory and observations give $H \approx 0\cdot003273$ (Jeffreys, 1970).

For the case of a model Earth of constant density, $(C-A)/C \approx \epsilon_a$ (§5.5.3), so that $H \approx \epsilon_a$. For the actual Earth, the values $H = 0\cdot003273$ and $\epsilon_a = 0\cdot003353$ (to be derived in §5.7) give $H \approx 0\cdot976\epsilon_a$. Thus the 'dynamical ellipticity' approximates to the actual surface ellipticity.

By (5.34) and (5.46),

$$y_a Ma^2 = I = C(1 - \tfrac{2}{3}H). \tag{5.47}$$

Since $a^3 = a_1^3(1-\epsilon_a)$, we then have

$$C/Ma_1^2 = y_a\{1 - \tfrac{2}{3}(\epsilon_a - H)\} \tag{5.48}$$

$$= \tfrac{2}{3}\{1 - \tfrac{2}{3}(\epsilon_a - H)\}\{1 - 0\cdot4\sqrt{(1+\eta_a)}\}, \tag{5.49}$$

by (5.45).

5.6 Numerical results on the hydrostatic theory

We have now arrived at the following five independent equations connecting the seven quantities J_2, C/Ma_1^2, $(C-A)/Ma_1^2$, η_a, ϵ_a, h, and H:

$$3J_2 = 2\epsilon_a - h - \epsilon_a^2 + 2h\epsilon_a/7 \tag{5.15}$$

$$2\eta_a = 5h/\epsilon_a - 4, \tag{5.29}$$

[Refs. on p. 58]

$$(C-A)/Ma_1^2 = J_2, \qquad (5.39)$$

$$(C-A)/C = H, \qquad (5.46)$$

$$C/Ma_1^2 = \tfrac{2}{3}\{1 - \tfrac{2}{3}(\epsilon_a - H)\}\{1 - 0.4\sqrt{(1 + \eta_a)}\}. \qquad (5.49)$$

When detailed numerical scrutiny is brought to bear, it is found that this set of equations covers the main requirements of the second-order hydrostatic theory. Lengthy examination (see e.g. Jeffreys, 1963) shows that the second-order terms disregarded in the equations other than (5.15) have much smaller effects than those of the second-order terms included in (5.15). Also, solving the set of equations by successive approximation yields values of ϵ_a and H such that $\tfrac{2}{3}(\epsilon_a - H) \approx 0.00006$. (This can be checked using the numerical results to follow.) Hence an error of less than one part in 10^4 is introduced when (5.49) is replaced by

$$C/Ma_1^2 = \tfrac{2}{3}\{1 - 0.4\sqrt{(1 + \eta_a)}\}. \qquad (5.50)$$

Prior to 1963, of the seven quantities J_2, C/Ma_1^2, etc., accurate observational evidence (derived independently of the above five equations) was available only on h and H. As seen in §§ 5.3.5 and 5.5.5, this evidence gave $h = 0.003450$ and $H = 0.003273$ to (about) four-figure accuracy. Apart from errors in the hydrostatic theory, the set of five equations is therefore just sufficient to yield correspondingly accurate values of the five quantities J_2, C/Ma_1^2, $(C-A)/Ma_1^2$, η_a, and ϵ_a.

5.6.1 *Evaluation of the surface ellipticity*

Eliminating J_2, C/Ma_1^2, $(C-A)/Ma_1^2$, and η_a from the five equations (5.15), (5.29), (5.39), (5.46), and (5.50) gives

$$H\{1 - 0.2\sqrt{(10h/\epsilon_a - 4)}\} = \epsilon_a - \tfrac{1}{2}h - \epsilon_a(\tfrac{1}{2}\epsilon_a - h/7). \qquad (5.51)$$

Substituting $h = 0.003450$ and $H = 0.003273$ into (5.51), and solving, yields $\epsilon_a = 0.003366 = 296.1^{-1}$.

5.6.2 *Moment of inertia*

Substituting $h = 0.003450$ and $\epsilon_a = 0.003366$ into (5.15) and (5.29), and solving, yields $J_2 = 0.001091$ and $\eta_a = 0.562$.

[Refs. on p. 58]

By (5.39), (5.46), and (5.47),

$$y_a = (C/Ma^2)(1 - \tfrac{2}{3}H)$$
$$= (C/Ma_1^2)\{1 + \tfrac{2}{3}(\epsilon_a - H)\}$$
$$= (J_2/H)\{1 + \tfrac{2}{3}(\epsilon_a - H)\}. \quad (5.52)$$

As seen in §5.6, the difference between the last factor in (5.52) and unity may be neglected, permitting us to write

$$y_a = J_2/H. \quad (5.53)$$

Substituting $J_2 = 0.001091$ and $H = 0.003273$ into (5.53) yields

$$I = y_a Ma^2 = 0.3333 Ma^2. \quad (5.54)$$

[The result $\eta_a = 0.562$ may be derived directly from (5.45) and (5.53).]

When account is taken of certain residual second-order corrections to (5.29) and (5.49) that have been disregarded in the foregoing account, the value of y_a yielded on the hydrostatic theory becomes slightly increased to 0.3335. It is obvious from the existence of mountains, ocean basins, etc., however, that appreciable stress-differences and therefore significant departures from hydrostatic conditions do occur in the outer part of the Earth. Consequently, the fourth digit in the estimate of y_a in (5.54) cannot in any case be considered reliable. As will be seen in §5.7, the extent of departure from hydrostatic conditions actually requires a change in the third digit.

An outstanding conclusion from (5.54) is that y_a has a value markedly less than 0.4 which would apply to a model spherical Earth of uniform density. The departures from the hydrostatic theory do not disturb this conclusion. Thus (5.54) supplied clear-cut evidence of a considerable central condensation of denser matter inside the Earth, constituting, in fact, a main part of the evidence which first suggested that the Earth contains a sizeable dense core.

More important still, knowledge of the value of y_a supplies a powerful equation of condition which has to be met in any attempt to calculate the density distribution inside the Earth. The value 0.3335 was used for this purpose for a long period prior to 1963 and, though now superseded, contributed to the determination of the first close approximation to the Earth's internal density distribution.

[*Refs. on p. 58*]

5.7 Use of artificial satellites

In deriving the expression (5.12) for K_2, use was made of (5.7) and (5.10). The form (5.10) involves only the external gravitational field, but (5.7) involves the hydrostatic theory. Hence (5.12) involves the hydrostatic theory through (5.7), and the same applies to the expression (5.15) for J_2.

Artificial satellites travelling around the Earth move through the Earth's external gravitational field, which is a principal factor in determining their orbits. In analysing the observed orbits, it has proved possible to separate resistance effects from the field effects and so to estimate J_2 completely independently of the hydrostatic theory. In this way, King-Hele, Cook, and Rees (1963) — see also A.H. Cook (1963) — obtained

$$J_2 = 0.0010828, \qquad (5.55)$$

an estimate which has to be preferred to the estimate 0.001091 of §5.6.2.

The derivation of (5.53) is also essentially independent of the hydrostatic theory. With $H = 0.003273$, (5.53) and (5.55) yield

$$I = y_a Ma^2 = 0.3308 Ma^2. \qquad (5.56)$$

The result (5.56), being independent of the hydrostatic theory, supersedes (5.54). The coefficient 0.3308 is considered likely to be accurate within about one part in 1000.

It is not possible to avoid the hydrostatic theory completely in redetermining ϵ_a, but an improved value can be derived by using (5.15) and (5.55) instead of (5.51). The result is

$$\epsilon_a = 0.003353 = 298.2^{-1}, \qquad (5.57)$$

which is close to the best current estimate (see §5.8). The revised estimate of η_a obtained using (5.45) and (5.56) is 0.586; this estimate is also subject to certain errors of the hydrostatic theory.

The extent of error in the hydrostatic theory, as indicated in the change from (5.54) to (5.56), came in 1963 as a surprise, and showed that stress-differences in the outer part of the Earth are greater than had been previously suspected. Jeffreys (1963) utilized the difference between (5.54) and (5.56) in investigating the order of magnitude of sustained deviatoric stresses (§17.1.3) in the outer part of the Earth. These stresses, although sizeable, were, however, found to be appreciably less than 10^8N/m^2; (this result is relevant to §7.6). See also Goldreich and Toomre (1969).

Observations of satellite orbits have incidentally (King-Hele and colleagues, 1965, 1967) yielded estimates of the even coefficients in (5.14) as far as J_8, and of the odd coefficients to J_{21}. One much-publicized result is that $J_3 \approx -2.4 \times 10^{-6}$. If accepted at face value, this would imply a 'pear-shaped' Earth (assuming the pear to be not Japanese) with an average distance from centre to surface some 15 m greater in the southern hemisphere than in the northern.

5.8 International reference systems

An 'international ellipsoid', based on work of Hayford (1909) and adopted in 1924 by the International Association of Geodesy for standard reference purposes, took

$$a_1 = 6 \cdot 378388 \times 10^6 \text{ m},$$
$$a_3 = 6 \cdot 356912 \times 10^6 \text{ m}, \quad (5.58)$$
$$a = 6 \cdot 371221 \times 10^6 \text{ m}.$$

and $\epsilon_a = 0 \cdot 00336700$.

The 'International Gravity Formula (1930)' gave [cf. equation (5.19)]

$$g_a = 9 \cdot 780490(1 + 0 \cdot 0052884 \sin^2\phi' - 0 \cdot 0000059 \sin^2 2\phi') \text{ m/s}^2, \quad (5.59)$$

where ϕ' is the geographic latitude ($\phi' \approx \phi + \epsilon_a \sin 2\phi$). A more recent proposal would replace (5.59) by

$$g_a = 9 \cdot 780318(1 + 0 \cdot 0053024 \sin^2\phi' - 0 \cdot 0000059 \sin^2 2\phi') \text{ m/s}^2. \quad (5.59a)$$

In 1964, the International Astronomical Union adopted for standard purposes:

$$a_1 = 6 \cdot 378160 \times 10^6 \text{ m},$$
$$J_2 = 0 \cdot 0010827, \quad (5.60)$$
$$GM = 3 \cdot 98603 \times 10^{14} \text{ m}^3 \text{ s}^{-2}.$$

Subsequently, the International Union of Geodesy and Geophysics adopted the values in (5.60) as part of the 'Geodetic Reference

[*Refs. on p. 58*]

System 1967'. A recently recommended standard value of ϵ_a is (to the first six significant figures)

$$\epsilon_a = 0\cdot00335292 = (298\cdot247)^{-1}. \qquad (5.61)$$

Taking a_1 and ϵ_a as in (5.60) and (5.61) gives

$$\left.\begin{array}{l} a = 6\cdot371023 \times 10^6 \text{ m}, \\ a_3 = 6\cdot356778 \times 10^6 \text{ m} \end{array}\right\} \qquad (5.62)$$

for the corresponding ellipsoid. The differences between the values of a_1, a_3, and a as given in (5.58), and in (5.60) and (5.62), indicate the degree of reliability to which a_1, a_3, and a have been determined. See also Marussi, Moritz, Rapp and Vicente (1974).

5.9 Ellipticities of internal surfaces of constant density

For a variety of geophysical purposes, for example, in evolving travel-times of bodily seismic waves in the Earth (see §9.5), values are required of the ellipticities ϵ of internal surfaces of constant density inside the Earth. The distribution of η inside the Earth can be estimated using (5.44) when knowledge of the distribution of density, and hence of y, is available. The distribution of ϵ is then derived by numerical integration of (5.28).

An early determination of the distributions of η and ϵ was made by Bullen (1936) and revised by Bullen and Haddon (1973) using a recent Earth model density distribution. The results, which are subject to certain of the errors of the hydrostatic theory, give at depths 0, 1000, 2000, 3000, 4000, and 6371 km below the Earth's surface: η = 0·59, 0·54, 0·48, 0·09, 0·05, and 0·00; and ϵ = 335, 305, 274, 254, 248, and 342 $\times 10^{-5}$.

The values of η thus derived fulfil the important purpose of providing a needed check on the high reliability of the Radau approximation in the interior of the Earth.

REFERENCES

Bullen, K.E. (1936). The variation of density and the ellipticities of strata of equal density within the Earth. *Mon. Not. R. Astr. Soc., Geophys. Suppl.*, 3, 395–401.

Bullen, K.E. and Haddon, R.A.W. (1973). The ellipticities of surfaces of equal density within the Earth. *Phys. Earth Planet. Interiors*, 7, 199–202.

REFERENCES

Clairaut, A.C. (1743). *Théorie de la Figure de la Terre*, Paris.
Cook, A.H. (1963). The contributions of satellites to the determination of the Earth's gravitational potential. *Space Sci. Reviews*, **2**, 355–437.
Cook, A.H. (1968). The polar flattening and gravity formula in the Geodetic Reference System 1967. *Geophys. J., R. Astr. Soc.*, **15**, 431–433.
Darwin, G.H. (1900). The theory of the figure of the Earth carried to the second order of small quantities. *Mon. Not. R. Astr. Soc.*, **60**, 82–124.
Goldreich, P. and Toomre, A. (1969). Some remarks on polar wandering. *J. Geophys. Res.*, **74**, 2555–2567.
Hayford, J.F. (1909–10). *The Figure of the Earth and Isostasy*; and *Supplementary Investigation*, U.S. Coast Geodetic Survey, Washington.
Jeffreys, Sir H. (1963). On the hydrostatic theory of the figure of the Earth. *Geophys. J., R. Astr. Soc.*, **8**, 196–202.
Jeffreys, Sir H. (1970). *The Earth* (5th Ed.), Cambridge University Press.
King-Hele, D.G., Cook, G.E. and Rees, Janice M. (1963). Determination of the even harmonics in the Earth's gravitational potential. *Geophys. J., R. Astr. Soc.*, **8**, 119–145.
King-Hele, D.G. and Cook, G.E. (1965). The even zonal harmonics of the Earth's gravitational potential. *Geophys. J., R. Astr. Soc.*, **10**, 17–29.
King-Hele, D.G., Cook, G.E. and Scott, D.W. (1967). Odd zonal harmonics in the geopotential determined from fourteen well-distributed satellite orbits. *Planet. Space Sci.*, **15**, 741–769.
Kovalevsky, K.J. (1966). Ellipsoide terrestre, U.A.I. *Bull. Astr.*, **1** (3), 19–21.
Marussi, A., Moritz, H., Rapp, R.H. and Vicente, R.O. (1974). Ellipsoidal density models and hydrostatic equilibrium. *Phys. Earth Planet. Interiors*, **9**, 4–6.
Radau, R.R. (1885). Sur la loi des densités à l'intérieur de la Terre. *C.R. Acad. Sci.* (Paris), C, 972.
Tisserand, F. (1891). *Traité de Mécanique Céleste*, Vol. II. Paris.

CHAPTER 6

Early models of the Earth's density variation

Early investigations of the Earth's moment of inertia I, summarized in §5.6, used equations, e.g. (5.27), from Clairaut's theory which involve the internal variation of the density ρ. The investigations were therefore accompanied by attention to the question of model representations of the variation of ρ and related properties of the Earth's interior. By the time I had become moderately well determined, it was possible to set down Earth density models that were useful for a variety of limited purposes. The early models had two parameters which could be determined from the known values of I and the radius a and mass M of the Earth. The present chapter is concerned with some of the early models and related theory, more particularly with results which have a bearing on later developments.

6.1 Earth models

6.1.1 Remarks on scientific inference

Scientific descriptions of the natural world – including theories and so-called 'laws' – are in terms of model representations of observational data. The representations have the character of 'mathematical models' if the term mathematical be widened to include discussions – verbal or symbolic, elegant or jargonistic – where some notice is taken of the rules of deductive logic. The more elegant models consist, in essence, of mathematical statements or equations, or sets of numerical tables. Newton's law of gravitation and Hooke's law are examples of elegant mathematical model representations. Inelegant models can, however, serve important purposes, especially at the crude beginning stages of a branch of science; in some contexts there is little point in making the equations of the model

[Refs. on p. 86]

explicit — indeed premature resort to formal mathematics can sometimes be a form of affectation.

On a given body of observational evidence, it is usually the case that some representations are preferable to others. Additional evidence may change the order of preference — the order of preference is a function of the evidence brought to bear. But however large the total quantity of observational evidence may be, it consists at bottom of a finite number of items (measurements, etc.) and there always remains an infinity of mathematical models compatible with it. Guiding principles in selecting an order of preference are that a model should (i) be compatible with the data within the estimated observational uncertainties; (ii) be as simple as possible; (iii) have properties which can lead to useful testing against new observations. The determination of an order of preference is complicated by degrees of imprecision in all observations and by influences partly or wholly unknown. Considerations of probability are brought to bear, explicitly or implicitly, in the determination.

The model equations, when made explicit, normally contain coefficients or parameters, n say in number, whose values are derived from the observational data. The requirement (ii) is largely met by preferring models in which n is judged to be near the minimum statistically required to fit the data.

In the early stages of an investigation, the data are often meagre. The requirement (i) can then usually be met by taking models in which n is quite small. As the investigation progresses, new data may require n to be increased, statistical judgment being again the basic deciding factor. A gifted investigator (for example, Newton with his inverse-square law) may sometimes succeed in drastically reducing n by producing a model much simpler than earlier models and compatible with all the data. (Sometimes 'simple' takes on a sophisticated meaning, e.g. in Relativity Theory.)

6.1.2 Inferences on the Earth's interior

The nether regions of the Earth are inaccessible in the ordinary sense. Before the time of Newton, when evidence about them was nearly totally lacking, it was not necessarily unreasonable to describe the Earth in terms of models involving say a Hell, or a subterranean monster shaking itself to cause earthquakes. The subsequent growth of evidence has lowered the plausibility of such models.

[Refs. on p. 86]

The inaccessibility of the nether regions still remains, however, and with it a tendency to deprecate conclusions about the interior of the Earth as 'inductive'. This attitude, which is implicit in many expositions of the so-called 'exact' sciences, is unsound. It implies that conclusions based on so-called direct visual evidence are not inductive — are presumably in a different category from all other scientific conclusions — and thereby ignores the model character of all scientific inference.

Inferences about the interior of the Earth, so far from being all inferior to those in the 'exact' sciences, range from those which are indeed flimsily based to inferences that are now as well established as commonly accepted results in standard physics. Differences in practice in the processes of drawing inferences about the natural world (or, for that matter, any field where evidence is intelligently discussed) lie, not in any distinction between inferences which are inductive and those which are not, but possibly in the degree to which the principles of scientific inference are consciously invoked in sorting out the better established conclusions from the worse. The principles themselves are quite general.

Let p denote a model representation of a body of data, and q a property deduced from p. It is an error to replace 'if p, then q' by 'because p, therefore q'. The effect is to identify a model with reality, ignoring assumptions underlying p, and assuming a uniqueness which model representations of properties of the natural world do not possess. The error is rather widespread, partly because in the 'exact' sciences the immediate effect often appears to be fairly harmless.

In the Earth context, the error is seldom harmless. Explicit discussion in terms of models and appreciation of the strengths as well as the limitations of models is constantly required. Failure to recognize this has been a source of unnecessary confusion in solid-Earth geophysics, and has sometimes given rise to cults of 'belief' and 'disbelief', with attendant 'conversions', especially in the more spectacular, speculative theories. (Even while such cults violate the canons of sound thinking, they not infrequently, however, inspire useful data gathering and other activity. Science, a human phenomenon, does not proceed by wholly rational steps.)

In this book there will be much reference to models. The first useful models on density inside the Earth were constructed after a, M, and I had been estimated to fair precision. These estimates

[Refs. on p. 86]

6.1] PHILOSOPHY OF EARTH MODELS

provided in effect two equations of condition. In accordance with the simplicity requirement, the most useful early models were therefore those with just two parameters. In fact two-parameter models persisted well into the twentieth century.

Mathematical convenience, always a guide in constructing models sometimes dictates procedures. This was strongly the case in eighteenth and nineteenth century geophysics, when new types of observational evidence were slow to emerge. The following case is of some general interest. Suppose that, in the course of constructing a model, a differential equation E containing the parameter h arises. Suppose that E is simply solvable only when $h = a$, and suppose further that the physical evidence on hand does not discriminate between a and any other value, b say, of h. Then mathematical convenience dictates the selection of a for the value of h in preference to b and carries with it a better chance of developing further tests of the model. The principles of scientific interferences are breached, however, if the model with $h = a$ is, in the absence of further evidence, considered to be nearer physical 'reality' than one with $h = b$. Confusion arises when mathematical beauty so dazzles the beholder that he identifies a model with physical reality (see e.g. some expositions of Relativity and Quantum Theory).

As the evidence grows, mathematical beauty and simplicity often have to give way to other considerations (though many still cling to the simple faith that the natural world is ultimately describable in simple terms). In the Earth context, it has transpired that, while the earlier models of density variation, etc., had simple mathematical forms, some recent models are expressed in terms of detailed numerical tables; the growing evidence since around the 1920s has demanded a big increase in the effective number of parameters required. The descriptions in the present chapter, being concerned with the earlier developments, are mainly couched in mathematical form. The details turn somewhat more numerical after Chapter 9.

6.1.3 *Provisional assumptions for Earth models*

Except where reference is made to deviations, Earth models will now be taken as spherically symmetrical and the hydrostatic relation, (6.13) below (see also §5.1), will continue to be assumed. Thus the density ρ, incompressibility k (§7.5.1), pressure p, gravitational intensity g, and related properties will be treated as functions

[*Refs. on p. 86*]

EARLY MODELS OF DENSITY VARIATION

of the distance r from the centre O or the depth z below the surface.

The subscripts a and 0 will denote values of variables at the surface and centre of the Earth, respectively.

6.2 Clairaut's equation and the density problem

Following Clairaut's derivation in 1743 of the equation (5.26) on the variation of the ellipticity ϵ with r, attempts were made to use the theory of the ellipticity, in conjunction with observational evidence on the constants h and H (§§5.3.1 and 5.5.5), to derive information on the variation of ρ with r. As shown in §5.6.2, the attempts led to a fairly reliable estimate of the moment of inertia coefficient y_a. But the attempts failed to give any additional information about ρ, and models with appreciably different density variations were found to give substantially the same value of ϵ_a. (This is illustrated in §6.7.1.3, where a model with $\rho(r)$ very different from that for the actual Earth is found to yield a value of ϵ_a correct within one part in 1000.)

Radau's approximation (§5.4.4) and the theory on moment of inertia later showed why the attempts failed. When h and H are given, η_a and ϵ_a are determined to good accuracy by the equations in §5.6. These equations also determine y_a but involve no further restriction on the variation of ρ.

The vigorous prosecution of the analytical theory of the ellipticity for more than a century after Clairaut laid, however, some important foundations bearing on density. In particular, the properties of a number of useful model density variations were examined, starting with a celebrated density law of Legendre and Laplace. Theory behind this law (see §6.4) has interesting connections with the modern approach to the problem of the Earth's density variation, and is given in the next section.

Throughout the chapter, use will be made of (5.21) and (5.22) which we rewrite as

$$\sigma r^3 = 3 \int_0^r \rho q^2 dq; \qquad \frac{d\sigma}{dr} = -\frac{3}{r}(\sigma - \rho). \qquad (6.1)$$

6.3 The Legendre-Laplace density law

6.3.1 *Legendre's derivation*

Legendre (1793) was interested in the problem of solving Clairaut's

[Refs. on p. 86]

equation (§ 5.4.2), which we shall take in the form

$$\frac{d^2\epsilon}{dr^2} + \frac{6\rho}{\sigma r}\frac{d\epsilon}{dr} - \frac{6\epsilon}{r^2} + \frac{6\rho\epsilon}{\sigma r^2} = 0, \qquad (6.2)$$

and sought a trial density law which would make (6.2) tractable.

In terms of the variable γ, where $\gamma = \epsilon\sigma r^3$, (6.2) becomes, on using (6.1),

$$\frac{d^2\gamma}{dr^2} - \left(\frac{6}{r^2} + \frac{3}{\sigma r}\frac{d\rho}{dr}\right)\gamma = 0. \qquad (6.3)$$

Legendre took

$$3(\sigma r)^{-1} d\rho/dr = -A^2, \qquad (6.4)$$

where A is a constant, as a trial density law, thereby reducing (6.3) to the comparatively simple form

$$\frac{d^2\gamma}{dr^2} - \left(\frac{6}{r^2} - A^2\right)\gamma = 0. \qquad (6.5)$$

He was able to solve both (6.4) and (6.5) in finite terms and thence derive a pair of model relations for ρ and ϵ in terms of r. [For the solution of (6.5), see Tisserand (1891), § 116.]

The differential equation (6.4) for the density variation becomes, on using (6.1),

$$\frac{d}{dr}\left(r^2 \frac{d\rho}{dr}\right) = -A^2 r^2 \rho, \qquad (6.6)$$

the general solution of which [as can be checked on substituting into (6.6)] is

$$\rho = \rho_0 (Ar)^{-1} \sin(Ar + \alpha),$$

where ρ_0 and α are integration constants. Legendre took $\alpha = 0$ in order to avoid having ρ infinite at $r = 0$, and so arrived at the model density law

$$\rho = \rho_0 (Ar)^{-1} \sin Ar, \qquad (6.7)$$

which contains the two parameters ρ_0 and A.

6.3.2 Laplace's derivation

Laplace (1825) ostensibly sought to arrive at a density law through plausible physical assumptions.

He assumed the hydrostatic relation $dp = -g\rho\, dr$, taking $g = Gm/r^2$

[Refs. on p. 86]

(§4.2.3) and $m = 4\pi r^3 \sigma/3$. By (6.1), these relations give

$$\frac{dp}{dr} = -4\pi G \rho r^{-2} \int_0^r \rho q^2 \, dq. \tag{6.8}$$

Laplace attributed density changes to compressibility, stated that "it is natural to think that liquids resist compression by as much more as they are compressed already", and then took

$$dp = C\rho \, d\rho, \tag{6.9}$$

where C is constant.

Eliminating dp from (6.8) and (6.9) gives

$$r^2 C \frac{d\rho}{dr} = -4\pi G \int_0^r \rho q^2 \, dq, \tag{6.10}$$

which happens to reproduce the form (6.6) and so yield the density law (6.7) again.

6.3.3 *Immediate comments*

From the point of view of scientific method, Legendre's procedure was possibly superior to Laplace's through being explicitly a model approach with no implication that (6.7) is more than one of many possible models compatible with the evidence. [Legendre looked upon (6.7) as an 'example'.]

Laplace's procedure had the defect that, in aspiring to something closer to reality (in itself commendable), it invoked in (6.9) a relation that cannot be sustained. The defect may, however, be only apparent since it has been suggested that Laplace's real reason for setting (6.9) down may have been his perception that the step would make Clairaut's equation integrable. However this may be, it is noteworthy that Laplace's approach as set down in §6.3.2 has in it the seeds of the modern approach, where key steps continue to be the use of (6.8) or its equivalent, and the search for a reliable connection between dp and $d\rho$.

The underlying assumptions in (6.7) will be examined in a more general way in §6.4, where some equations needed prominently in later developments are introduced.

[Refs. on p. 86]

6.4 Background theory in density determination

The relations
$$g = Gm/r^2, \quad (6.11)$$
$$dm = 4\pi r^2 \rho\, dr, \quad (6.12)$$

which assume only the inverse-square law and spherical symmetry, are the most firmly established in the whole theory of the Earth's density variation. The hydrostatic relation
$$dp/dz = -dp/dr = g\rho \quad (6.13)$$
is also reliable within fairly closely assessable limits (see §§ 5.6.2 and 5.7).

These three relations yield Laplace's relation (6.8). We now proceed more generally than Laplace.

6.4.1 *Effect of compression*

For the present, we take model conditions which ignore deviations from an adiabatic temperature gradient in the Earth, and also ignore changes of chemical composition and of phase. Thus we take ρ as depending only on p, in which case (see **B**, §2.3.3, and §7.5.1 below)
$$d\rho/dp = \rho/k. \quad (6.14)$$

The equations (6.11)–(6.14) yield
$$d\rho/dr = -Gm\rho/(r^2\phi), \quad (6.15)$$
where
$$\phi = k/\rho, \quad (6.16)$$
as an expression for the density gradient in the stated conditions. From (6.12), (6.15), and (6.16), we deduce
$$\frac{d}{dr}\left\{r^2 k \rho^{-2} \frac{d\rho}{dr}\right\} = -4\pi G r^2 \rho. \quad (6.17)$$

In the above stated conditions, a functional relation between any two of ρ, k, and p provides an equation of state for the material concerned. Given such a relation, k can, through (6.14) if necessary, be connected with ρ. Then (6.17) provides a second-order differential equation for the variation of ρ with r.

Comparison of (6.10) and (6.17) shows that Laplace in effect

[Refs. on p. 86]

assumed $k = C\rho^2$ as equation of state. This equation also comes immediately from (6.9) and (6.14).

6.4.2 Emden equation on density

The model equation of state

$$dk/d\rho = n, \tag{6.18}$$

where n is constant, has proved to be serviceable in a number of discussions on the Earth's interior (see e.g. §11.2.7), and leads to the following results.

Eliminating p between (6.14) and (6.18) gives $k = C\rho^n$, where C is constant. Substituting this expression for k into (6.17) gives

$$\frac{d}{dr}\left\{r^2 \rho^{n-2} \frac{d\rho}{dr}\right\} = -A^2 r^2 \rho, \tag{6.19}$$

where $A = \sqrt{(4\pi G/C)}$ = constant. Comparison of (6.19) with (6.6) and (6.10) shows again that the Legendre-Laplace model is the particular case $n = 2$.

On putting $\rho^{n-1} = (n-1)\theta$, so that $\rho^{n-2}d\rho = d\theta$, and putting $n - 1 = \nu^{-1}$, (6.19) becomes

$$d(r^2 d\theta/dr)/dr = -A^2 \nu^{-\nu} r^2 \theta^\nu.$$

Putting $r = A^{-1}\nu^{\frac{1}{2}\nu}\xi$ then gives

$$\frac{d^2\theta}{d\xi^2} + \frac{2}{\xi}\frac{d\theta}{d\xi} + \theta^\nu = 0. \tag{6.20}$$

The equations (6.19) and (6.20) are forms of Emden's equation, solutions of which have been much investigated in astrophysics and geophysics. (See also §11.5.2.)

[The relation (6.20) was first given by Ritter. The constant ν, called the polytropic index, entered astrophysics through the Lane-Ritter assumptions $p \propto T^{\nu+1}$ and $\rho \propto T^\nu$, which correspond to a perfect gas in convective equilibrium. These assumptions yield $p^\nu \propto \rho^{\nu+1}$, which (apart from an integration constant) is also deducible from equations in the above geophysical approach.]

6.4.3 Solutions of Emden equations

The Emden equations (6.19) and (6.20) admit of simple solution

[Refs. on p. 86]

only in the cases $\nu = 5$ and $\nu = 1$, i.e. $n = 1 \cdot 2$ and $n = 2$. [When $\nu = 0$, (6.20) has another simple solution, but the connection between (6.19) and (6.20) then involves a singularity. For other values of ν and n, the equations can be solved through heavier analysis (see e.g. Fowler, 1930) or resort to electronic computers (Lyttleton, 1965).]

When $\nu = 5$, substitution shows that (6.20) is satisfied by $\theta = c(1 + \frac{1}{3}c^4\xi^2)^{-\frac{1}{2}}$, where c is constant. The corresponding solution of (6.19), with $n = 1 \cdot 2$, is readily found to be

$$\rho = \rho_0(1 + B^2 r^2)^{-\frac{5}{2}}, \tag{6.21}$$

where $B^2 = \rho_0^{4/5} A^2/15$. The solution (6.21) satisfies immediate physical requirements in giving ρ bounded and $d\rho/dr \leq 0$ for all r.

When $\nu = 1$, substitution shows that $\theta = c\xi^{-1}\sin\xi$ satisfies (6.20), the corresponding solution of (6.19), with $n = 2$, being

$$\rho = \rho_0(Ar)^{-1}\sin Ar. \tag{6.7}$$

This gives ρ bounded and $d\rho/dr \leq 0$ for all r in the range $0 \leq Ar \leq \pi/2$.

6.4.4 Comments

The models (6.7) and (6.21) both accord with the requirements of scientific method outlined in §§6.1.1 and 6.1.2. In the absence of much evidence on dk/dp, they were to be preferred to other solutions of (6.19) and (6.20) on grounds of mathematical convenience. The presence of just two parameters $(\rho_0, A; \rho_0, B)$ was appropriate to fitting the very limited observational data available at the time on the Earth's density variation. Through their simplicity, (6.7) and (6.21) had, moreover, the potentiality of leading readily to further observational test.

It is interesting that Legendre and Laplace should both have stumbled, in apparently different ways, upon one of the two models (6.7) and (6.21). It is also interesting that, in stumbling upon (6.7) rather than (6.21), they happened to pick on the model with the greater value of dk/dp, though the value (2 units) is still too low on modern estimates (see Chapter 11).

It needs to be stressed that, even with n not specified, (6.19) is still a highly restricted model relation. Underlying assumptions include (6.14) and (6.18), the limitations of which have not always been adequately appreciated even by some modern writers.

[Refs. on p. 86]

6.5 Other early model density laws

6.5.1 Roche's law

Roche (1848) followed Laplace in using (6.8), and then took in place of (6.9)

$$d p = (C\rho + C'\rho^2)d\rho, \tag{6.22}$$

thus obtaining

$$\frac{d}{dr}\left\{r^2(C + C'\rho)\frac{d\rho}{dr}\right\} = -4\pi G r^2 \rho. \tag{6.23}$$

Without seeking a general solution, he showed that (6.23) is satisfied by

$$\rho = \rho_0(1 - Kr^2), \tag{6.24}$$

where $\rho_0 = 3C/2C'$ and $K = 4\pi G/15C$.

Roche's law (6.24) provides another useful two-parameter density law, and is in fact still used for model purposes over limited ranges of depth in the Earth. But the chief importance of (6.24) rests, not on Roche's derivation through (6.22) and (6.23), but on the fact that (6.24) gives the mathematically simplest two-parameter model for which $d\rho/dr$ is zero at $r = 0$.

6.5.2 Laws of Lipschitz and Lévy

In wrestling with the mathematics of the ellipticity theory, Lipschitz proposed the law $\rho = \rho_0(1 - Kr^\lambda)$, which is a particular case of a more complex law proposed by Lévy. Such laws hold interest today only because of the mathematical detail available on their implications. Since they contain more than two parameters, they were unnecessarily elaborate in relation to the observational data available at the time.

6.5.3 Darwin's law

Darwin (1884) proposed the two-parameter model $\rho = Ar^{-n}$, where $0 < n < 3$. (With $n < 0$, $d\rho/dz$ would be negative; $n = 0$ is the case of uniform density; with $n = 3$, m would be infinite; with $n > 3$, m would be negative.) Let σ denote the mean density of the matter inside the sphere of radius r. By (6.1) and (6.11), Darwin's model has the algebraically simple properties

[Refs. on p. 86]

$$\sigma = \rho(1 - \tfrac{1}{3}n)^{-1}, \quad (6.25)$$

$$m = 4\pi r^3 \rho (3-n)^{-1}, \quad (6.26)$$

$$g = 4\pi G r \rho (3-n)^{-1}. \quad (6.27)$$

By (6.25), ρ/σ is constant throughout the model.

When the hydrostatic relation (6.13) is assumed, the model gives p also as a simple algebraic function of r, namely,

$$p = \frac{4\pi G A^2}{2(n-1)(3-n)}\left\{\frac{1}{r^{2n-2}} - \frac{1}{a^{2n-2}}\right\} \quad (n \neq 1), \quad (6.28a)$$

$$p = 2\pi G A^2 \ln(a/r) \quad (n = 1). \quad (6.28b)$$

The model has some other analytical advantages as well as some remarkable properties connected with the moment of inertia coefficient y and the internal variation of the ellipticities ϵ of surfaces of constant density (see §§ 6.6.3 and 6.7.1.3).

Simple expressions for k and dk/dp can also be derived. Writing $k = \eta' \rho dp/d\rho$ [a generalization of (6.14) which allows for departure from chemical homogeneity (see §11.5.2) – η' corresponds to the parameter η of §11.5.2], and using (6.28a), we obtain

$$k = \frac{4\pi \eta' G A^2}{n(3-n)r^{2n-2}}. \quad (6.29)$$

Treating η' as constant for the sake of simplicity, we then obtain

$$dk/dp = 2\eta'(1 - n^{-1}). \quad (6.30)$$

For $1 < n < 3$, (6.30) gives $dk/dp < 4\eta'/3$. (For $0 < n < 1$, (6.30) would give η' negative for positive dk/dp.) Since it is now known that $dk/dp = O(3)$ inside much of the Earth's deeper interior, the model would thus require η' when positive to be greater than 2, and so would involve significant departures from chemical homogeneity. (See also §6.6.3.)

The model has the formal limitation that it gives ρ (also p and k if $n > 1$) tending to infinity as $r \to 0$. In spite of this limitation, the analytical properties make the law $\rho = A r^{-n}$ particularly valuable in throwing light on certain mathematical and numerical aspects of the Earth's density variation, especially in connection with the coefficient y and the ellipticity ϵ.

[Refs. on p. 86]

6.5.4 Early models with core

Thomson and Tait (1879) contemplated an Earth model consisting of a core of uniform density ρ_1 and radius a_1 surrounded by a shell of uniform density ρ_2 (Fig. 6.1).

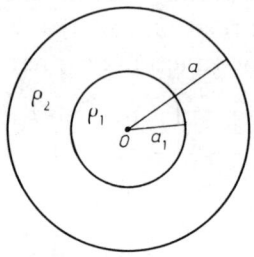

Fig. 6.1

Radau (1885) and Wiechert (1897) worked out numerical details for such models, taking ρ_2 as known (equal to the average density of rocks found near the Earth's surface) and a_1 and ρ_1 as parameters. Properties of these models, referred to as the Wiechert type, are considered in §§6.6.4 and 6.7.1.4.

6.6 Numerical results for early models

Any acceptable density distribution for the Earth must satisfy the known values of the radius a, mean density σ_a, and moment of inertia coefficient y_a. With a, σ_a, and y_a given, the relations (5.37) and (5.38), namely

$$a^3 \sigma_a = 3 \int_0^a \rho q^2 \, dq, \qquad (6.31)$$

$$y_a a^5 \sigma_a = 2 \int_0^a \rho q^4 \, dq, \qquad (6.32)$$

enable the parameters in a two-parameter density law to be uniquely determined.

6.6.1 The Legendre-Laplace model

For the Legendre-Laplace model (6.7), the equations (6.31) and (6.32) give

$$\sigma_a = 3(Aa)^{-3} \rho_0 (\sin Aa - Aa \cos Aa), \qquad (6.33)$$

$$y_a \sigma_a = 2(Aa)^{-5} \rho_0 \{(3A^2 a^2 - 6) \sin Aa - (A^3 a^3 - 6Aa) \cos Aa\}, \tag{6.34}$$

which are readily solvable for ρ_0 and Aa by assuming trial values of Aa. Taking $\sigma_a = 5 \cdot 52 \text{ g/cm}^3$ and $y_a = 0 \cdot 3308$ gives

$$Aa = 2 \cdot 533, \quad \rho_a = 2 \cdot 55 \text{ g/cm}^3, \quad \rho_0 = 11 \cdot 29 \text{ g/cm}^3. \tag{6.35}$$

6.6.2 The Roche model

Roche's model (6.24) gives

$$\sigma_a = \rho_0(1 - 0 \cdot 6 K a^2), \tag{6.36}$$

$$y_a \sigma_a = \rho_0(0 \cdot 4 - 2 K a^2 / 7). \tag{6.37}$$

With $\sigma_a = 5 \cdot 52 \text{ g/cm}^3$ and $y_a = 0 \cdot 3308$, (6.36) and (6.37) give

$$Ka^2 = 0 \cdot 793, \quad \rho_a = 2 \cdot 18 \text{ g/cm}^3, \quad \rho_0 = 10 \cdot 53 \text{ g/cm}^3. \tag{6.38}$$

6.6.3 The Darwin model

Let y be the moment of inertia coefficient for the matter inside any internal sphere of radius r and centre O. Then (5.36) applies. Using (6.25), we then find that the Darwin model gives

$$y = 2(3 - n)/\{3(5 - n)\}, \tag{6.39}$$

so that y is the same for all r. (It is easy to show that the Darwin model is the unique model which has this property.)

Putting $y_a = 0 \cdot 3308$ in (6.39) gives $n = 1 \cdot 030$; then taking $\sigma_a = 5 \cdot 52 \text{ g/cm}^3$ gives $\rho_a = 3 \cdot 62 \text{ g/cm}^3$ by (6.25).

By (6.30), these numerical data also give $dk/dp = 0 \cdot 06 \eta'$ and so would require η' to be impracticably high to meet current evidence on dk/dp in the Earth (Chapter 11). Thus the model is not useful for inferences involving k. (A postscript in Darwin's paper refers to a warning by Kelvin which, on slightly different grounds, led Darwin to discount the reliability of the model in respect of k.)

Some further numerical aspects of the Darwin model are considered in §6.7.1.3.

6.6.4 Models with uniform shell and core

For models specified as in §6.5.4, (6.31) and (6.32) give

$$\sigma_a = \rho_2\{1 + \mu(a_1/a)^3\}, \qquad (6.40)$$

$$y_a\sigma_a = 0\cdot 4\rho_2\{1 + \mu(a_1/a)^5\}, \qquad (6.41)$$

where $\mu = (\rho_1 - \rho_2)/\rho_2$.

Radau took the shell density ρ_2 as $2\cdot 7\,\text{g/cm}^3$ and obtained $7\cdot 47\,\text{g/cm}^3$ and $0\cdot 844a$ for the core density ρ_1 and core radius a_1, respectively. These results were incidental to Radau's examination (see §6.7.2.2) of the minimum central density of the Earth.

Wiechert took $\rho_2 = 3\cdot 2\,\text{g/cm}^3$ and obtained $\rho_1 = 8\cdot 21\,\text{g/cm}^3$ and $a_1 = 0\cdot 779a$. Jeffreys (1926), taking account of seismological evidence, amended a_1 to $0\cdot 545a$, used more recent estimates of σ_a and y_a, and gave the revised values $\rho_1 = 12\cdot 04$ and $\rho_2 = 4\cdot 27\,\text{g/cm}^3$ (later corrected to $12\cdot 32, 4\cdot 22\,\text{g/cm}^3$).

6.7 Some further nineteenth century results

Nineteenth century work on the Earth's interior was characterized by intense application of mathematical analysis in an endeavour to squeeze the maximum return from limited observational data. Some of the formal theorems derived are still important. Other results, for example, attempts to set limits to the central density ρ_0, though now by-passed, are of historical interest. The following sub-sections touch on a few selected details.

6.7.1 *Range of variation of Radau's parameter* η

The utility of Radau's approximation (§5.4.4) rests on the assumption that η, defined by

$$\eta = (r/\epsilon)\mathrm{d}\epsilon/\mathrm{d}r, \qquad (6.42)$$

does not depart significantly outside the range $0 \leqslant \eta \leqslant 0\cdot 6$ anywhere inside the Earth. Much attention was therefore devoted to investigating limits to η inside the Earth.

It has already been seen (§§5.4.3, 5.6.2, and 5.7) that $\eta_0 = 0$ (for regular forms of density variation — but see also §6.7.1.3) and $\eta_a \approx 0\cdot 58$. (Nineteenth century studies generally gave values of η_a within $0\cdot 03$ units of $0\cdot 58$.) Also $\eta \geqslant 0$ for all r (§5.4.3). Thus the condition for Radau's approximation to apply is that η must throughout the Earth lie (very nearly) between the central and surface values. A sufficient condition for this is that $\mathrm{d}\eta/\mathrm{d}r \geqslant 0$ for all r.

[*Refs. on p. 86*]

6.7.1.1 Upper bound to η

From (6.42), it can be deduced that

$$r^4 \epsilon^{-1} d(\epsilon r^{-3})/dr = \eta - 3. \quad (6.43)$$

Since $d(\epsilon r^{-3})/dr \leq 0$ (§5.4.3), it follows that $\eta \leq 3$ for all r, a result due to Poincaré (1888).

Callandreau (1885) derived a reduced upper bound to η. Divide (5.24) through by r^5 and differentiate. Then, using (5.22),

$$\sigma(d\epsilon/dr) - 3\epsilon\sigma/r + 3r^{-6} \int_0^r \{\rho d(\epsilon r^5)/dr\} dr = 0. \quad (6.44)$$

Integrating by parts in the third term leads to

$$\eta = 3 - 3\sigma^{-1} \{\rho + (\epsilon r^5)^{-1} \int_\rho^{\rho_0} \epsilon r^5 d\rho\}, \quad (6.45)$$

which gives Callandreau's result

$$\eta \leq 3(1 - \rho/\sigma) \leq 3(1 - \rho_a/\rho_0). \quad (6.46)$$

6.7.1.2 Models with continuous density variation

Most of the simpler representations of the Earth's density distribution had ρ continuous, $d\rho/dr \leq 0$, and $d^2\rho/dr^2 \leq 0$, for all r. This set of conditions applies in particular to the Legendre-Laplace and Roche models. Callandreau (1885) showed that when these conditions hold, η increases monotonely with r, so that the requirement $0 \leq \eta \leq \eta_a$ is satisfied. Callandreau's derivation is given by Tisserand (1891, §112). The following alternative derivation is of some general interest.

We first rewrite Clairaut's equation in the form

$$r d\eta/dr = 6(1 - \tau)(1 + \eta) - (\eta^2 + 5\eta), \quad (6.47)$$

where $\tau = \rho/\sigma$. The form (6.47) is derived using (5.26) and (6.1). The writer has found it very useful for a variety of purposes. Differentiating $\tau = \rho/\sigma$ and using (6.1) gives

$$\tau^{-1} d\tau/dr = \rho^{-1} d\rho/dr - \sigma^{-1} d\sigma/dr$$
$$= \rho^{-1} d\rho/dr + 3r^{-1}(1 - \tau).$$

Differentiating again gives, for any r at which $d\tau/dr = 0$,

$$\tau^{-1}d^2\tau/dr^2 = \rho^{-1}d^2\rho/dr^2 - \rho^{-2}(d\rho/dr)^2 - 3r^{-2}(1-\tau). \quad (6.48)$$

Given that $d\rho/dr \leqslant 0$, then, for all r, we have $\tau \leqslant \tau_0 = 1$. Given further that $d^2\rho/dr^2 \leqslant 0$, equation (6.48) would then require that $d^2\tau/dr^2 \leqslant 0$. Thus τ cannot pass through a minimum with respect to r. Also, $\tau_a \leqslant 1$. Hence $d\tau/dr \leqslant 0$ for all r.

Differentiating (6.47) gives, for any r at which $d\eta/dr = 0$,

$$rd^2\eta/dr^2 = -6(1+\eta)d\tau/dr \geqslant 0 \quad (6.49)$$

since $d\tau/dr \leqslant 0$. Hence η cannot pass through a maximum. Since $\eta_0 \leqslant \eta_a$, it follows that $d\eta/dr \geqslant 0$ for all r.

6.7.1.3 *The Darwin model*

The Darwin law $\rho = Ar^{-n}$ gives $d^2\rho/dr^2 > 0$ (since $n > 0$) and so does not meet all of Callandreau's conditions. But the law does meet the requirement for Radau's approximation and is particularly interesting as exhibiting an extreme case which does so.

Possibly the most interesting property of the law (Darwin does not appear to have noticed this property) is that it is the unique law which gives η constant throughout the model. This is seen on taking Clairaut's equation in the form (6.47) and noting that any model which has η constant must have τ constant, from which Darwin's law follows. (In order to show the converse, that the law entails $\eta = $ constant, it is necessary to invoke the physical requirement that ϵ must not be infinite at $r = 0$.) By (6.42), another simple property of Darwin's model is thus $\epsilon = Cr^\eta$, where η is constant.

The constant value of η in the model is not zero, since by (6.47) this would entail $\tau = 1$ and therefore $n = 0$ (which has been excluded). Thus the singularity in the model at $r = 0$ is associated with a deviation from the result $\eta = 0$ at $r = 0$ which holds for regular models (§5.4.3). However, this deviation is accompanied by a property acutely relevant to the theory of the Radau approximation (§5.4.4). For, among the class of models for which η_a is the maximum permissible for this purpose, the Darwin model, through having η constant, has the maximum permissible η at all depths. Thus the model provides an extreme case for applicability of the Radau theory and, because of the algebraic simplicity of its properties, gives added insight into the Radau theory.

[*Refs. on p. 86*]

A related interesting point, which also does not seem to have been noticed in Darwin's time, is that the main numerical properties of the Darwin model can be readily derived without any recourse to the Radau theory.

For example, given observational values of h and H (§§5.3.1 and 5.5.5), values of η_a, ϵ_a, y_a, and n can be deduced from the Darwin law $\rho = Ar^{-n}$ as follows. (For simplicity, the derivation will here be limited to the first-order ellipticity theory.) The equations (5.15), (5.39), (5.46), and (5.47) (none of which depends on the Radau theory) yield

$$3Hy_a = 2\epsilon_a - h. \qquad (6.50)$$

Since η is constant in the Darwin model, Clairaut's equation in the form (6.47) gives

$$2n(1 + \eta) = 5\eta + \eta^2. \qquad (6.51)$$

The equation (5.29), also independent of the Radau theory, is

$$2\eta_a = 5h/\epsilon_a - 4. \qquad (6.52)$$

The four equations (6.39), (6.50), (6.51), and (6.52) connect the six quantities $\eta\,(= \eta_a)$, ϵ_a, $y\,(= y_a)$, n, h, and H.

With $h = 0.003450$ and $H = 0.003273$, the equations yield $n = 1.011$ and

$$y_a = 0.3324, \quad \eta_a = 0.570, \quad \epsilon_a = 0.003356. \qquad (6.53)$$

Taking $\sigma_a = 5.52 \text{ g/cm}^3$ then gives $\rho_a = 3.66 \text{ g/cm}^3$.

It is interesting to compare the results (6.53) with the results in §§5.6.1, 5.6.2, and 5.7. It is mildly astonishing that the results (6.53), based on the assumption of Darwin's law, and derived by comparatively simple algebra, happen to be closer to estimates of y_a and ϵ_a that were derived after 1963 with the help of artificial satellite data than were estimates regarded as the best prior to 1963. Since modern estimates of the Earth's density distribution depart substantially from Darwin's law for particular sizeable ranges of depth, the results (6.53) thus demonstrate strikingly that the theory of the Earth's ellipticity and moment of inertia cannot of itself enable fine inferences to be drawn about the density distribution.

6.7.1.4 Wiechert-type models

Throughout the core of a Wiechert-type model (§6.6.4), $\sigma = \rho$ and

(6.47) then gives $\eta = 0$ since $\eta_0 = 0$.

Inside the shell, (6.1) and (6.47) give

$$d\sigma/dr = -3r^{-1}(\sigma - \rho_2), \qquad (6.54)$$

$$r\, d\eta/dr = 6(1 - \rho_2/\sigma)(1 + \eta) - (\eta^2 + 5\eta). \qquad (6.55)$$

By (6.54), σ and hence $(1 - \rho_2/\sigma)$ are decreasing functions of r, while $(\eta^2 + 5\eta)/(\eta + 1)$ is an increasing function of η. Hence, by (6.55), a necessary and sufficient condition for η to be monotonic throughout the shell is

$$6(1 - \rho_2/\sigma_a) \geqslant \{(\eta^2 + 5\eta)/(\eta + 1)\}_a,$$

i.e. $\rho_2/\sigma_a \leqslant 0{\cdot}663$, taking $\eta_a = 0{\cdot}57$.

This condition is satisfied in the original Wiechert model, where $\rho_2/\sigma_a = 3{\cdot}2/5{\cdot}57 = 0{\cdot}575$, but not in the model as revised by Jeffreys, where $\rho_2/\sigma_a = 4{\cdot}22/5{\cdot}52 = 0{\cdot}764$. [Using (6.40) and (6.41), it is easy to show that, when $\eta_a = 0{\cdot}57$, the condition is satisfied or not according as $a_1/a \gtreqless \sqrt{(7{\cdot}42y_a - 1{\cdot}97)}$, the value of which is $0{\cdot}70$ when $y_a = 0{\cdot}3308$. (Cf. Jeffreys, 1924.) In the original and revised Wiechert models, $a_1/a = 0{\cdot}78$ and $0{\cdot}545$, respectively.]

6.7.1.5 *Summary*

The results in §§ 6.7.1.2, 6.7.1.3, and 6.7.1.4 show that, for a wide class of trial density distributions in the Earth, Radau's approximation is satisfactory. The one exception among the models considered is the revised Wiechert model, for which, however, Jeffreys pointed out that allowance for continuous increase of density with depth inside the shell and core would act in the direction of reducing η for values of r where it exceeded η_a.

Thus an outcome of nineteenth century investigations was that it became highly probable that $\eta < 0{\cdot}6$ throughout the Earth, but that the result could not be formally deduced on the then available evidence. This in fact remained the position until after 1930. Investigations proceeded on the assumption that Radau's approximation is reliable for the Earth, but confirmation had to await the detailed determination of the density distribution. Modern density determinations have given $d\tau/dr < 0$, and hence, by the argument in §6.7.1.2, $d\eta/dr > 0$, for all r.

[*Refs.* on p. 86]

6.7.2 Limits to central density

The following early attempts to set limits to the central density ρ_0 are of some historical interest.

6.7.2.1 Deduction from Clairaut's work

The equation (5.25) on the variation of ϵ in the Earth is

$$-2\sigma\epsilon - \sigma r d\epsilon/dr + 3\int_r^a \rho(d\epsilon/dr)dr + 5h\sigma_a/2 = 0. \qquad (6.56)$$

Putting, in particular, $r = 0$ in (6.56) gives

$$-2\rho_0\epsilon_0 + 3\int_{\epsilon_0}^{\epsilon_a} \rho d\epsilon + 5h\sigma_a/2 = 0. \qquad (6.57)$$

Since $d\epsilon/dr > 0$ for all r, (6.57) gives

$$\rho_0 > 5h\sigma_a/4\epsilon_0 > 5h\sigma_a/4\epsilon_a. \qquad (6.58)$$

Inserting numerical values on the right side of (6.58), Tisserand (1891) derived $\rho_0 > 7.07$ g/cm^3.

6.7.2.2 Results of Stieltjes and Radau

Stieltjes and Radau (see Tisserand, 1891) gave a slightly increased lower limit to ρ_0, assuming a value for the surface density ρ_a. Their argument is essentially as follows.

Integrating (6.31) and (6.32) by parts gives

$$\sigma_a = \rho_a + a^{-3}\int_{\rho_a}^{\rho_0} r^3 d\rho, \qquad (6.59)$$

$$2.5 y_a \sigma_a = \rho_a + a^{-5}\int_{\rho_a}^{\rho_0} r^5 d\rho. \qquad (6.60)$$

Also

$$\rho_0 = \rho_a + \int_{\rho_a}^{\rho_0} d\rho. \qquad (6.61)$$

The inequality

$$\left(\frac{\int_{\rho_a}^{\rho_0} r^3 d\rho}{\int_{\rho_a}^{\rho_0} d\rho}\right)^{\frac{1}{3}} \leqslant \left(\frac{\int_{\rho_a}^{\rho_0} r^5 d\rho}{\int_{\rho_a}^{\rho_0} d\rho}\right)^{\frac{1}{5}} \qquad (6.62)$$

then gives

$$\{(\sigma_a - \rho_a)/(\rho_0 - \rho_a)\}^{\frac{1}{3}} \leqslant \{(2{\cdot}5 y_a \sigma_a - \rho_a)/(\rho_0 - \rho_a)\}^{\frac{1}{5}},$$

whence

$$\rho_0 \geqslant \rho_a + (\sigma_a - \rho_a)^{\frac{5}{2}} (2{\cdot}5 y_a \sigma_a - \rho_a)^{-\frac{3}{2}}. \quad (6.63)$$

Tisserand (1891), using the then available values of σ_a and y_a, applied (6.63) to obtain $\rho_0 \geqslant 7{\cdot}3, 7{\cdot}4, 7{\cdot}6$ g/cm^3, corresponding to the trial values $\rho_a = 2{\cdot}0, 2{\cdot}5, 3{\cdot}0$ g/cm^3, respectively.

Using (6.40) and (6.41), it is easy to show that for Wiechert-type models (§§ 6.5.4 and 6.6.4), equation (6.63) applies with the equality sign. In a model of this type, Radau assumed $\rho_a = 2{\cdot}7$ g/cm^3 and derived $\rho_0 = 7{\cdot}47$ g/cm^3; by virtue of (6.63), this value would be a lower limit for the Earth.

The calculations made by Stieltjes gave, in addition, $\rho_a \geqslant 3{\cdot}1$ and $\rho_0 \leqslant 12{\cdot}94$ g/cm^3. But this was on the assumption that $d^2\rho/dr^2 \leqslant 0$ for all r.

6.7.3 Saigey's theorem on g

Since $m = (4\pi/3)r^3\sigma$, (6.11) gives

$$g = (4\pi/3)Gr\sigma, \quad (6.64)$$

and hence, using (6.1),

$$dg/dr = (4\pi/3)G(\sigma + r d\sigma/dr)$$

$$= 4\pi G(\rho - 2\sigma/3). \quad (6.65)$$

Taking $\sigma_a = 5{\cdot}52$ g/cm^3, (6.65) gives $(dg/dr)_a < 0$ unless ρ_a exceeds $3{\cdot}68$ g/cm^3. Assuming that the appropriate value to take for ρ_a is less than $3{\cdot}68$ g/cm^3, Saigey (see Tisserand, 1891) thus inferred that g rises to a maximum somewhere below the Earth's surface.

At the Earth's centre, (6.64) shows g to be zero and, since $\sigma_0 = \rho_0$, (6.65) gives $(dg/dr)_0 = (4\pi/3)G\rho_0 > 0$. [The conclusions on g_0 and $(dg/dr)_0$ do not hold in general with density laws like Darwin's which have a singularity at $r = 0$.]

Tisserand (1891) found that Roche's law gave g a maximum (equal to $1{\cdot}05 g_a$) near $r = 0{\cdot}85 a$, and sought to compare values of $(dg/dr)_a$ as given on Roche's law with observational results of Airy and von Sterneck (§ 2.3) in mines.

Since Saigey's theorem has been known for nearly a century, it is surprising to find books on dynamics which still assume g to diminish from the surface downwards. [In a model Earth of constant density, (6.64) gives $g \propto r$, so that g does of course diminish monotonely with depth in this highly unrealistic case.]

6.7.4 Early results on pressure

By (6.13) and (6.64),

$$dp = -(4\pi/3)G\sigma\rho r\, dr. \qquad (6.66)$$

For a model Earth of constant density, (6.66) yields for the central pressure p_0

$$p_0 = \tfrac{2}{3}\pi G a^2 \sigma_a^2. \qquad (6.67)$$

Substituting known values of G, a, and σ_a into (6.67) gives $p_0 \approx 1.7 \times 10^{11}\,\text{N/m}^2$, an early result which indicated the order of magnitude of the pressure in the Earth's deep interior.

With Roche's law $\rho = \rho_0(1 - Kr^2)$, we find

$$p_0 = \tfrac{2}{3}\pi G a^2 \rho_0^2 (1 - 0.8 Ka^2 + 0.2 K^2 a^4). \qquad (6.68)$$

Taking $\rho_0 = 10.53\,\text{g/cm}^3$ and $Ka^2 = 0.793$, as found in §6.6.2, then gives $p_0 \approx 3.1 \times 10^{11}\,\text{N/m}^2$.

These results may be compared with modern values of p_0 given in §§ 10.7.2, 12.7.2, and 16.4.

6.7.5 Estimates of incompressibility

Given a model density relation, it is possible, using (6.11)–(6.14), to infer values of the incompressibility k in the Earth's interior. Calculations made last century gave $k_a = 0.4$ and $0.2 \times 10^{11}\,\text{N/m}^2$ with the Legendre-Laplace and Roche models, respectively, and so gave a rough indication of the order of magnitude of k near the Earth's surface. These results, however, assumed (6.14) and therefore uniform chemical composition, and so were much less reliable than the early results on pressure.

6.8 Early evidence on the Earth's rigidity

As will be seen in following chapters, the determination of density

[Refs. on p. 86]

in the Earth is linked not only with the incompressibility but also with the rigidity μ. The first important calculation on the Earth's rigidity was made by Kelvin in 1863. In describing Kelvin's calculation, it is convenient to define certain dimensionless numbers h and k introduced by Love at a later stage. (These symbols h and k are not connected with the h and k used earlier in the chapter.)

6.8.1 *Love's numbers*

The material of the Earth is deformed not only by the Earth's gravitational field, but also by the attractions of extra-terrestrial bodies, predominantly the Sun and Moon, as well as by the centrifugal forces arising from the Earth's axial rotation. To a good first approximation, the total resulting 'disturbing' potential U can be treated as a spherical harmonic of the second degree, referred to axes fixed in the Earth. The deformation associated with U causes a contribution, U' say, to the overall gravitational potential at any point.

Love's number k is defined as the value of U'/U at the (displaced) Earth's surface. Love's number h is defined as the value of gu/U at the surface, where u is the radial displacement arising from the potential $U(1 + k)$, and g is undisturbed gravity. For a totally rigid Earth, h and k would be zero. Thus knowledge of the value of h and k can throw light on the internal rigidity. (Another dimensionless number l, introduced by Lambert and Shida, has also been used in discussing the deformation and relates to non-radial components of the displacement.) See also Munk and MacDonald (1960).

The motion of the Sun and Moon relative to the Earth causes U to change with diurnal and other periods. Thus an effect of U is to raise tides in the ocean and also (bodily tides) in the solid Earth. In treating the bodily tides, it is customary to use an 'equilibrium theory', i.e. to neglect acceleration terms in the equations of motion. The equilibrium theory is sufficiently accurate for bodily tides since the periods of the deforming forces are appreciably greater than the periods of the Earth's free elastic vibrations (which are of order one hour or less — see §14.5.3). The theory is not reliable for oceanic tides, the free ocean periods being greater than those for the solid Earth; but a formally defined 'equilibrium oceanic tide' is introduced for mathematical convenience.

With this understanding, h may be regarded as the ratio of the height of the bodily tide to the height U/g of the corresponding equilibrium oceanic tide.

Various methods (see Jeffreys, 1970, §§7.03−7.051) have been applied to determining h and k observationally, and yield $h \approx 0{\cdot}60$ and $k \approx 0{\cdot}30$.

6.8.2 Kelvin's calculation

In investigating the deformation of a homogeneous incompressible elastic sphere under body forces, Kelvin obtained a result equivalent to

$$15g\rho a/(19\mu + 2g\rho a) = 3h = 5k, \qquad (6.69)$$

where a, ρ, μ, and g denote the radius, density, rigidity, and surface gravity. Applying (6.69) to the Earth, and taking $a = 6{\cdot}4 \times 10^6$ m, $\rho = 5{\cdot}5$ g/cm^3, $g = 9{\cdot}80$ m/s^2 and $h = 0{\cdot}6$, gives $\mu = 1{\cdot}15 \times 10^{11}$ N/m^2. Taking $k = 0{\cdot}3$ instead of $h = 0{\cdot}6$ gives $\mu = 1{\cdot}45 \times 10^{11}$ N/m^2.

The rigidity of steel at atmospheric pressure is of the order of $0{\cdot}8 \times 10^{11}$ N/m^2, so that Kelvin's calculation implied that the Earth is on the whole substantially more rigid than steel. This was Kelvin's conclusion in 1863 obtained using observations of fortnightly tides to determine h. The conclusion came as a great surprise, being contrary to widespread contemporary opinion that most of the interior of the Earth is molten. But the conclusion has since been well substantiated.

Kelvin's investigation is particularly important to the problem of the Earth's density variation because it supplied the basis for a later inference (§10.2) that the rigidity in most of the Earth's central core is small or zero. This is a vital piece of evidence required in estimating densities in the core.

6.9 Early twentieth century Earth models

We conclude the chapter with an account of some attempts earlier this century to arrive at more detailed knowledge of the Earth's density distribution than in the models so far discussed. The symbol k will now revert to denoting the incompressibility.

6.9.1 Three-layer models

From Wiechert's time, practically all work on Earth models allowed

[Refs. on p. 86]

for a core and aimed at models closer to reality than Wiechert's.

In 1914, Klussmann constructed several models, each with three layers of constant densities and assigned thicknesses. Density values ranging from 3·0 to 3·6 g/cm^3 were postulated in an 'upper mantle', and the values of M and I were used to determine ρ in the two layers below. The yielded densities were between 7·2 and 5·4 g/cm^3 in the 'lower mantle', and between 8·3 and 9·6 g/cm^3 in the core. (See Gutenberg, 1959.)

In 1925, Haalck sought to allow for variation of density inside layers. He postulated a linear variation down to a depth of 1200 km and a different linear variation from there to the core boundary at 2900 km depth. He took ρ continuous outside the core and constant inside. Using values of M and I, he obtained a model in which ρ varies from 2·7 to 6·4 g/cm^3 through the two highest layers and is equal to 9·7 g/cm^3 throughout the core. In the meantime, however, Williamson and Adams had brought seismological evidence to bear on density gradients in the Earth.

6.9.2 Contributions of Williamson and Adams

As will be seen (§9.6.2), seismology supplies fairly reliable values of the function ϕ ($= k/\rho$) in terms of r. Williamson and Adams (1923) took the important step of substituting values of ϕ thus obtained into (6.15) and so [subject to the limitations of (6.15)] making direct estimates of density gradients inside the Earth.

Their procedures enabled them to show that compressibility could account for a total density increase of about 2 g/cm^3 between the surface and a depth of 3400 km, but that compressibility alone cannot bring about a sufficiently high density in the deeper interior to meet the known mass M of the Earth. Williamson and Adams thus supplied the first direct evidence that there is a substantial change of chemical composition (or phase change) in the Earth's deeper interior — that the average core material is in a different chemical or physical state from the average shell material.

In deriving this result, they took account of the values of M and I and certain other evidence to infer limits to the density ρ_b just below the crust (the 'crust' is defined in §10.2; see also §10.4.3). They inferred that $\rho_b > 3\cdot25$ g/cm^3 (otherwise the central density would be improbably high) and that $\rho_b < 3\cdot45$ g/cm^3 (otherwise $d\rho/dr > 0$ in some part of the Earth), and took $\rho_b \approx 3\cdot3$ g/cm^3 in their calculations. Although later seismological evidence has not supported certain of the assumptions they made in estimating ρ_b, it is of great interest that the preferred estimate of ρ_b has remained close to $3 \cdot 3 \text{ g/cm}^3$. Thus their use of (6.15) to draw their important conclusion on differences between the shell and core has been well substantiated.

In taking $\rho_b = 3 \cdot 3 \text{ g/cm}^3$, Williamson and Adams brought supporting evidence to bear from rock measurements in conjunction with seismological data on the outer layers of the Earth. They inferred the main constituent just below the crust to be ultrabasic rock, and proposed an olivine-like composition for the shell.

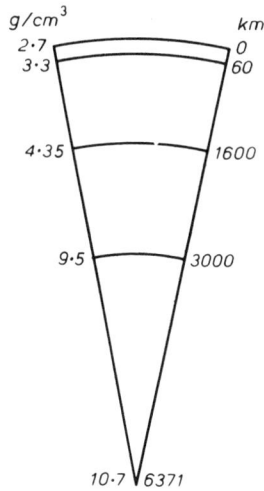

Fig. 6.2. *The Williamson-Adams density model*

Williamson and Adams constructed an Earth model consistent with the values of M and I, using the available seismological data and taking ρ as continuous. Their model (Fig. 6.2) consisted of an outermost layer (thickness $\Delta r = 60 \text{ km}$, ρ increasing from $2 \cdot 7$ to $3 \cdot 3 \text{ g/cm}^3$), a chemically homogeneous layer ($\Delta r = 1540 \text{ km}$, $3 \cdot 3 < \rho < 4 \cdot 35 \text{ g/cm}^3$), a layer of varying chemical composition ($\Delta r = 1400 \text{ km}$, $4 \cdot 35 < \rho < 9 \cdot 5 \text{ g/cm}^3$), and a central core ($\Delta r = 3370 \text{ km}$, $9 \cdot 5 < \rho \leqslant 10 \cdot 7 \text{ g/cm}^3$). The assumption of continuous density, however, resulted in ρ being overestimated by up to 4 g/cm^3 inside the region above the core.

[*Refs. on p. 86*]

EARLY MODELS OF DENSITY VARIATION

REFERENCES

Callandreau, O. (1885). *C.R. Acad. Sci.* (Paris), **100**, 1024.

Darwin, G.H. (1884). On the figure of equilibrium of a planet of heterogeneous density. *Proc. R. Soc.,* **36**, 158–166.

Fowler, R.H. (1930). The solutions of Emden's and similar differential equations. *Mon. Not. R. Astr. Soc.,* **91**, 63–91.

Gutenberg, B. (1959). *Physics of the Earth's Interior*, Academic Press, New York.

Haalck, H. (1925). Ueber die Lagerung der Massen in Innern der Erde und deren Elastizitatskonstanten auf Grund der neuesten Ergebnisse. *Z. angew. Geophys.,* **1**, 257–280.

Jeffreys, H. (1924). On the Radau transformation in the theory of the figure of the Earth. *Mon. Not. R. Astr. Soc., Geophys. Suppl.,* **1**, 121–124.

Jeffreys, H. (1926). The rigidity of the Earth's central core. *Mon. Not. R. Astr. Soc., Geophys. Suppl.,* **1**, 371–383.

Jeffreys, Sir. H. (1970). *The Earth* (5th Ed.), Cambridge University Press.

Klussmann, W. (1915). Ueber das Innere der Erde. *Gerl. Beitr. Geophys.,* **14**, 1–38.

Laplace, P.S. de (1825). *Mécanique Céleste,* Vol. V. Paris.

Legendre, A.-M. (1973). *Suite des Recherches sur la Figure des Planètes.* Paris.

Lyttleton, R.A. (1965). The phase-change hypothesis of the structure of the Earth. *Proc. R. Soc.,* A, **287**, 47–493.

Munk, W.H. and MacDonald, G.J.F. (1960). *The Rotation of the Earth.* Cambridge University Press.

Poincaré, H. (1888). *C.R. Acad. Sci.* (Paris), p.107.

Radau, R.R. (1885). See Tisserand, *loc. cit.*, p.227.

Roche, E. (1848). Mémoire sur la loi de densité à l'intérieur de la Terre. *Acad. Sci. Montpellier.*

Thomson, Sir W. (later Lord Kelvin) and Tait, P.G. (1879). *Treatise on Natural Philosophy,* Cambridge University Press.

Tisserand, F. (1891). *Traité de Mécanique Céleste,* Vol. II. Paris.

Wiechert, E. (1897). Ueber die Massenverteilung in Innern der Erde. *Nachr. Ges. Wiss. Göttingen, Math.-Phys. Klasse,* 221–243.

Williamson, E.D. and Adams, L.H. (1923). Density distribution in the Earth. *J. Wash. Acad. Sci.,* **13**, 413–428.

CHAPTER 7

Representation of elasticity in the Earth

Reference has already been made in Chapter 6 to the involvement of the incompressibility k and the rigidity μ in determining the Earth's density variation. Before developing further theory connecting k and μ with the density ρ, it is desirable, however, to look into questions of the interpretation of k and μ and certain other aspects of deformation in the Earth's interior, where stresses reach values beyond the ordinary laboratory range.

Attention has again to be given to scientific method, and a first task is to select suitable model equations to describe the observations of interest. A glance at the available samples of Earth materials reveals fantastic complexity, orderly and disorderly, on the small scale. But the observations of interest to our subject matter relate mainly to larger-scale properties – for example, to average values of k, μ, etc., over ranges of distance inside the Earth that may be measured in kilometres. In representing observations on that scale, it is sufficient to work in terms of Earth models consisting of a limited number of internal regions inside each of which the matter is treated as continuously distributed.

Observations of the transient deformations which accompany seismic waves (see Chapter 8) contribute prominently to the determination of values of ρ, k, and μ inside the Earth. The theory of the present chapter provides, among other things, necessary background for discussing such deformations.

Properties of stress and strain that are of wide generality will be discussed first, and the question of equations connecting them then examined. Questions of physical interpretation will be discussed at some length. Earlier parts of the theory will be set down only in outline. (References are made to sections in B which give derivations of results to be quoted without proof.)

[*References on p. 107*]

7.1 Stress

7.1.1 Stress tensor

Let P be an internal point of a material medium whose mass is continuously distributed, and let A and B refer to the material adjacent to the two sides of a small plane area δS containing P. Referred to cartesian axes Ox_i ($i = 1, 2, 3$), let l_i be the direction-cosines of the outward normal to δS on the side A.

In general, a set Z of forces is exerted across δS by B on A, and an equal and opposite set by A on B. Let δF_{ij} ($j = 1, 2, 3$) be the component parallel to Ox_j of the resultant of the forces Z. The limit of $\delta F_{ij}/\delta S$ ($j = 1, 2, 3$), as $\delta S \to 0$, gives the components of *stress* at P corresponding to the direction l_i.

The stress at P for any direction l_i can be deduced from knowledge of the three components of the stress at P for any three different directions. For this purpose the nine-component *stress tensor* p_{ij} ($i, j = 1, 2, 3$) for the stress at P is introduced (see **B**, §2.1.1); p_{23} denotes the component, parallel to Ox_3, of stress across a small face perpendicular to Ox_2; and similarly with other members of the p_{ij}. In terms of p_{ij}, the stress for the direction l_i is $\Sigma l_j p_{ij}$. (Summations when not otherwise specified are with respect to the repeated subscript, j in this case, and run from 1 to 3.)

The stress tensor is symmetrical (**B**, §2.1.2); i.e. $p_{ij} = p_{ji}$ for all i, j. Thus only six of the nine components p_{ij} are independent.

7.1.2 Principal stresses and axes

For any point P inside the medium, there exists a set of three mutually perpendicular directions such that the stress acts normally across a small plane area normal to any one of these directions (**B**, §2.1.4). The *principal axes of stress* at P are axes through P with these directions. When the axes are selected as principal axes at P, then p_{23}, p_{31}, p_{12} are zero and p_{11}, p_{22}, p_{33} — sometimes to be then denoted as p_1, p_2, p_3 — are called the *principal stresses* at P.

It can be shown that Σp_{kk} is invariant (see **B**, §2.2.2); i.e. $p_{11} + p_{22} + p_{33}$ is independent of the orientation of the set of axes Ox_i. Hence $\frac{1}{3}\Sigma p_{kk}$ (whether referred to principal axes or not) is equal to the mean of the three principal stresses. It is convenient to denote this mean by $-p$.

[*References on p. 107*]

7.1.3 Deviatoric stress

The *deviatoric stress tensor* P_{ij} at P is defined by

$$P_{ij} = p_{ij} - \tfrac{1}{3}\Sigma p_{kk}\delta_{ij} = p_{ij} + p\delta_{ij}, \qquad (7.1)$$

where δ_{ij} (the Kronecker delta) is defined to have its components equal to unity when $i = j$, and to zero when $i \neq j$.

The stress components p_{ij} at P are therefore fully specified by knowledge of the mean of the principal stresses and the five independent deviatoric components [five since (7.1) gives $P_{ij} = P_{ji}$ and $P_{11} + P_{22} + P_{33} = 0$].

7.1.4 Hydrostatic stress

A characteristic of a fluid at rest is that the deviatoric stress is everywhere zero, so that the stress tensor at any point then has the form $-p\delta_{ij}$. In this case, p is identical with the pressure as ordinarily defined.

In the particular case where the deviatoric stress is zero throughout a solid, the stress at any point again has the form $-p\delta_{ij}$, and is referred to as a *hydrostatic stress*.

7.2 Strain and rotation

7.2.1 Strain tensor

In a small, assumed continuous, movement of the particles of the medium from a prior configuration, let u_i be the components of displacement which the particle now at the point P, of coordinates x_i, has undergone. Neglecting higher order terms, i.e. limiting consideration to the *infinitesimal* (or *linear*) strain theory, the displacement of the particle at the neighbouring point $x_i + \delta x_i$ will have been $u_i + \Sigma(\partial u_i/\partial x_j)\delta x_j$. The relative displacement of the two particles will therefore have been

$$\Sigma e_{ij}\delta x_j - \Sigma \xi_{ij}\delta x_j, \qquad (7.2)$$

where

$$e_{ij} = \tfrac{1}{2}\left(\frac{\partial u_j}{\partial x_i} + \frac{\partial u_i}{\partial x_j}\right), \qquad (7.3)$$

[References on p. 107]

$$\xi_{ij} = \tfrac{1}{2}\left(\frac{\partial u_j}{\partial x_i} - \frac{\partial u_i}{\partial x_j}\right). \tag{7.4}$$

The contribution $\Sigma e_{ij}\delta x_j$ to the relative displacement, when not zero, is always associated with deformation in the neighbourhood of P (B, §2.2), and e_{ij} is called the *strain tensor* at P. (The term $-\Sigma \xi_{ij}\delta x_j$ does not contribute to deformation — see §7.2.5.)

Associated with the strain tensor, there are *principal axes of strain* at any point P and corresponding *principal strains* e_1, e_2, and e_3, just as in the case of the stress tensor, although the principal axes of stress and strain do not necessarily coincide. Also Σe_{kk} is invariant.

7.2.2 Dilatation and compression

The *dilatation* θ at P is defined as the limit, as $V \to 0$, of the proportionate increase of volume V in the neighbourhood of P during the deformation. It is also equal to the limit of the proportionate decrease in density at P. The *compression* is $-\theta$. The dilatation θ is expressible as Σe_{kk} or $\Sigma(\partial u_k/\partial x_k)$ (B, §2.2.3).

7.2.3 Deviatoric strain

The deviatoric strain tensor E_{ij} is defined analogously to P_{ij}. Thus

$$E_{ij} = e_{ij} - \tfrac{1}{3}\Sigma e_{kk}\delta_{ij} = e_{ij} - \tfrac{1}{3}\theta \delta_{ij}. \tag{7.5}$$

As with P_{ij}, only five of the components E_{ij} are independent.

7.2.4 Linear and finite-strain theory

In contexts where higher order terms in $\partial u_i/\partial x_j$ cannot be neglected, *finite-strain* theory has to be used and additional terms have to be considered in equations like (7.3).

In the interior of the Earth, the strains in the undisturbed state reach quite large values (see §11.1). But in treating the ordinary problems of seismic wave transmission, it is sufficient to apply the linear theory to the additional strains accompanying the wave transmission. This will be done in the seismological theory to follow, where u_i, e_{ij}, p_{ij}, etc. will relate to the excesses over the displacements, strains, stresses, etc. already present in the undisturbed state.

[References on p. 107]

At the same time, finite-strain theory does assist in some other aspects of the Earth's density problem and further reference to the theory will be made in §§7.8.2 and 7.9.7.

7.2.5 Rotation

It can be shown (B, §2.2.1) that the contribution $-\Sigma \xi_{ij}\delta x_j$ to the relative displacement (7.2) gives in the neighbourhood of P a pure rotation, as of a rigid body, about some axis through P.

The array ξ_{ij} is called the *rotation tensor* at P. By (7.4), ξ_{ij} is antisymmetrical; i.e. $\xi_{ij} = -\xi_{ji}$ for all i,j. Thus $\xi_{11} = \xi_{22} = \xi_{33} = 0$, and ξ_{ij} has only three independent components, which are equal to half the components of the vector curl **u**, where $\mathbf{u} = (u_1, u_2, u_3)$.

7.3 Model stress-strain relations

Changes of stress in any material are accompanied in general by changes of deformation. A first step in deformation theory is to arrive empirically at a suitable set of model relations connecting the p_{ij} and e_{ij}.

7.3.1 *The perfect-elasticity isotropic model*

One much used set of model stress-strain relations is a generalization of Hooke's law which (on the linear theory) connects the p_{ij} and e_{ij} by linear equations of the form

$$p_{kl} = \Sigma\Sigma A_{ijkl} e_{ij}, \qquad (7.6)$$

the coefficients A_{ijkl} depending on the composition of the material at P and on the thermodynamical conditions under which the strain takes place. Only 36 of the coefficients could be independent because of the symmetry of p_{ij} and e_{ij}. Under 'reversible' thermodynamical conditions, the number of independent coefficients is reducible to 21. In these conditions the stress-strain relations (7.6) are associated with *perfect elasticity*, there being in this case no dissipation of energy during change of strain.

An *isotropic* material is a model material whose response to stress is independent of orientation. In the case of a perfectly elastic isotropic material, the 21 coefficients are reducible to 2 (B, §2.3.1). (Crystals are examples of materials which cannot be

[*References on p. 107*]

treated as isotropic, and description of their response to stress requires more than 2 coefficients.)

The set of stress-strain relations for the perfect-elasticity isotropic model may (B, §2.3.3) be written as

$$p_{ij} = (k - 2\mu/3)\theta\delta_{ij} + 2\mu e_{ij}. \tag{7.7}$$

In (7.7), the coefficients k and μ are a pair of parameters whose values specify the elastic properties of the model at the point P; the physical interpretation of k and μ is given in §7.5. In formal mathematical work, the pair λ and μ, called the *Lamé parameters*, where

$$\lambda = k - 2\mu/3, \tag{7.8}$$

is commonly used, but λ does not have an immediate physical interpretation.

An important set of relations equivalent to (7.7), readily derivable from (7.7) using (7.1) and (7.5), is

$$p = -k\theta, \qquad P_{ij} = 2\mu E_{ij}. \tag{7.9,10}$$

With the perfect-elasticity isotropic model, the principal axes of stress and strain coincide since, for $i \ne j$, the components p_{ij} and e_{ij} vanish for the same axes.

7.3.2 Perfect fluid

The particular case of (7.7) or (7.9,10) in which μ is everywhere zero (whether the material is at rest or not) defines a *perfect fluid*. In this case, the deviatoric components of stress are all zero. The equation (7.9) and knowledge of the value of the single parameter k then specify the complete elastic behaviour.

7.3.3 Model relations for anelasticity

There is no narrow limit to the possible stress-strain relations for materials, and the choice of a model depends (as always) on the observational evidence, the context of interest, and the criteria stated in §6.1.

The following are examples of the simplest model representations, all isotropic, which have been found serviceable in particular (through usually limited) contexts where *anelasticity* is significant, i.e. where the elasticity cannot be treated as perfect. Each of

[References on p. 107]

these models includes (7.9) unchanged, and includes, in place of (7.10), one of the following sets of equations:

$$P_{ij} = 2\nu dE_{ij}/dt, \qquad (7.11)$$

$$P_{ij} = 2\mu E_{ij} + 2\nu dE_{ij}/dt, \qquad (7.12)$$

$$P_{ij} + \tau dP_{ij}/dt = 2\nu dE_{ij}/dt, \qquad (7.13)$$

$$P_{ij} + \tau dP_{ij}/dt = 2\mu E_{ij} + 2\nu dE_{ij}/dt, \qquad (7.14)$$

where ν and τ are parameters additional to k and μ.

Along with (7.9), the equation (7.11) gives a model representation for a fluid of viscosity ν; (7.12) and (7.13) give, respectively, the Kelvin (or Voigt) and Maxwell models for an imperfectly elastic solid; (7.14) gives a model representing some observed features of elastic afterworking. (For further details see B, §§2.5.1–2.5.4.)

When P_{ij} is periodic, the representations (7.12), (7.13), and (7.14) show different characteristics according to the magnitude T of the predominant period. For example, when T/τ is small (rapidly changing stress), (7.13) approximates to the perfectly elastic behaviour (7.10) with $\mu = \nu/\tau$. When T/τ is large (slowly changing stress), (7.13) approximates to the viscous fluid behaviour (7.11). These results illustrate an important property, namely that the periods of the stresses involved can be a crucial feature in determining the extent to which and manner in which anelasticity manifests itself. Also, the actual suitability of a particular model set of stress-strain relations for a particular context is liable to depend, among other things, on the stress periods.

The results in the last paragraph show that the model relations (7.11)–(7.14) can serve to illustrate important aspects of anelasticity. The relations are often also useful in indicating orders of magnitude of some dissipation effects during deformation. But, while thus useful for certain broad purposes, they are all too simple to represent sets of observations in reliable quantitative detail.

The investigation of more realistic representations of anelasticity is a large active branch of science in its own right, drawing evidence both on the microscopic scale (e.g. from solid-state and lattice theory) and on the macroscopic scale (e.g. from experiments on metals and rocks carried out over wide ranges of pressure and temperature).

[*References on p. 107*]

7.4 Stress-strain relations for the Earth

Since the time of Hooke (1678), the perfect elasticity model has been found serviceable in describing the behaviour of many, but by no means all, materials under ordinary laboratory conditions. In accordance with the simplicity rule (§6.1), it is therefore appropriate to use this as a trial initial model for the Earth, to be tested against observations as they come. It is also appropriate to start with the isotropic case since there is a likelihood that orientation effects will, in the context of present interest, be largely averaged out. Thus investigations of the elasticity of the Earth suitably began by taking (7.7), or (7.9,10), as a trial model.

The utility of this model was first clearly indicated when Oldham (1899, 1900) identified P, S and surface waves on seismograms in agreement with theory developed from (7.7). Subsequent studies have shown that (7.7) and (7.9,10) give a remarkably close representation of the Earth's elasticity as far as the principal seismic observations go. When model representations such as (7.12) and (7.14) are tried, the indicated values of terms involving ν and τ are found to be very small compared with those involving k and (in solid regions) μ.

Thus where seismic periods (which may reach the order of an hour) are concerned, (7.7) and (7.9,10) provide an extremely satisfactory model for most purposes; k and μ are in fact coefficients of dominant importance in the study of the present state of the interior of the Earth. The model is also serviceable for stress changes of somewhat longer period, including tidal periods, and is assumed in interpreting evidence such as Kelvin's (§6.8.2) on the Earth's rigidity — evidence which is also needed in determining the Earth's density variation.

At the same time, the discussion in §7.3.3 on simple anelasticity models indicates that increasing the periods of duration of stress can increase the relative importance of imperfections of elasticity. Thus where periods on the geological time-scale are involved, it cannot be assumed that (7.7) is adequate. Indeed the evidence indicates [see e.g. Jeffreys (1964)] that substantially more elaborate models are then required. Hence it has to be appreciated that values of k and μ (more particularly μ) given later in the book apply to the response of the interior of the Earth to stresses whose periods are comparatively short.

[References on p. 107]

Although the literature on anelasticity is vast, the formulation of suitable models to represent the long term behaviour of the materials of the Earth's interior is still in its infancy. An indication of some of the developments may be read in work of Knopoff (1964), Jeffreys (1970), Jackson and Anderson (1970), and Stacey (1970).

Even in the case of seismic wave periods, observations of attenuation make it necessary to take account of anelasticity in some problems. The observations are frequently described in terms of the quantity Q, defined (there are alternative definitions) so that the attenuation factor is $e^{-2\pi/Q}$ for one wavelength. Inside the Earth's mantle (§10.2) the estimated Q ranges from the order of 100–200 near the top to 1000–2000 deeper down. In the fluid outer core, Q appears to be about 6000; in the inner core, 500–1000. [These estimates have been made ignoring the dependence of Q on the wave period, but are useful as a general guide. For further detail see Fedotov (1963), D.L. Anderson, Ben-Menahem, and Archambeau (1965), Jeffreys (1970) p. 333.]

7.5 Interpretation of coefficients in perfect elasticity

The relevance of (7.7) or (7.9,10) to a large class of problems on the interior of the Earth being well established, we proceed to the question of the physical interpretation of k and μ, and some related coefficients.

7.5.1 Incompressibility (bulk modulus)

By (7.9), k^{-1} is the ratio of the compression to the pressure, and so measures the *compressibility* of the material at the point $P(x_i)$ concerned; k is the *incompressibility*.

The order of magnitude of k discriminates between gases and liquids. At ordinaty temperatures and pressures, a fluid whose behaviour is represented by (7.9) is called a *gas* if k is of the order of 10^5 N/m^2 and a *liquid* if k is substantially greater (say 10^9 N/m^2 or more).

From the definition of θ (§7.2.2), it is deducible from (7.9) that

$$\partial \rho / \partial p = \rho/k \qquad (7.15)$$

[*References on p. 107*]

[cf. (6.14)]. The partial derivative notation is used in (7.15) since ρ depends in general on several variables, for example temperature and chemical composition, in addition to p. This additional dependence can be important in the Earth context, where ρ is investigated as a function of position and where pressure, temperature, and composition may vary with position independently of one another. The full set of variables on which ρ depends and features of the dependence are examined in some detail in §10.1.

7.5.2 *Rigidity*

The coefficient μ in (7.7) or (7.10) is called the *rigidity* at P, and is thus an index of the strain produced by an assigned deviatoric stress.

The order of magnitude of μ discriminates between fluids and solids in contexts where the representation (7.10) is suitable. For fluids, μ is negligibly small (zero for an ideal fluid). For most metals and rocks under normal conditions, μ is of the order of 10^9 to 10^{11} N/m². A perfectly elastic material is called a *solid* when μ is not negligible. Sometimes a material is called a fluid when the evidence shows only that μ does not exceed 10^9 N/m²; this is because it is difficult to detect smaller values of μ in practice.

An ideal material having $\mu = \infty$ at all points would resist deformation under any finite deviatoric stress. If also $k = \infty$ everywhere, the material would be an ideal rigid body.

It is to be noted that incompressibility and rigidity have just the same meanings for the material of the Earth's deep interior as they do for any ordinary laboratory material which can be treated as perfectly elastic. In both cases the elastic properties follow as consequences of the suitability of (7.7) in representing the pertinent observations.

Where periods on the geological time-scale are involved in the Earth and in other contexts where (7.7) is not adequate, it needs to be noted that the rigidity is not necessarily given by μ. For example, the model relation (7.14) gives a rigidity approximating to μ under slowly changing stress, but to ν/τ under rapidly changing stress (see **B**, §2.5.3).

[*References on p. 107*]

7.5.3 Young's modulus and Poisson's ratio

By (7.9) and (7.10), $\theta = -p/k$ and $E_{ij} = P_{ij}/2\mu$. Using (7.1) and (7.5), simple algebra gives

$$e_{ij} = (2\mu)^{-1} p_{ij} + \{(2\mu)^{-1} - (3k)^{-1}\} p \delta_{ij}.$$

Putting

$$3k = E/(1 - 2\sigma), \quad 2\mu = E/(1 + \sigma) \quad (7.16,17)$$

then gives

$$e_{ij} = (1 + \sigma)E^{-1} p_{ij} + 3\sigma E^{-1} p \delta_{ij}, \quad (7.18)$$

which is a usual form for the expression of the strain components e_{ij} in terms of the stress components p_{ij}.

Young's modulus E and *Poisson's ratio* σ constitute an alternative pair of coefficients to k and μ in expressing the stress-strain relations of a perfectly elastic isotropic material. The coefficients E and σ are commonly interpreted in terms of the deformation of a homogeneous cylindrical wire subjected to uniform normal stresses, p_1 say, at the ends. If e_1, e_2, and e_3 are the principal strains produced (e_1 is longitudinal, and e_2 and e_3 are lateral, with $e_2 = e_3$), it is deducible from (7.18) that $E = p_1/e_1$ and $\sigma = -e_2/e_1$.

By simple algebra from (7.16, 17),

$$E = 9k\mu/(3k + \mu), \quad \sigma = (3k - 2\mu)/(6k + 2\mu).$$

$$(7.19, 20)$$

Thus when the values of the incompressibility and rigidity are known, the values of Young's modulus and Poisson's ratio can be immediately deduced.

If it is assumed that the base configuration of the material (i.e. the configuration $e_{ij} = 0$) is elastically stable, then k, μ, and E cannot be negative. Otherwise there is no theoretical restriction on the ranges of values of k, μ, and E. It then follows from (7.20) that $-1 \leqslant \sigma \leqslant 0.5$. With observed materials, σ is normally positive. The value of σ increases as μ/k decreases and is 0.5 for a perfect fluid.

7.6 Strength of a material

Under limited deviatoric stress, (7.7) serves as a suitable model for many laboratory solids. This ceases to be the case when, under increasing stress, a certain function F of the deviatoric stress components reaches at some point P inside the solid a value called the *strength* of the material at P. When the strength is exceeded, continued application of stress causes dissipation of elastic energy, and flow [e.g. if (7.13) should now apply] or fracture may occur.

Let p_1, p_2, and p_3 be the principal stresses at P in descending order of magnitude, and P_1, P_2, and P_3 the corresponding members of P_{ij}. Functions that have been proposed for F include the *stress-difference* F_S, where

$$F_S = p_1 - p_3 = P_1 - P_3, \qquad (7.21)$$

and the *Mises function* F_M, where

$$F_M = \sqrt{\{(p_1 - p_3)^2 + (p_1 - p_2)^2 + (p_2 - p_3)^2\}}. \qquad (7.22)$$

Since $P_1 + P_2 + P_3 = 0$ (§7.1.3), we can also write

$$F_M = \sqrt{\{3(P_1^2 + P_2^2 + P_3^2)\}}, \qquad (7.23)$$

By simple algebra, the difference ratio $(F_M - F_S)/F_M$ is found to lie between 0·18 and 0·30 for all values of the P_i. This difference is small in relation to the order of accuracy needed in considering the strengths of materials inside the Earth, and so either function (F_S or F_M) is generally suitable for the Earth context.

When, under accumulating stress, the function F_S or F_M reaches a value equal to the strength of the material, the tendency to flow or fracture prevents the value of the function from increasing much further. Thus the strength sets a limit to the stress-differences and deviatoric stresses that can be sustained. With observed materials the strength rarely exceeds the order of 10^8 N/m² and it is surmised that values of P_1, P_2, and P_3 beyond 10^9 N/m² are unlikely to exist anywhere below the Earth's surface; by the properties of tensors, the same applies to the whole set of P_{ij} referred to axes that are not necessarily principal axes. Relevant geophysical observations (see e.g. §5.7) tend to confirm the surmise.

It may be remarked that in assessing the strength of a material a question of model representation involving the time-scale can arise. Thus on the geological time-scale, some investigators have asserted that the strength of the Earth is effectively zero — that under a sufficiently long sustained deviatoric stress, however small, flow or fracture will occur. For our immediate purpose, however, the main result required is the assessment of the maximum order of magnitude of the P_{ij} likely to occur in the interior of the Earth.

[*References on p. 107*]

7.7 The terms 'fluid' and 'solid'

The classification into solid and fluid as given in §7.5.2 will be retained for the materials of the whole interior of the Earth. The classification is in terms of the magnitude of the rigidity μ, and avails so long as the model relations (7.7) and (7.9,10) are appropriate. This means that the terms fluid and solid as used in this book relate to the response of materials of the Earth's interior to stresses with periods not greatly exceeding the seismic and tidal periods. The determination of density in the Earth does not use much evidence involving longer periods.

Where periods on the geological time-scale are involved, stress-strain relations may involve time-dependent terms (cf. (7.11)–(7.14)) and the question of defining fluid and solid has to be re-opened. On this question, different criteria have been used, some depending on the physical property of rigidity (not necessarily the coefficient μ – see §7.5.2) and some on the strength. (See **B**, §2.5.6.) Confusion is avoided by restricting the terms fluid and solid to contexts where the representation (7.7) is suitable.

7.8 Pressure and finite strain

7.8.1 *The term 'pressure'*

In the interior of the Earth, the term pressure is used for both solid and fluid regions. For a solid region, the *pressure* is minus the mean of the three principal stresses and so corresponds to p as defined in §7.1.2 and used in §7.1.3 and subsequently.

Changes in the density ρ depend on changes in p through equations such as (7.9) and (7.15), but do not depend on changes in the P_{ij}. Hence in solid as well as fluid regions of the Earth, the determination of the density variation is closely connected with the variation of p.

An important point is that, throughout most of the Earth, the deviatoric stress components are all negligible compared with p. As indicated in §7.6, the P_{ij} are not likely to exceed the order of 10^9 N/m². On the other hand, p reaches this order at a depth of about 35 km in the Earth, and nearly 4×10^{11} N/m² at the centre (see §10.7.2). Since, by (7.1),

$$p_{ij} = -p\delta_{ij} + P_{ij}, \tag{7.24}$$

[*References on p. 107*]

it follows that inside much of the Earth $p_{ij} \approx -p\delta_{ij}$; thus the internal stress is predominently hydrostatic. This provides some justification for assuming hydrostatic conditions in investigating the figure of the Earth (Chapter 5).

7.8.2 *Strain under high pressure*

Some elementary formulae relating to finite strain in the Earth will now be derived for a purpose needed in §7.9.7.

Let x_i and $x_i + dx_i$ be the positions of two neighbouring particles P and Q inside a body which has undergone finite strain, and let u_i and $(u_i + du_i)$ be the displacements of P and Q from the configuration of zero stress (Fig. 7.1). The change χ in the square of the distance between P and Q is given by

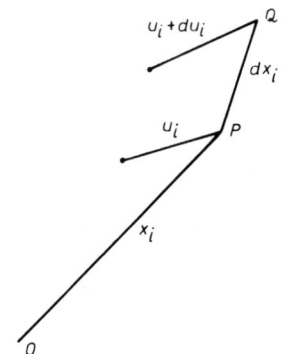

Fig. 7.1

$$\chi = \Sigma(dx_i)^2 - \Sigma(dx_i - du_i)^2$$
$$= 2\Sigma(dx_i du_i) - \Sigma(du_i)^2.$$

Since $du_i = \Sigma(\partial u_i/\partial x_j)dx_j$, we obtain, on simplifying and using (7.3),

$$\chi = 2\Sigma\Sigma\epsilon_{ij}dx_i\,dx_j, \tag{7.25}$$

where

$$\epsilon_{ij} = e_{ij} - \tfrac{1}{2}\Sigma\left(\frac{\partial u_m}{\partial x_i}\frac{\partial u_m}{\partial x_j}\right). \tag{7.26}$$

The *finite-strain tensor* ϵ_{ij} of course reduces to e_{ij} when second powers of $\partial u_i/\partial x_j$ are neglected.

[*References on p. 107*]

Consider now conditions in the undisturbed Earth (seismic and tidal disturbances are not here being considered) at depths greater than 35 km. As seen in §7.8.1, the stress components p_{ij} are predominantly non-deviatoric and the same applies to the corresponding finite-strain components ϵ_{ij}. Thus the strain is predominantly compressional and to good accuracy $\partial u_i/\partial x_j$ may be taken to have the form $q\delta_{ij}$, where q is scalar. Then $q = \partial u_1/\partial x_1 = \partial u_2/\partial x_2 = \partial u_3/\partial x_3 = \frac{1}{3}\theta$, where θ continues to be defined by the formula in §7.2.2, and ϵ_{ij} has the form $\epsilon\delta_{ij}$. By (7.26), putting $j = i$ and summing for i, we then have

$$3\epsilon = \theta - \tfrac{1}{6}\theta^2,$$
$$1 - 2\epsilon = (1 - \tfrac{1}{3}\theta)^2. \tag{7.27}$$

Since ϵ is negative when the pressure is positive, we shall write $\epsilon = -f$. It has been customary to call f the 'compression' in applications of finite-strain theory to the Earth, though it needs to be noted that this is a different usage from that in §7.2.2. Let V denote an element of volume containing P, and let the subscript zero denote values in the unstrained state. Then, using (7.27), we have

$$\begin{aligned}\rho/\rho_0 = V_0/V &= (1 - \partial u_1/\partial x_1)^3 \\ &= (1 - \tfrac{1}{3}\theta)^3 \\ &= (1 + 2f)^{3/2}. \end{aligned} \tag{7.28}$$

7.9 Thermodynamical conditions

It was pointed out in §7.3.1 that coefficients such as k and μ depend on the thermodynamical conditions under which the strain takes place. It is necessary to examine this dependence so that the coefficients can be interpreted without ambiguity in a particular context. The following simplified discussion brings out the main results which bear on the determination of the Earth's density distribution.

The discussion in §7.9.1 does not require detailed specification of the stress-strain relations, but after that we shall limit consideration to the particular case of perfect elasticity. Linear stress-strain relations are assumed except where otherwise indicated.

[References on p. 107]

7.9.1 *Expression for internal energy*

During a change of strain de_{ij}, let dW' be the work done by the external forces and dQ the heat received. Let dT and du be the consequent increases in the kinetic and internal energies, and dW the work done against the internal forces. In §§7.9.1–7.9.5.3, all of W', Q, T, u, and W are understood as being reckoned per unit volume and applying to the material in the neighbourhood of the particle at P.

By the principle of mechanical energy and the first law of thermodynamics,

$$dW' - dW = dT, \qquad dW' + dQ = dT + du. \quad (7.29,30)$$

By the contraction property of tensors (B, §2.3.3), $\Sigma\Sigma p_{ij}de_{ij}$ is invariant and so equal to $\Sigma p_i de_i$, where p_i and e_i are the principal stresses and strains at P. Hence

$$\begin{aligned}
du &= dQ + dW \\
&= dQ + \Sigma p_i de_i \\
&= dQ + \Sigma\Sigma p_{ij}de_{ij} \\
&= dQ - pd\theta + \Sigma\Sigma P_{ij}dE_{ij}, \quad (7.31)
\end{aligned}$$

on using (7.1) and (7.5).

7.9.2 *Thermodynamical variables for fluids and solids*

The thermodynamical state of a perfect fluid of given mass and uniform composition is sufficiently, and is commonly, specified by the (absolute) temperature τ and volume V, taken as independent variables. For our immediate context (to §7.9.5.3) it is more suitable to use the dilatation θ in place of V. (Changes in V and θ are connected by $d\theta = dV/V$.) A variable such as the pressure p depends on τ and θ through the equation of state for the particular fluid, and may be taken as an alternative independent variable when convenient.

For a solid conforming to the stress-strain relations (7.9,10), the specification is similar except that the additional variables E_{ij} (with the P_{ij} as a possible alternative set) are required. A full set of independent thermodynamical variables is now τ, θ, and five of the E_{ij} (see §§7.1.3 and 7.2.3). The pressure p is dependent on variables in this set. By (7.9), when τ and θ are kept fixed, p also stays fixed, independently of any change in the deviatoric strain E_{ij}. Thus p may depend on θ and τ, but not on the E_{ij}. Similarly, by (7.10), the P_{ij} may depend on the E_{ij} and τ, but not on θ. (The dependences on τ would be through the coefficients k and μ.)

Let τ change by $d\tau$ during the deformation given by (7.9) and (7.10). Then (7.9) and (7.10) cannot both involve $d\tau$, since eliminating $d\tau$ would then yield relations connecting p and θ with P_{ij} and E_{ij} in a way which would violate the independence of (7.9) and (7.10).

It is necessary to select (7.9) as the relation involving $d\tau$ since the results must agree with the particular case of a fluid. Thus (7.10) does not involve $d\tau$.

7.9.3 Isothermal and adiabatic changes

For reversible changes, which include isothermal and adiabatic changes, the change ds in the *entropy* s per unit volume is given by

$$\tau^{-1} dQ = ds. \tag{7.32}$$

By the second law of thermodynamics, ds is a perfect differential in the variables specifying the state. In treating reversible changes, s may be used as an alternative independent variable, e.g. as an alternative to either τ or θ.

Isothermal and *adiabatic* (or *isentropic*) changes are defined as changes in which τ and Q, respectively, are kept fixed; thus, for small changes, $d\tau = 0$ and $dQ = 0$, respectively. By (7.32), $ds = 0$ for small adiabatic changes. Isothermal and adiabatic changes are reversible.

Representations in terms of isothermal or adiabatic conditions ignore practical complications connected with possible heat transfer by conduction – a non-reversible process. The complications have only minor effects in the applications we shall be making, including the usual applications to seismic wave transmission, and will here be ignored.

7.9.4 Isothermal and adiabatic rigidity

It was seen in §7.9.2 that (7.10) does not involve the temperature change $d\tau$. It follows that the coefficient μ in (7.10) is the same for all reversible thermodynamical changes.

[References on p. 107]

In particular, the isothermal rigidity and the adiabatic rigidity are equal.

7.9.5 Isothermal and adiabatic incompressibility

The separation established in §7.9.2 between the dependences of p and the P_{ij} means that the relation between the isothermal and adiabatic incompressibilities, k' and k say, for a solid conforming to (7.9) and (7.10) has the same form as for a fluid conforming to (7.9). A derivation of the relation will now be given.

7.9.5.1 Definitions

Following are the definitions of k', k, the specific heat c at constant strain, and the coefficient γ of cubical expansion at constant pressure, subscripts indicating variables which are kept fixed:

$$k' = -(\partial p/\partial \theta)_\tau, \qquad k = -(\partial p/\partial \theta)_{s, E_{ij}}, \qquad (7.33, 34)$$

$$c = \rho^{-1}(\partial Q/\partial \tau)_{\theta, E_{ij}}, \qquad \gamma = (\partial \theta/\partial \tau)_p. \qquad (7.35, 36)$$

7.9.5.2 Auxiliary formulae

By (7.31) and (7.32),

$$du = \tau ds - p d\theta + \Sigma\Sigma P_{ij} dE_{ij}. \qquad (7.37)$$

By the first law of thermodynamics, du is a perfect differential. Hence

$$(\partial p/\partial s)_{\theta, E_{ij}} = -(\partial \tau/\partial \theta)_{s, E_{ij}}. \qquad (7.38)$$

By (7.32) and (7.35),

$$c\rho = \tau(\partial s/\partial \tau)_{\theta, E_{ij}}. \qquad (7.39)$$

Looking at p as a function of τ and θ, with τ as a function of θ, s, and E_{ij}, we can write

$$\left(\frac{\partial p}{\partial \theta}\right)_{s, E_{ij}} = \left(\frac{\partial p}{\partial \tau}\right)_\theta \left(\frac{\partial \tau}{\partial \theta}\right)_{s, E_{ij}} + \left(\frac{\partial p}{\partial \theta}\right)_\tau. \qquad (7.40)$$

Use will also be made of the relations

$$\left(\frac{\partial y}{\partial z}\right)_x \left(\frac{\partial z}{\partial x}\right)_y \left(\frac{\partial x}{\partial y}\right)_z = -1, \qquad \left(\frac{\partial x}{\partial y}\right)_z = \left(\frac{\partial y}{\partial x}\right)_z^{-1}, \qquad (7.41)$$

[References on p. 107]

7.9.5.3 Derivation of relation between k and k'

By the definitions (7.33) and (7.34), we have, using (7.40),

$$k - k' = -\left(\frac{\partial p}{\partial \tau}\right)_\theta \left(\frac{\partial \tau}{\partial \theta}\right)_{s, E_{ij}}$$

$$= \left(\frac{\partial p}{\partial \tau}\right)_\theta \left(\frac{\partial p}{\partial s}\right)_{\theta, E_{ij}} \qquad \text{[using (7.38)]}$$

$$= \frac{\tau\{(\partial p/\partial \tau)_\theta\}^2}{\tau(\partial s/\partial \tau)_{\theta, E_{ij}}}$$

$$= \frac{\tau\{(\partial p/\partial \theta)_\tau (\partial \theta/\partial \tau)_p\}^2}{\rho c},$$

using (7.39) and (7.41). Hence, by (7.33) and (7.36), we have the required relation

$$k - k' = \tau k'^2 \gamma^2 / \rho c. \qquad (7.42)$$

7.9.6 Helmholtz free energy

We now let U and S be the internal energy and entropy in volume V of a material, instead of reckoning per unit volume as in §§7.9.1-7.9.5.3, and limit consideration to the case of hydrostatic stress. In place of (7.37), we then have

$$dU = \tau dS - p dV. \qquad (7.43)$$

The *Helmholtz free energy* Ψ is defined as $(U - \tau S)$. Hence, by (7.43),

$$d\Psi = -S d\tau - p dV. \qquad (7.44)$$

For the particular case of an isothermal change, (7.44) becomes

$$(d\Psi)_\tau = -p dV. \qquad (7.45)$$

We shall need the following result:

$$9k'_0 V_0 = \lim(\partial \Psi/f \partial f)_\tau, \qquad (7.46)$$

where k'_0 is the isothermal incompressibility at zero pressure, V_0 and f are as in §7.8.2, and the limit is for $p \to 0$ (and therefore

[*References on p. 107*]

$\theta \to 0$ and $f \to 0$). The result (7.46) is obtained by writing

$$k_0' = -\lim (p\partial V/\theta \partial V)_\tau = \lim (\partial \Psi/\theta \partial V)_\tau$$

and then using (7.28), which gives $\theta \approx -3f$ and $dV \approx -3V_0 df$.

7.9.7 Murnaghan-Birch theory of finite strain

Birch (1952) sought to adapt Murnaghan's (1951) theory of finite strain to obtain results applicable to the Earth's interior. He assumed hydrostatic stress, and assumed that Ψ can be expressed as a Taylor expansion

$$\Psi = af^2 + bf^3 + cf^4 + ..., \qquad (7.47)$$

where the coefficients $a, b, c, ...$ depend on τ. Then (7.46) gives

$$9k_0' V_0 = 2a. \qquad (7.48)$$

From (7.28), we derive

$$3V_0 df/dV = -(1 + 2f)^{5/2}.$$

For a first approximation in which terms of (7.47) apart from af^2 are ignored, (7.45) and (7.48) then give

$$p = -(\partial \Psi/\partial f)_\tau df/dV$$
$$= 3k_0' f(1 + 2f)^{5/2}. \qquad (7.49)$$

The second approximation is

$$p = 3k_0' f(1 + 2f)^{5/2}(1 + 3bf/2a). \qquad (7.50)$$

By (7.28), (7.49) becomes, in terms of ρ,

$$p = (3k_0'/2)\{(\rho/\rho_0)^{7/3} - (\rho/\rho_0)^{5/3}\}. \qquad (7.51)$$

The equations (7.49)–(7.51) and related equations are commonly referred to as the Murnaghan-Birch equations of state.

Birch (1952) tested (7.51) against experimental results of Bridgman (1948) for the alkali metals at pressures up to 10^{10} N/m², and inferred that it is within the experimental errors to take $b/a, c/a, ...,$ zero. He then postulated that (7.51) is a suitable equation of state to take for the materials of the Earth's deep interior.

Using (7.15), we derive from (7.28) and (7.49)

$$k' = k_0'(1 + 2f)^{5/2}(1 + 7f), \qquad (7.52)$$

$$\left(\frac{\partial k'}{\partial p}\right)_T = \frac{12 + 49f}{3 + 21f}, \qquad (7.53)$$

$$(k'/\rho) = (k'/\rho)_0 (1 + 9f + 14f^2). \qquad (7.54)$$

The relations (7.52) and (7.53) serve useful purposes in indicating the expected orders of magnitude of k and dk/dp in the Earth's deep interior, while (7.54) indicates the expected rate of increase of an important function ($\phi = k/\rho = \alpha^2 - 4\beta^2/3$) of the seismic P and S velocities α and β (see §§8.3.1, 8.3.2, and 10.1.2) in homogeneous regions of the Earth.

The above theory also leads to estimates of f throughout the Earth (see §11.1). Neglecting possible complications due to phase changes (§10.1.4), the estimates give f about 0·13 in the vicinity of the Earth's mantle-core boundary and about 0·2 near the Earth's centre.

Applications to the deep interior of the Earth of theory on incompressibility, including finite-strain theory, are made in Chapter 11. In §15.1, equations of state for the Earth are further discussed.

REFERENCES

Anderson, D.L., Ben-Menahem, A. and Archambeau, C.B. (1965). Attenuation of seismic energy in the upper mantle. *J. Geophys. Res.*, 70, 1441–1448.

Birch, F. (1952). Elasticity and constitution of the Earth's interior. *J. Geophys. Res.*, 57, 227–286.

Bridgman, P.W. (1948). The compression of 39 substances to 100 000 kg./cm.^{2}. *Proc. Amer. Acad. Arts Sci.*, 76, 55–70.

Fedotov, S.A. (1963). *Izv. Akad. Nauk. U.S.S.R., Ser. Geofiz.*, 509–520.

Hooke, R. (1678). *De Potentiâ Restitutiva*. London.

Jackson, D.D. and Anderson, D.L. (1970). Physical mechanisms of seismic-wave attenuation. *Rev. Geophys. Space Phys.*, 8, 1–63.

Jeffreys, Sir H. (1964). How soft is the Earth? *Q.J. Roy. Astr. Soc.*, 5, 10–22.

Jeffreys, Sir H. (1970). *The Earth* (5th Ed.), Cambridge University Press.

Knopoff, L. (1964). *Q. Rev. Geophys.*, 2, 625–660.

Murnaghan, F.D. (1951). *Finite deformation of an elastic solid*, Wiley, New York.

Oldham, R.D. (1899). *Mem. Geol. Surv.* (India), p. 29.

Oldham, R.D. (1900). On the propagation of earthquake motion to great distances. *Phil. Trans.*, A, 194, 135–174.

Stacey, F.D. (1970). *Physics of the Earth*, Wiley, New York.

CHAPTER 8

Seismic wave transmission

As seen in Chapter 6, the total observational evidence available by the end of the nineteenth century could supply a severely limited number of equations of condition on the distribution of the density ρ in the Earth. Models giving widely different values of ρ at particular depths were permissible.

The advent of the Milne seismograph in 1892 heralded a dramatic increase in the sources of information. In due course, data from seismology led to well-determined values of k/ρ and μ/ρ ($k =$ incompressibility, $\mu =$ rigidity) throughout a large part of the Earth's interior and made it possible to narrow estimates of the density distribution very greatly. The seismological evidence emerged at first very slowly, with intermittent important advances such as Oldham's specific detection of the Earth's core in 1906, Gutenberg's fairly close location of the core boundary in 1914, and the application in 1923 by Williamson and Adams (§6.9.2) of seismic data in estimating density gradients. During the period 1930–40, the development of seismology and related studies was particularly marked and reached the stage where fairly close limits could be reliably set to the density throughout much of the Earth.

The present chapter is concerned with those parts of the basic theory of seismic wave transmission through which evidence is provided on the density and related properties. Much of the theory will be given only in outline, and reference will continue to be made to the text-book (**B**) on seismology for fuller details.

8.1 Earthquakes and other sources of seismic waves

In the present context, we are interested in earthquakes as generators of seismic waves. For any one earthquake, the main seismic waves

[*Refs. on p. 127*]

issue during a short time-interval from a confined 'focal region' inside the Earth. The time-interval rarely exceeds a few seconds and the dimensions of this region are small compared with the radius of the Earth. With most earthquakes, including nearly all large ones, the energy released is commonly regarded as coming principally from elastic strain energy previously accumulated. (Other mechanisms are possible in some cases.) The energy ranges from about 10^5 J in the smallest recorded earthquakes to about 10^{17} J in the largest.

For many mathematical model purposes, the energy in an earthquake is treated as issuing from an impulsive point source, or *focus*, below the Earth's surface. The *epicentre* is the point of the surface vertically above the focus. Departures from this model condition contribute to uncertainties in interpreting seismic wave observations and are, in the first instance, dealt with statistically. The focal depths of earthquakes range from near-zero to about 700 km.

The energy even in moderate earthquakes is sufficient to send, through all parts of the Earth's interior, waves that can be recorded on emerging at the surface. The waves communicate, via seismograms written at some 10^3 observatories distributed over the Earth's surface, evidence on the detailed structure below.

8.1.1 *Nuclear and chemical explosions*

Sizeable artificial explosions also generate seismic waves. In advancing knowledge of the structure of the Earth's interior, artificial explosions have some advantages over natural earthquakes in that they approximate more closely to the above model source conditions and, particularly important, their source locations and origin times can be closely known.

Chemical explosions are not normally large enough to provide information on conditions very far below the surface, but have been powerfully used in unravelling the structure down to some 50 km depth in particular geographical regions.

With nuclear explosions (B, Chapter 16), the released energy is commonly expressed in equivalent kilotons (kt) or megatons (Mt) of TNT explosive (1 kt = 10^3 tons, 1 Mt = 10^6 tons), 1 kt corresponding to 4×10^{12} J. In general only a small proportion of released nuclear energy (1 per cent or less) goes into seismic waves. Although the seismic wave energy in the largest nuclear explosions is consequently well short of that in the largest natural earthquakes, records

[Refs. on p. 127]

of nuclear explosions have communicated information on all regions of the Earth's interior, including the deepest. The information, moreover, is much more precise than that derived through natural earthquakes, though it is subject to the limitation that the source locations of nuclear explosions to date have been fairly narrowly restricted. Thus nuclear explosions have so far filled only a supplementary role, albeit an invaluable one, to natural earthquakes in unravelling the structure of the Earth's interior.

8.2 Equations of motion of seismic disturbances

The theory used in interpreting the information written on seismograms will now be outlined, so far as needed for the density problem. The theory here given will relate mainly to wave transmission through the Earth at points not too close to the source. It is sufficient for our purpose (see §7.4) to use the model relations (7.7) or (7.9,10). The passage of a group of waves through the material in the neighbourhood of a point P (x_i) is in general sufficiently rapid to preclude much transfer of heat. Hence it is the adiabatic incompressibility k (see §7.9.5) that is essentially involved.

8.2.1 *General form of equations*

At time t, let u_i be the components of displacement at P, let p_{ij} be the components of stress, and U the gravitational potential. We take u_i, p_{ij}, and U to be zero in the relative equilibrium configuration existing prior to the passage of the waves. (Thus the changes in U of interest here are associated with the disturbance to density and gravity as the waves pass.) Let ρ be the density at P.

By considering the forces on a small parallelepiped of matter containing P, it can be shown (see B, §2.1.5) that, to sufficient accuracy,

$$\rho\left(\frac{\partial^2 u_i}{\partial t^2}\right) = \sum \left(\frac{\partial p_{ij}}{\partial x_j}\right) + \rho\left(\frac{\partial U}{\partial x_i}\right). \tag{8.1}$$

8.2.2 *Perfect-elasticity isotropic case*

The equation (8.1) holds for any continuously deformable material, irrespective of the particular stress-strain relations. We now assume the stress-strain relations (7.7,9,10), and take the components $\partial u_i/\partial x_j$ (and hence θ, e_{ij}, and ξ_{ij}) to be all zero in the equilibrium

[Refs. on p. 127]

configuration.

From the definitions of p, p_{ij}, P_{ij}, θ, e_{ij}, E_{ij}, and ξ_{ij} (§§7.1–7.2.5) and relations among (7.1–10), we derive

$$p_{ij} = k\theta\delta_{ij} + 2\mu E_{ij}, \tag{8.2}$$

$$\sum \left(\frac{\partial E_{ij}}{\partial x_j}\right) = \sum \left\{\frac{\partial}{\partial x_j}\left(\frac{\partial u_j}{\partial x_i} - \xi_{ij} - \tfrac{1}{3}\theta\delta_{ij}\right)\right\}$$

$$= \frac{2}{3}\frac{\partial\theta}{\partial x_i} - \sum \left(\frac{\partial \xi_{ij}}{\partial x_j}\right). \tag{8.3}$$

The last line of (8.3) follows since $\sum \partial^2 u_j/\partial x_j \partial x_i = \partial \sum (\partial u_j/\partial x_j)/\partial x_i = \partial\theta/\partial x_i$ and $\sum(\delta_{ij}\partial/\partial x_j) = \partial/\partial x_i$.

Substituting from (8.2) into (8.1) and using (8.3) gives

$$\rho\frac{\partial^2 u_i}{\partial t^2} = (k + \tfrac{4}{3}\mu)\frac{\partial\theta}{\partial x_i} - \sum\left(2\mu\frac{\partial\xi_{ij}}{\partial x_j}\right) + \rho\frac{\partial U}{\partial x_i} + \theta\frac{\partial k}{\partial x_i} + 2\sum\left(E_{ij}\frac{\partial\mu}{\partial x_j}\right). \tag{8.4}$$

8.2.3 Simplification

In a completely uniform medium, the gradients $\partial\rho/\partial x_i$, $\partial k/\partial x_i$, and $\partial\mu/\partial x_i$ are zero. In the Earth, these gradients are of course not zero, but in (8.4) and equations derived from (8.4), it is found (after numerical examination) that the terms containing them are in general small compared with other terms. The effect of taking these gradients into account is to introduce a little dispersion (see §8.5.4) into the associated wave transmission, but the effect is minor (§8.6) in relation to our main problem of determining values of ρ, k, and μ in the Earth.

Thus it is sufficient for present purposes to use the simplified equations

$$\frac{\partial^2 u_i}{\partial t^2} = \alpha^2 \frac{\partial\theta}{\partial x_i} - 2\beta^2 \sum\left(\frac{\partial\xi_{ij}}{\partial x_j}\right) + \frac{\partial U}{\partial x_i}, \tag{8.5a}$$

or, in vector form,

$$\frac{\partial^2 \mathbf{u}}{\partial t^2} = \alpha^2 \operatorname{grad} \theta - \beta^2 \operatorname{curl} \boldsymbol{\xi} + \operatorname{grad} U, \tag{8.5b}$$

where

$$\alpha^2 \rho = k + 4\mu/3, \quad \beta^2 \rho = \mu, \tag{8.6,7}$$

and $\boldsymbol{\xi} = \text{curl } \mathbf{u}$ (see §7.2.5).

Further, in differentiating terms in (8.5) with respect to x_i, it is generally sufficient to treat α and β as constants.

8.3 Bodily seismic waves

The equations (8.5) are satisfied by several types of wave form. The waves include bodily waves which travel three-dimensionally through the medium, and surface waves which travel in the vicinity of a guiding boundary with amplitudes steadily diminishing in general with distance from the boundary. The present section is concerned with bodily waves.

In a medium to which (8.5) applies, two distinct types of bodily waves — the primary (P) and the secondary (S) — can in general be transmitted. This is because the expressions on the right side of (8.5b) are either gradients or curls of vectors.

8.3.1 P waves

If q is any scalar and $\mathbf{v}$ any vector function of position, standard vector theory gives div grad $q = \nabla^2 q$ and div curl $\mathbf{v} = 0$. Also div $\mathbf{u} = \Sigma(\partial u_i/\partial x_i) = \theta$.

Taking the divergence of the vector equation (8.5b) [or applying the operation $\partial/\partial x_i$ to (8.5b)], we therefore obtain

$$\frac{\partial^2 \theta}{\partial t^2} = \alpha^2 \nabla^2 \theta + \nabla^2 U$$

$$= \alpha^2 \nabla^2 \theta + 4\pi G \rho \theta, \qquad (8.8)$$

using Poisson's equation (4.8); U is the gravitational potential associated with the disturbance in density, equal to $-\rho\theta$ by §7.2.2; G is the gravitation constant.

Using a representative value of ρ and taking representative earthquake source conditions, Jeffreys (1931) inferred that the influence of the last term in (8.8) would normally be unobservable on a seismogram.

To high accuracy, we then have

$$\frac{\partial^2 \theta}{\partial t^2} = \alpha^2 \nabla^2 \theta, \qquad (8.9)$$

which (on the various mathematical model conditions taken) is the equation for the transmission of P bodily waves inside regions of the Earth where ρ, k, and μ are treated as continuous. The equation (8.9), which is a particular case of the classical three-dimensional wave equation, is associated with a dilatational (irrotational) disturbance travelling with speed $\alpha = \sqrt{\{(k + 4\mu/3)/\rho\}}$.

8.3.2 S waves

Given that q is any scalar function of position, vector theory further gives curl grad $q = 0$, and curl curl $\mathbf{v}$ = grad div $\mathbf{v} - \nabla^2\mathbf{v}$. Thus curl curl $\mathbf{v} = -\nabla^2\mathbf{v}$ in the particular case where $\mathbf{v}$ is expressible as the curl of a vector.

Taking the curl of equation (8.5b) hence yields

$$\frac{\partial^2 \boldsymbol{\xi}}{\partial t^2} = \beta^2 \nabla^2 \boldsymbol{\xi}, \tag{8.10}$$

the usual equation for transmission of S bodily waves. This is again a particular case of the classical wave equation and is associated with a rotational (equivoluminal) disturbance travelling with speed $\beta = \sqrt{(\mu/\rho)}$.

By (8.7), the S wave speed is zero for a perfect fluid. Thus the detection of S as well as P waves in a region of the Earth provides evidence of solidity in the sense defined in §7.7. Failure to detect S waves where P waves are observed generally suggests fluidity.

8.3.3 Further simplification

The term involving U in (8.5) was seen to be unimportant in P wave transmission, and it disappeared in the derivation of the S wave equation (8.10). We shall therefore simplify further by taking in place of (8.5), unless stated otherwise,

$$\frac{\partial^2 u_i}{\partial t^2} = \alpha^2 \frac{\partial \theta}{\partial x_i} - 2\beta^2 \sum \left(\frac{\partial \xi_{ij}}{\partial x_j}\right) \tag{8.11a}$$

or

$$\frac{\partial^2 \mathbf{u}}{\partial t^2} = \alpha^2 \operatorname{grad} \theta - \beta^2 \operatorname{curl} \boldsymbol{\xi}, \tag{8.11b}$$

in which the term in U has been dropped. Equations equivalent to

(8.11) are commonly taken as the starting point in routine seismology theory.

[The gravity term grad U of (8.5) can have a significant effect in some surface wave problems where the periods are very long. Reference to this term will be made again in §13.4.]

8.3.4 *Plane P and S waves*

The form of a plane wave involving the components u_i of displacement and advancing with speed v in the direction l_i is expressible as

$$u_i = F_i\{\Sigma(l_j x_j) - vt\}, \qquad (8.12)$$

where the F_i ($i = 1, 2, 3$) are functions of the expression in curled brackets.

Substituting (8.12) as a trial solution into (8.11a) is found [see e.g. Bullen (1954), p. 25] to yield a cubic equation in v^2, with roots α^2, β^2, β^2. The case $v = \alpha$ gives longitudinal waves, the u_i having the direction given by l_i; these are P-type waves, the plane-wave case of dilatational waves. The two cases $v = \beta$ give two sets of transverse waves polarized, respectively, in two planes which are at right angles to one another and are parallel to the direction of wave advance; these are S-type waves, the plane-wave case of rotational waves.

It follows that P and S waves are the only bodily plane waves possible under the stated conditions, and that S waves may be plane polarized. The plane-wave case is relevant to our density problem which rests to a considerable extent on observations of seismic waves sufficiently distant from the source to be treated as having plane wave-fronts.

Polarized S waves in which the disturbed particles all move horizontally are referred to in seismology as SH; those in which the particles all move in vertical planes, SV.

8.4 Scalar and vector potentials

In conditions that are sufficiently general for the present context, vector theory enables the displacement **u** to be expressed in the form

$$\mathbf{u} = \text{grad } \phi - \text{curl } \boldsymbol{\psi}, \qquad (8.13)$$

where div $\boldsymbol{\psi} = 0$; ϕ and $\boldsymbol{\psi}$ are called scalar and vector potentials. Taking the divergence and curl of (8.13) yields, respectively,

[*Refs. on p.* 127]

8.5] SCALAR AND VECTOR POTENTIALS

$$\theta = \nabla^2 \phi, \quad \boldsymbol{\xi} = \nabla^2 \boldsymbol{\psi}. \qquad (8.14, 15)$$

If we set down the relations

$$\frac{\partial^2 \phi}{\partial t^2} = \alpha^2 \nabla^2 \phi, \quad \frac{\partial^2 \boldsymbol{\psi}}{\partial t^2} = \beta^2 \nabla^2 \boldsymbol{\psi}, \qquad (8.16, 17)$$

it is immediately obvious, using (8.14,15), that (8.11b) is satisfied. For many purposes it is convenient to use (8.13), (8.16), and (8.17) in place of (8.11) in deriving properties of seismic waves. (It is only the derivatives of ϕ and $\boldsymbol{\psi}$ that are physically significant.)

An advantage of introducing ϕ and $\boldsymbol{\psi}$ is that they enable wave equations (8.16) and (8.17) to be formulated in which the dilatational and rotational disturbances are expressed in separate equations from the outset. In particular, (8.16) and (8.17) (on applying the operation ∇^2 to both sides) immediately yield (8.9) and (8.10), and hence the P and S wave velocities. The separation is particularly useful in investigating surface waves, where the mathematical analysis is more complicated than in §8.3.

8.5 Surface seismic waves

8.5.1 *Preliminaries*

In surface waves, the maximum motion is in the vicinity of a boundary surface which acts as a wave guide. (The distance from the boundary to where the displacements cease to be physically significant depends on the wave lengths — see Chapter 13.) With seismic surface waves, the dominating boundary of interest is usually the outer surface of the Earth since this is where seismic observations are gathered.

Because of the complexity of the mathematics, the procedure in treating the theory of seismic surface waves is to assume severely restricted mathematical model conditions to start with, introducing modifications stage by stage in endeavours to improve the fit with observation. The initial model is a horizontally stratified 'flat Earth'. Axes Ox_i will be taken so that O is in the surface and Ox_3 points vertically downward. The surface waves will be taken as advancing in the direction Ox_1, the wave-fronts being planes parallel to $Ox_2 x_3$ (Fig. 8.1), and the displacements being the same at any instant at all points of any line parallel to Ox_2.

[*Refs. on p. 127*]

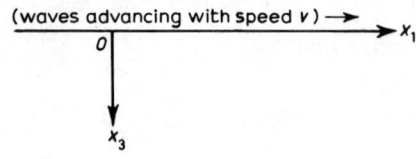

Fig. 8.1

Since derivatives with respect to x_2 are zero in these conditions, u_2 is equal to the component of $-\text{curl }\boldsymbol{\psi}$ parallel to Ox_2, and (8.13) gives

$$u_1 = \frac{\partial \phi}{\partial x_1} + \frac{\partial \psi}{\partial x_3}, \quad u_3 = \frac{\partial \phi}{\partial x_3} - \frac{\partial \psi}{\partial x_1}, \qquad (8.18,19)$$

where ψ denotes the component of $\boldsymbol{\psi}$ parallel to Ox_2. Then (8.16,17) become

$$\frac{\partial^2 \phi}{\partial t^2} = \alpha^2 \nabla^2 \phi, \quad \frac{\partial^2 \psi}{\partial t^2} = \beta^2 \nabla^2 \psi, \quad \frac{\partial^2 u_2}{\partial t^2} = \beta^2 \nabla^2 u_2, \qquad (8.16,17a,17b)$$

where $\nabla^2 = \partial^2/\partial x_1^2 + \partial^2/\partial x_3^2$.

In seeking to solve (8.16,17) we take trial solutions of sinoidal (monochromatic) form, namely

$$\phi, \psi, u_2 = (f, g, h)\exp\{\iota\kappa(x_1 - vt)\}. \qquad (8.20)$$

In (8.20), $2\pi/\kappa$ is the *wave length*, (f, g, h) are functions of x_3, and v is the *wave velocity*; κ is sometimes called the *wave number*; $2\pi/\kappa v$ is the *wave period*. On the right sides of equations like (8.20), it is understood that real parts are to be taken.

It is convenient to introduce r and s, defined by

$$\alpha^2(1+r^2) = \beta^2(1+s^2) = v^2; \qquad (8.21)$$

r and s when real are taken to be positive, and when imaginary to have the form $a\iota$, where a is positive.

Substituting from (8.20) into (8.16,17) gives

$$\frac{d^2(f, g, h)}{dx_3^2} + \kappa^2(fr^2, gs^2, hs^2) = 0, \qquad (8.22)$$

[*Refs. on p. 127*]

whence
$$f = A\exp(\iota\kappa rx_3) + D\exp(-\iota\kappa rx_3), \tag{8.23}$$
and two similar equations, where A, D, etc., are constants. Thus (8.20) become

$$\phi = A\exp\{\iota\kappa(rx_3 + x_1 - vt)\} + D\exp\{\iota\kappa(-rx_3 + x_1 - vt)\}, \tag{8.24a}$$

$$\psi = B\exp\{\iota\kappa(sx_3 + x_1 - vt)\} + E\exp\{\iota\kappa(-sx_3 + x_1 - vt)\}, \tag{8.24b}$$

$$u_2 = C\exp\{\iota\kappa(sx_3 + x_1 - vt)\} + F\exp\{\iota\kappa(-sx_3 + x_1 - vt)\}. \tag{8.24c}$$

8.5.2 Rayleigh waves

Rayleigh (1885) examined the simple case of surface wave transmission over a uniform flat Earth model.

In this case, the wave amplitudes are required to diminish steadily with the distance x_3 below the surface. Hence D, E, and F in (8.24) must all be zero and r and s must be imaginary. One immediate consequence is that the velocity v of Rayleigh waves is less than β.

The boundary conditions require the stress components p_{31}, p_{32}, and p_{33} to be zero at the surface. By (7.3), (8.14), (8.18), and (8.19),

$$\theta = \frac{\partial^2\phi}{\partial x_1^2} + \frac{\partial^2\phi}{\partial x_3^2}, \quad e_{33} = \frac{\partial^2\phi}{\partial x_3^2} - \frac{\partial^2\psi}{\partial x_1 \partial x_3}, \tag{8.25,26}$$

$$2e_{31} = \frac{2\partial^2\phi}{\partial x_1 \partial x_3} - \frac{\partial^2\psi}{\partial x_1^2} + \frac{\partial^2\psi}{\partial x_3^2}, \quad 2e_{32} = \frac{\partial u_2}{\partial x_3}, \tag{8.27,28}$$

derivatives with respect to x_2 being zero. By (7.7), the boundary conditions therefore require that, at $x_3 = 0$,

$$2\frac{\partial^2\phi}{\partial x_1 \partial x_3} - \frac{\partial^2\psi}{\partial x_1^2} + \frac{\partial^2\psi}{\partial x_3^2} = 0, \quad \frac{\partial u_2}{\partial x_3} = 0, \tag{8.29,30}$$

$$(\alpha^2 - 2\beta^2)\frac{\partial^2\phi}{\partial x_1^2} + \alpha^2\frac{\partial^2\phi}{\partial x_3^2} - 2\beta^2\frac{\partial^2\psi}{\partial x_1 \partial x_3} = 0, \tag{8.31}$$

(8.6) and (8.7) being used in obtaining (8.31).

Substituting from (8.24c) (with $F = 0$) into (8.30) gives $C = 0$. This gives the important result that there cannot be SH Rayleigh waves.

Substituting from (8.24a,b) (with $D = E = 0$) into (8.29) and (8.31), putting $x_3 = 0$ and cancelling the factor $\kappa^2 \exp\{\iota\kappa(x_1 - vt)\}$,

[Refs. on p. 127]

gives
$$2rA + (s^2 - 1)B = 0, \tag{8.32}$$
$$\{\alpha^2(1 + r^2) - 2\beta^2\}A - 2\beta^2 sB = 0. \tag{8.33}$$

The equations (8.32) and (8.33) show that if A is not zero, B is in general also not zero, and vice versa. Hence Rayleigh waves involve both ϕ and ψ, and are therefore associated with combined P and SV motion.

Eliminating A and B from (8.32) and (8.33), and using (8.21), we obtain for the speed v of Rayleigh waves,

$$\{2 - (v/\beta)^2\}^4 = 16(1 - v^2/\alpha^2)(1 - v^2/\beta^2), \tag{8.34}$$

which reduces to a cubic equation in v^2. It can be shown (B, §5.2) that (8.34) always yields one and only one positive value of v^2 and that this value is less than β^2. Hence, in the Rayleigh model conditions, a P-SV disturbance can always be transmitted along the surface.

Specimen calculations for solids frequently assume $\alpha^2 = 3\beta^2$. (This is equivalent to assuming Poisson's relation $\sigma = \frac{1}{4}$, a moderately good approximation for many solids.) In this case, the Rayleigh wave speed is found from (8.34) to be $v^2 = (2 - 2/\sqrt{3})\beta^2$ and the motion of the surface particles is given by

$$u_1 \approx 0.42a \cos\{\kappa(x_1 - vt)\}, \quad u_3 \approx 0.62a \sin\{\kappa(x_1 - vt)\}, \tag{8.35}$$

where a is constant. According to (8.35), each particle describes a vertical ellipse, the motion being retrograde in relation to the direction of wave advance.

For the more general case in which the waves are not sinoidal, the solution is obtained by superposing the results for sinoidal waves with different wave lengths $2\pi/\kappa$. The equation for the surface particle motion is then of course more complicated than (8.35).

An important property of the simple Rayleigh wave theory is that κ does not enter the equation (8.34) for v. The Rayleigh wave speed is therefore independent of the wave length, and the result of superposing waves of different κ is a displacement pattern which keeps a constant shape as it advances. In other words, ordinary Rayleigh waves are not subject to dispersion (see §8.5.4).

8.5.3 Love waves

In two important respects, the simple Rayleigh model is seriously inadequate as a description of the usual seismic surface wave observations. First, SH surface waves are usually observed as prominently as the P and SV. Secondly, the observed surface waves — all of P, SV, and SH — are subject to marked dispersion. Love (1911) showed that both inadequacies could be corrected by including in the model a uniform layer L', of constant thickness H' say, immediately below the free surface, this layer L' resting in welded contact on the semi-infinite Rayleigh-type medium, L say, below. Following Love, we investigate in this section only the u_2 component of displacement. Thus we here seek solutions of the form (8.24c), not (8.24a) or (8.24b).

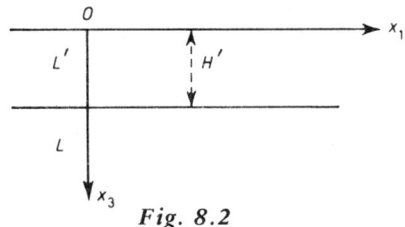

Fig. 8.2

We take axes as in §8.5.1, with L' lying between the planes $x_3 = 0$, $x_3 = H'$, and L lying below the plane $x_3 = H'$ (Fig. 8.2). For L, unprimed symbols will be retained, and (8.24c) used as it stands; for L', (8.24c), modified by having the primed symbols μ', β', r', s', C', and F', will be used. (The wave velocity v and wave length $2\pi/\kappa$, being specific for a given sinoidal wave, are the same in both L and L'; hence there is no need to use primed symbols for v and κ in L'.)

The surface-wave stipulation requires u_2 to diminish with depth inside L, but this does not apply inside L'. Thus s has still to be imaginary, but not s'. Since s is imaginary, F has to be zero; but C' and F' are both non-zero. Thus the appropriate equations for the displacements in L' and L are, respectively,

$$u_2 = C' \exp\{\iota\kappa(s'x_3 + x_1 - vt)\} + F' \exp\{\iota\kappa(-s'x_3 + x_1 - vt)\}, \quad (8.36)$$
$$u_2 = C \exp\{\iota\kappa(sx_3 + x_1 - vt)\}. \quad (8.37)$$

The further boundary requirements are that $p_{32} = 0$ at $x_3 = 0$, and that u_2 and p_{32} are continuous at $x_3 = H'$. Since $p_{32} = 2\mu e_{32} = \mu \partial u_2/\partial x_3$, these requirements give

$$C' - F' = 0, \quad (8.38)$$
$$C' \exp(\iota\kappa s'H') + F' \exp(-\iota\kappa s'H') = C \exp(\iota\kappa sH'), \quad (8.39)$$

[Refs. on p. 127]

$$\mu's'\{C' \exp(\iota\kappa s'H') - F' \exp(-\iota\kappa s'H')\} = \mu s C \exp(\iota\kappa s H'). \tag{8.40}$$

Eliminating C, C', and F' between (8.38), (8.39), and (8.40) gives

$$\mu\iota s + \mu's' \tan(\kappa s'H') = 0. \tag{8.41}$$

Substituting for s from (8.21), and similarly for s', into (8.41) gives

$$\mu(1 - v^2/\beta^2)^{\frac{1}{2}} - \mu'(v^2/\beta'^2 - 1)^{\frac{1}{2}} \tan\{\kappa H'(v^2/\beta'^2 - 1)^{\frac{1}{2}}\} = 0, \tag{8.42}$$

which is the equation for the velocity v of Love waves.

Provided $\beta' < \beta$, which is in accord with general observed S velocity trends in the outer part of the Earth, (8.42) yields a real value of v. (The value lies in the range $\beta' < v < \beta$.) Hence the Love model conditions allow the transmission of SH surface waves, thus meeting one of the observed features of seismic surface waves.

Further, the equation (8.42), unlike (8.34) for Rayleigh waves, gives v dependent on the wave length $2\pi/\kappa$. Hence (see §8.5.4) Love waves are subject to dispersion, thus meeting the second observational feature.

By (8.36) and (8.38), the motion inside L' is given by

$$u_2 = 2C' \cos(\kappa s' x_3) \cos\{\kappa(x_1 - vt)\}. \tag{8.43}$$

In (8.37) and (8.43), C and C' are functions of κ.

Further theory on Love waves is given in §13.1. See also B, Chapter 5, for additional detail on both Rayleigh and Love waves.)

8.5.4 *Group velocity*

Consider a disturbance, for example that due to an earthquake, which originates inside a limited region R during a limited time-interval and travels outward from R. Take an origin O inside R and take the zero of time inside the initial time-interval. Let t denote any later time, and x the distance from O.

In general, the single sinoidal forms assumed for u_2 etc. in §§ 8.5.1–8.5.3 would not by themselves represent such a disturbance adequately. But by Fourier theory it is normally sufficient to treat a a more general disturbance as the result of superposing an indefinite number of sinoidal constituents, the wave lengths $2\pi/\kappa$ (and hence also the periods $2\pi/\kappa v$) being different from one constituent to another.

Where, as in (8.42), the formal assumption of sinoidal waves yields

[Refs. on p. 127]

an equation in which the wave or 'phase' velocity v depends on κ, the individual sinoidal constituents of a general disturbance travel with different speeds. In such a case, as time goes on, the disturbance therefore becomes increasingly spread or *dispersed*, and the distributions of displacement inside limited ranges of the distance x approximate increasingly to sinoidal form.

The main characteristics of dispersion are exhibited by considering the case of one-dimensional wave transmission through a uniform medium and analysing asymptotic approximations to the displacement, $y(x,t)$ say, for large x and t. It can be shown (see B, §3.3.5.1) that in general

$$y(x,t) \sim \phi(\kappa)\{\tfrac{1}{2}\pi|d^2(\kappa v)/d\kappa^2|t\}^{-\frac{1}{2}} \cos(\kappa x - \kappa v t + \epsilon(\kappa) \pm \tfrac{1}{4}\pi), \quad (8.44)$$

where $\phi(\kappa)$ depends on the initial conditions, ϵ (which also depends on κ — see Brune, Nafe, and Alsop, 1961) is a phase angle, and κ and v are connected with x and t by the relation

$$x/t = d(\kappa v)/d\kappa = V(\kappa), \text{ say.} \quad (8.45)$$

An alternative approximation (B, §8.8), useful in some contexts, is

$$y(x,t) \sim \psi(\kappa v)\{\tfrac{1}{2}\pi|d^2\kappa/d(\kappa v)^2|x\}^{-\frac{1}{2}} \cos(\kappa x - \kappa v t + \epsilon \pm \tfrac{1}{4}\pi), \quad (8.46)$$

where $\psi(\kappa v)$ corresponds to $\phi(\kappa)$.

In (8.44) and (8.46), the upper or lower sign is taken according as $d^2(\kappa v)/d\kappa^2$, i.e. $dV/d\kappa$, is negative or positive. Both (8.44) and (8.46) apply fairly generally, although they fail in certain exceptional conditions (B, §3.3.5.4) which can arise in some important practical contexts. The following discussion is relevant to the more usual contexts in which (8.44) and (8.46) hold.

If x and t are sufficiently large, (8.44) shows that, for given t, the displacement is approximately sinoidal in the vicinity of any particular point x. There is thus a cluster of waves, all with nearly the same wave length $2\pi/\kappa$, corresponding to any given (x,t).

Further, (8.45) shows that, if x and t are continuously increased in such manner that x/t is kept constant, then the function $V(\kappa)$ is equal to that constant. Expressed otherwise, a group of waves with wave lengths always clustering about a given value of $2\pi/\kappa$ travels away from O at a speed equal to the *group velocity* $V(\kappa)$. For different wave lengths $2\pi/\kappa$, there are different groups, each travelling with its own group velocity. By (8.45),

[*Refs. on p. 127*]

$$V = \mathrm{d}(\kappa v)/\mathrm{d}\kappa = v + \kappa \mathrm{d}v/\mathrm{d}\kappa. \tag{8.47}$$

By (8.47), the group velocity V in general differs from the phase velocity v; but $V = v$ when v is independent of κ (as with Rayleigh waves).

8.5.5 *Extensions of Rayleigh and Love wave theory*

The theory of simple Rayleigh waves relates to a flat half-space throughout which the density is uniform, and so could not (even apart from other limitations) itself contribute information on changes in density etc. inside the Earth. Nevertheless the theory is important as a step towards investigating *P-SV* waves in more complicated models.

The next least complicated model is the Love model structure of §8.5.3. For this structure, Love (1911) himself supplied theory for *P-SV* waves but only for the particular case of incompressible media. The analysis was generalized to include compressible media, notably by Sezawa (1927), Stoneley (1928), Lee (1932), and Jeffreys (1935). The *P-SV* waves involve the displacement components u_1 and u_3, and the algebraic detail is appreciably more complicated than for u_2 since u_1 and u_3 are involved jointly. There are now six boundary conditions in place of the three given by (8.38)–(8.40), and six constants in place of C, C', and F' of §8.5.3. Eliminating the six constants gives a single equation analogous to (8.42), but now including the P velocities α and α', as well as the S velocities β and β', for the two media L and L'. Examination of the roots of the equation shows that *P-SV* waves exist for the Love-type structure and the equation for the wave velocity v again involves κ, so that the waves, like the *SH* waves, are dispersed. (See also B, §5.4.)

The assumption of a simple Love model structure served usefully in early seismological investigations of the Earth's crust. At later stages the studies were extended to include models with two and three surface layers resting on a uniform semi-infinite substratum. These and further theoretical developments are referred to in §13.2.

The theoretical developments have accompanied developments in observational technique that have greatly extended the range of seismic periods that can be reliably measured. In fact, observations of the whole seismic spectrum, including periods much greater than the order of a minute (an earlier observational limit), have now become available. As a consequence, surface-wave observations can now supply useful information on the structure of the Earth down to depths of several hundred kilometres.

When depths of this order are involved, the model conditions of §§8.5–8.5.3 have to be modified in further respects. Allowances need to be included for such features as the Earth's curvature and for continuous variation of ρ, k, and μ inside particular layers. The additional features all contribute to the observed wave dispersion. These points will be elaborated in §§13.2, and 13.4.

With periods exceeding the order of three minutes, surface wave observations merge with observations of the Earth's free oscillations (Chapter 14), a recent additional source of information on the Earth's density distribution.

8.5.6 *Dispersion curves*

In the process of deriving information about the Earth's interior from observations of surface waves, use is made of *dispersion curves*. These curves connect any two of a set of variables which include V, v, κ, and related variables such as the period. (Sometimes it is convenient to include in the set a variable which involves a feature of the Earth's structure, e.g. the variable $\kappa H'$ in the simple Love case.)

Observations of a single seismogram yield, among other things, a numerical relation between the periods $2\pi/\kappa v$ of groups of surface waves and the corresponding group velocities V for a particular displacement component. An observational dispersion curve can be constructed from these observations. A corresponding theoretical dispersion curve may also be constructed. For this purpose, some model Earth structure is assumed and numerical values taken from the model are substituted into the wave-velocity equation. For example, in the simple Love case, numerical values of μ/μ', β, and β' may be substituted into (8.42), yielding a relation between $\kappa H'$ and v/β'.

Comparison of corresponding observational and theoretical dispersion curves assists in the process of utilizing surface-wave observations (see §13.1).

8.6 Refraction and reflexion of bodily seismic waves

In deriving the equations of §8.2.3, variation of ρ, k, and μ (and hence also of α and β) inside the medium was neglected. Continuous variation of these quantities causes bodily seismic waves to be

[*Refs. on p. 127*]

continuously refracted. At boundaries where α or β is discontinuous, the waves are in general refracted and reflected, with discontinuous changes in the directions of wave advance.

Terms such as continuous and discontinuous relate of course only to mathematical models. When, within distances of order Δx along normals to a level-surface S, there are proportionate changes in ρ, k, or μ that are large compared with the ratio of Δx to the wave lengths of interest, it is commonly suitable to treat S as a surface of mathematical discontinuity. At such a surface, incident bodily waves are in general reflected as well as refracted. But with shorter waves the discontinuous model may be unsuitable and there may be no detectable reflexion. Additional questions of model representation can arise wherever diffraction effects (§9.1.1) are significant.

Another effect of continuous variation of ρ, k, and μ is to introduce into equations such as (8.5)–(8.7), (8.11), etc., terms which depend on $\partial \rho/\partial x_i$, $\partial k/\partial x_i$, and $\partial \mu/\partial x_i$. Such terms introduce dispersion effects for bodily waves as well as surface waves, but the corresponding observed dispersion is generally very slight in the case of bodily waves and will here be ignored.

In interpreting bodily-wave observations, it is sufficient for the main requirements of our density problem to consider fairly simple model representations as in the following subsections.

8.6.1 *Refraction and reflexion at a plane boundary*

When a train of P or S bodily waves meets a boundary surface across which one or more of ρ, k, and μ is taken to change discontinuously, there are in general four derived wave trains – refracted P and S, and reflected P and S.

The general character of refraction and reflexion is exhibited in Fig. 8.3 for the case of SV waves incident against a plane boundary separating two different uniform media L and L'; the boundary has been conventionally taken as horizontal. It is convenient to take axes as in Fig. 8.3, to use notation as in §8.5 and to represent the displacement again in terms of ϕ, ψ, and u_2.

Incident plane waves approaching the boundary may be represented [cf. (8.12)] by one of the forms

$$\phi = F_0(x_1 \cos e_0 - x_3 \sin e_0 - \alpha t), \qquad (8.48)$$

$$\psi = G_0(x_1 \cos f_0 - x_3 \sin f_0 - \beta t), \qquad (8.49)$$

[*Refs. on p. 127*]

$$u_2 = H_0(x_1 \cos f_0 - x_3 \sin f_0 - \beta t), \qquad (8.50)$$

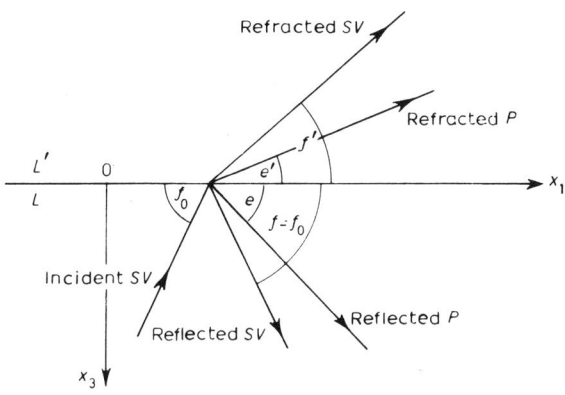

Fig. 8.3

corresponding respectively to plane P, SV, or SH waves approaching the boundary with velocities α, β, and β in directions making angles e_0, f_0, and f_0 with the boundary. In (8.48)–(8.50), F_0, G_0, and H_0 are functions of the expressions in brackets. The angle e_0 or f_0 is called the *angle of emergence;* its complement is the *angle of incidence*. Forms similar to (8.48)–(8.50) correspond to the reflected and refracted waves [see e.g. (8.51) and (8.52) for the case where the incident waves are SH]. In Fig. 8.2, e and f relate to the reflected P and SV waves; e' and f' to the refracted P and SV waves.

Given one of (8.48)–(8.50) as representing an incident wave train, the properties of the derived waves are obtained through applying the boundary conditions: at $x_3 = 0$, the stress components p_{31}, p_{32}, and p_{33} are to be continuous, and also the displacement components u_1, u_2, and u_3 (assuming welded contact).

Incident P waves are found to give rise in general to refracted and reflected P and SV waves, but not SH; incident SV waves to P and SV, but not SH; incident SH waves to SH, but not P or SV. (For details, see B, Chapter 6.)

For given α, β, α', and β', the theory also gives details on how the incident energy is partitioned among the derived waves for a specified angle of emergence. For certain particular values of the angle of emergence, there may be no refracted waves (total reflexion), no reflected waves, or refraction or reflexion confined to the P or S

[*Refs. on p.* 127]

type. For some ranges of values of e_0 or f_0, the energies in particular derived waves may be abnormally small.

8.6.2 *Generalization of Snell's law*

The angles e_0, e, e', f_0, f, and f' (see Fig. 8.3) are connected by relations that correspond to a generalization of Snell's law in optics. This will be shown in the particular case of *SH* waves, taking the model conditions of §8.6.1 and representing the incident waves by (8.50).

The refracted and reflected waves are appropriately represented by

$$u_2 = H'(x_1 \cos f - x_3 \sin f' - \beta't), \qquad (8.51)$$

$$u_2 = H(x_1 \cos f + x_3 \sin f - \beta t), \qquad (8.52)$$

respectively. The motion inside L' is completely given by (8.51), while that inside L is given by

$$u_2 = H_0(x_1 \cos f_0 - x_3 \sin f_0 - \beta t) + H(x_1 \cos f + x_3 \sin f - \beta t). \qquad (8.53)$$

Continuity of u_2 at $x_3 = 0$ gives

$$H_0(x_1 \cos f_0 - \beta t) + H(x_1 \cos f - \beta t) = H'(x_1 \cos f' - \beta't) \qquad (8.54)$$

for all x_1, t. Since $p_{32} = \mu \partial u_2/\partial x_3$, continuity of p_{32} at $x_3 = 0$ gives

$$-(\mu \sin f_0)H_0^*(x_1 \cos f_0 - \beta t) + (\mu \sin f)H^*(x_1 \cos f - \beta t)$$
$$= -(\mu' \sin f')H'^*(x_1 \cos f' - \beta't) \qquad (8.55)$$

for all x_1 and t, where a star * here denotes the derivative of the function concerned.

It is possible to satisfy (8.54) and (8.55) for all x_1 and t only if the arguments of the various H functions are in a constant ratio, i.e. if $(\cos f_0)/\beta = (\cos f)/\beta = (\cos f')/\beta'$; thus incidentally, $f = f_0$.

For incident *P* or *SV* waves, it can be shown similarly that, in similar notation,

$$\frac{\cos e_0}{\alpha} \left(\text{or } \frac{\cos f_0}{\beta} \right) = \frac{\cos e}{\alpha} = \frac{\cos f}{\beta} = \frac{\cos e'}{\alpha'} = \frac{\cos f'}{\beta'}, \qquad (8.56)$$

the generalized Snell's law for the model case taken.

8.6.3 *Continuous refraction*

The relations (8.56) can be applied in simple model cases to

[Refs. on p. 127]

determine the changes of wave direction when there is continuous refraction due to continuous change in α or β. For example, when the surfaces of constant α and β are parallel planes, $(\cos e)/\alpha$ is constant along the path of continuously refracted P waves. In other cases, modified forms of (8.56) need to be used; this applies in particular in a spherically symmetrical Earth model, where the curvature of the surfaces of constant α and β causes a modification, to be dealt with in §9.1.2.

REFERENCES

Brune, J.N., Nafe, J.E. and Alsop, L.E. (1961). The polar phase shift of surface waves on a sphere. *Bull. Seismol. Soc. Amer.*, 51, 247–257.

Bullen, K.E. (1954). *Seismology.* London: Methuen.

Gutenberg, B. (1914). Über Erdbebenwellen. VIIA. Beobachtungen an Registrierungen von Fernbeben in Göttingen und Folgerungen über die konstitution des Erdkörpers. *Nachr. Ges. Wiss. Göttingen*, Math.–Phys. Klasse, pp. 1–52 and 125–176.

Jeffreys, H. (1931). On the cause of oscillatory movement in seismograms. *Mon. Not. R. Astr. Soc., Geophys. Suppl.*, 2, 407–416.

Jeffreys H. (1935). The surface waves of earthquakes. *Mon. Not. R. Astr. Soc., Geophys. Suppl.*, 3, 253–261.

Lee, A.W. (1932). The effect of geological structure upon microseismic disturbance. *Mon. Not. R. Astr. Soc., Geophys. Suppl.*, 3, 83–105.

Love, A.E.H. (1911). *Some Problems of Geodynamics.* Cambridge University Press.

Oldham, R.D. (1906). Constitution of the interior of the Earth as revealed by earthquakes. *Quart. J. Geol. Soc.*, 62, 456–475.

Rayleigh, Lord (Strutt, J.W.) (1885). On waves propagated along the plane surface of an elastic solid. *Proc. Lond. Math. Soc.*, 17, 4–11.

Sezawa, K. (1927). Dispersion of elastic waves propagated on the surface of stratified bodies and on curved surfaces. *Bull. Eq. Res. Inst.* (Tokyo), 3, 1–18.

Stoneley, R. (1928). A Rayleigh wave problem. *Proc. Leeds Phil. Soc.*, 1, 217–225.

CHAPTER 9

First approximation to seismic P and S velocity distributions in the Earth

Theory given in Chapter 8 on bodily P and S seismic waves will now be applied to extracting information about the Earth's interior from messages written on seismograms by waves from events such as natural earthquakes and artificial explosions.

To good approximation, the energy in bodily seismic waves can be regarded as transmitted along seismic 'rays'. The rays start from an earthquake focus or explosion source, pass through the interior of the Earth, and emerge at the surface. The record (seismogram) of the bodily waves from an event normally shows a series of *phases*, each of which is associated with travel along a particular class (or family) of rays. Detail on the concept of seismic *ray* and on seismic phases is given in §§9.1 and 9.4.

The times of onset of various phases can often be measured to fairly high precision on seismograms, and the measurements constitute the primary observational evidence on the values at specific depths inside the Earth of certain functions of important physical variables.

The first essential step toward determining these values is to derive from the observed times of onset empirical tables which give the times of travel of particular bodily waves along the rays. The task of evolving satisfactory travel-time tables is interwoven with that of calculating the times of origin and the source locations of the natural earthquakes used for the purpose. Much of the present chapter (§§9.2–9.5) is concerned with this task. One of several complications is the existence of abnormalities in the P and S velocity variations inside some parts of the Earth. These abnormalities need to be discussed in some detail (§9.3) both in interpreting the observational data and in assessing the uncertainties of the final numerical results.

[*Refs. on p. 150*]

The second step is to derive values of the P and S velocities α and β inside the Earth (§9.6) from the travel-time tables. As will be seen, the two steps cannot be wholly separated because of the complications due to abnormal velocity.

The year 1940 marked an important stage where, after a long process of evolution, a reliable first approximation to the distributions of α and β had become available inside much of the Earth, along with useful estimates of the uncertainties Correspondingly reliable first approximations to k/ρ and μ/ρ, where ρ, k, and μ denote density, incompressibility, and rigidity, likewise became available. Information on k/ρ and μ/ρ constitutes an important section of the evidence needed to determine the Earth's density distribution to good precision.

Except where stated, the following theory will relate to a spherically symmetrical Earth model. Thus α and β will generally be treated as functions of the distance r from the Earth's centre or the depth z below the surface. Reference will be made, however, in §§9.2 and 9.5 to deviations from spherical symmetry due to the Earth's ellipticity of figure. Other deviations are discussed in §§12.1.1, 13.6.5, 15.7 and 16.1.2.

9.1 Seismic rays

9.1.1 *General characteristics*

The theory in §§8.3–8.3.3 shows that P and S waves are transmissible from a source of disturbance in a perfectly elastic isotropic medium with speeds α and β depending on the density ρ and elastic properties k and μ. If the medium were uniform and conditions spherically symmetrical with respect to the source, the wave fronts would be spheres advancing outward with constant speed.

A similar situation applies in the transmission of optical waves, the context which gave rise to the concept of ray. In the optical context with a uniform medium, the wave energy is commonly regarded as being transmitted, to good approximation, radially outward inside a host of cones of small solid angle, each cone having its vertex at the source. For sufficiently small solid angles and at distances from the source large compared with the source dimensions, the cones approximate to 'pencils'; the energy is looked upon as being transmitted along these pencils, or along rectilinear rays which are the axes of the indefinitely thin pencils.

[*Refs. on p. 150*]

The classical wave equation governs optical wave transmission, and, subject to certain restrictions, the ray approximation is formally derivable from it. The classical wave equation [in the forms (8.9,10)] also governs P and S seismic wave transmission. Hence the ray concept can be equally well used with seismic waves, subject to restrictions similar to those in the optical case.

When the medium is not uniform, α and β are in general variable functions of position, and the wave fronts are no longer spheres. The ray concept may, however, still be used. The rays still cut wave fronts normally, but are no longer straight lines. This is the case generally in seismology: changes in α and β, continuous and discontinuous, cause the directions of rays to change continuously and discontinuously, respectively (see §§ 8.6–8.6.3). The changes in direction are determined by equations such as (8.56) and the extension given by (9.3) below.

The ray theory is usually satisfactory at distances from the source that are large compared with the source dimensions. The theory is usually very inadequate close to an earthquake source as a consequence of deviations from symmetry in the initial conditions and the source neighbourhood. The theory can also be inadequate wherever the velocity varies substantially inside distances comparable with the wave length. Small-scale irregularities can cause deviations called *scattering* (see § 15.6), resulting in some loss of resolving power when the ray theory is applied. In general, waves are said to be *diffracted* when deviations from the ray theory have to be considered. Diffraction effects can also be important with waves approaching boundaries near grazing incidence. It is possible (e.g. with certain seismic waves which meet the Earth's mantle-core boundary) to have part of the energy in waves to which the ray theory is applicable, and part in scattered or diffracted waves.

In the present chapter, except where mentioned, ray theory will be used. The main theory of the chapter is therefore applicable only to restricted data, for example, data gathered not too close to epicentres. In parts of the Earth where ρ, k, and μ are changing fairly rapidly, the use of ray theory requires restrictions on the wave lengths in the observations used. Other restrictions may be involved near boundaries. It transpires, however, that the applicability of ray theory is fairly wide since the main observed bodily seismic waves have wave lengths of the order of 10 km or less and, for much of the Earth's interior, changes in ρ, k, and μ are mostly fairly small

[Refs. on p. 150]

over distances of this order. [Inside the Earth's crust (§ 10.3.1), and especially where seismic prospecting methods are used, variations of property with distance are more rapid and the lengths of the waves principally studied are correspondingly smaller.]

9.1.2 Parameter of a seismic ray

Let FJ (Fig. 9.1) be a particular seismic ray running from an earthquake focus F to a point J of the Earth's surface, and let L be the lowest point of the ray. Let i be the inclination of the ray to the vertical at any point P of coordinates (r, θ); thus i is the complement of the angle of emergence e (§ 8.6.1).

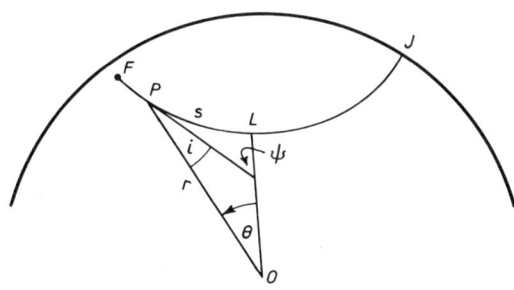

Fig. 9.1

The equations (8.56) govern the refraction and reflexion of P and S seismic rays at plane boundaries. Thus in a horizontally stratified (flat) Earth model, $v^{-1} \sin i, = q$ say, would be constant along the ray, where v is the wave velocity α or β.

Now take a spherically symmetrical Earth model. In general, along any ray, the direction θ of the vertical now changes as r changes. Thus q must now be treated as dependent on r as well as on v, and its rate of change along a ray is expressible as

$$\frac{dq}{dr} = \left(\frac{\partial q}{\partial r}\right)_v + \left(\frac{\partial q}{\partial v}\right)_r \frac{dv}{dr}; \qquad (9.1)$$

the subscripts r and v here indicate variables kept fixed during differentiations.

The factor $(\partial q/\partial v)_r$ in (9.1) is zero since q has been shown to be constant in the flat Earth case. Also, if v is unvaried, a ray runs straight, with $i + \theta = \frac{1}{2}\pi$, as in Fig. 9.2. Hence (9.1) gives

$$\frac{d(v^{-1} \sin i)}{dr} = \left\{\frac{\partial}{\partial r}(v^{-1} \sin i)\right\}_v$$

$$= (v^{-1} \cos i)\left(\frac{\partial i}{\partial r}\right)_v$$

$$= -(v^{-1} \cos i)\frac{d\theta}{dr}$$

$$= -r^{-1}v^{-1} \sin i, \qquad (9.2)$$

since $\tan i = r d\theta/dr$. Integrating (9.2) gives

$$rv^{-1} \sin i = p, \qquad (9.3)$$

where p, called the *ray parameter*, is constant along the particular ray.

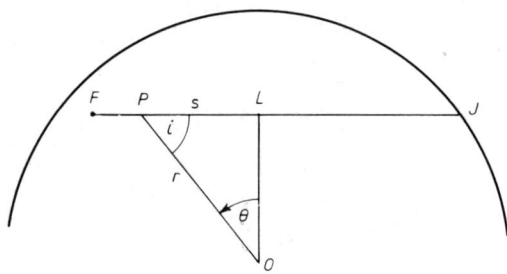

Fig. 9.2

9.1.3 Use of Fermat's principle

According to *Fermat's principle* (which applies to seismic as well as optical rays), the travel time T along a ray is stationary compared with the time along neighbouring paths with the same end points. Fermat's principle can be used to give the following alternative derivation of (9.3).

Let s be the arc length LP (Fig. 9.1). Differentiation with respect to θ will be indicated by a prime. Since $(ds)^2 = r^2(d\theta)^2 + (dr)^2$, we have

$$T = \int v^{-1} ds = \int \phi d\theta, \quad \text{where} \quad \phi = v^{-1}\sqrt{(r^2 + r'^2)}. \quad (9.4)$$

By the calculus of variations, the condition for T to be stationary is $\partial\phi/\partial r = d(\partial\phi/\partial r')/d\theta$, a first integral of which is known to be $\phi = r'\partial\phi/\partial r' + p$, where p is constant. Substituting for ϕ from (9.4) and using $\sin i = r(r^2 + r'^2)^{-\frac{1}{2}}$ gives (9.3) again.

9.1.4 Families of rays

We shall be concerned with *families* of rays which have the following properties: all the rays of any one family start from a focus F at an assigned level (corresponding to the *focal depth*) and finish at another assigned level (usually the Earth's surface in our applications); the P or S character after leaving F is specified; all the rays encounter the same surfaces of discontinuity, and the P or S and refracted or reflected character after every such encounter is specified. A ray will be referred to as being of a specified type when the family to which it belongs is specified.

Let FJ be a ray of specified type, let Δ be its angular length, i.e. the angle subtended by FJ at the Earth's centre O, let T be the time of travel of the seismic waves along FJ, and let p be the parameter of the ray. For a spherically symmetrical Earth model, T depends on Δ and on the depths of F and J below the surface, but not otherwise on the locations of F and J.

By (9.3),
$$p = r_f v_f^{-1} \sin i_f = r_a v_a^{-1} \sin i_a, \tag{9.5}$$

where the subscripts f and a indicate values at the level of F and at the surface, respectively. The angles i_f and i_a determine the steepness at the focus and at the surface, and vary from ray to ray of the family. Thus (9.5) exhibits the fact that the parameter p, while constant along a single ray, varies from ray to ray.

When considering properties of a family, we shall let the steepness at F increase continuously from ray to ray, and so let i_f and p decrease continuously. In general, Δ and T will then be continuous functions of p. (Abnormal velocity variation with depth can cause exceptions — see §9.3.)

A simple formula connects p with Δ and T. Let FJ and FJ' be neighbouring rays (Fig. 9.3), and let $d\Delta$ and dT be the differences in Δ and T for the two rays. Let JN be normal to FJ'. Then $\angle NJJ' \approx i_a$, $NJ' \approx v_a dT$, and $JJ' \approx r_a d\Delta$. Hence $\sin i_a = (v/r)_a dT/d\Delta$, and (9.5) gives
$$p = dT/d\Delta. \tag{9.6}$$

[Refs. on p. 150]

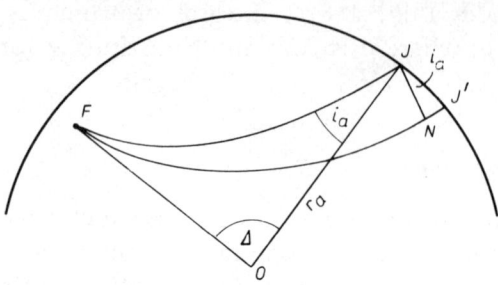

Fig. 9.3

In practice, the dependence of T on Δ for a family of seismic rays is usually expressed in the form of numerical tables. Representations in terms of formal functional relations are, however, used for a variety of purposes. (See, e.g., B, §§ 7.3.1, 7.5.4, on the form $T = A \sin B\Delta$.) The details are also often exhibited in the form of (T, Δ) curves.

9.1.5 *The variables* η, ξ, *and* ζ

It is convenient to introduce η (not the η used in any of Chapters 5, 6, and 11), ξ, ζ, as functions of r defined by

$$\eta = rv^{-1}; \quad \xi = 2(1-\zeta)^{-1} = 2 d \ln r / d \ln \eta. \qquad (9.7, 8)$$

Thus

$$\zeta = (r/v) dv/dr. \qquad (9.9)$$

The relation (9.3) can now be written as

$$\eta \sin i = p. \qquad (9.10)$$

The subscript p will indicate values of functions of r at the lowest point L of the ray whose parameter is p. Since $i_p = \tfrac{1}{2}\pi$, we have $\eta_p = p$.

9.1.6 *Curvature of a ray*

The upward curvature κ at the point P of a seismic ray is equal to $-d\psi/ds$ where $\psi = i + \theta$ (Fig. 9.1). By (9.10) and (9.8), $-\cot i\, di = d \ln \eta = (1-\zeta) d \ln r$ along the ray. Using $\sin i = r d\theta/ds$ and $\tan i = r d\theta/dr$, we then have

$$-r\kappa = r di/ds + r d\theta/ds = -(1-\zeta) \sin i + \sin i.$$

[Refs. on p. 150]

Hence, using (9.9),
$$\kappa = -r^{-1}\zeta \sin i = -(v^{-1}\sin i)dv/dr. \tag{9.11}$$

9.1.7 Expressions connecting Δ and T with r and v

For points on a seismic ray of parameter p, we have, using (9.3),
$$d\theta/dr = r^{-1}\tan i = pvr^{-1}(r^2 - p^2v^2)^{-\frac{1}{2}},$$
$$ds/dr = \sec i = r(r^2 - p^2v^2)^{-\frac{1}{2}}.$$

Using (9.7), we then obtain
$$\Delta = \int d\theta = p\int r^{-1}(\eta^2 - p^2)^{-\frac{1}{2}}dr, \tag{9.12}$$
$$T = \int v^{-1}ds = \int \eta^2 r^{-1}(\eta^2 - p^2)^{-\frac{1}{2}}dr, \tag{9.13}$$

where $\int$ denotes the sum of two integrals ranging upward from the lowest point L on the two sides of L.

When v, and therefore η, is a given function of r, (9.12) and (9.13) in general enable Δ and T to be derived in terms of p. Thus (9.12) and (9.13) supply a linkage between (v,r) and (T,Δ) relations. In general, the divergence of the integrands at L is a source of difficulty only in connection with certain types of abnormal velocity variation to be discussed in §9.3.

9.2 Effect of the Earth's ellipticity on seismic travel times

The Earth's ellipticity, though small enough to be neglected in the first approximation, nevertheless produces observable effects on seismic travel times. A first-order theory, outlined below, is sufficient in practice to meet requirements.

Let X and Y be spherically and spheroidally symmetrical Earth models related as follows. Let $r = q$ be an internal sphere of X over which the seismic (P or S) velocity v has the (constant) value v_q. Following (4.35), let
$$r = q(1 + \epsilon p_2) = q\{1 + \epsilon(\tfrac{1}{3} - \sin^2\phi)\} \tag{9.14}$$

be the internal surface Σ of Y over which v has this same constant value v_q. In (9.14), q and ϵ are the mean radius and ellipticity of Σ; r and ϕ are the radius vector and geocentric latitude of any point on Σ; and p_2 is as defined in (4.27).

Let FJ be a seismic ray of specified type in Y, with F at the focus

[Refs. on p. 150]

and J at the outside surface. Let ϕ_F, z_F, Z, and Δ be the geocentric latitude of F, the depth of F below the outside surface, the azimuth of J from F, and the geocentric angular length of FJ, respectively. In X, a corresponding ray is taken which has the same z_F and Y. Let $(T + \delta T)$ and T be the travel times along the two rays, respectively.

Then it can be shown (see **B**, § 10.7) that

$$\delta T = H + K + p^{-1} \int \epsilon p_2 \eta^3 (dv/dq) d\theta, \qquad (9.15)$$

where H and K are the values of $\epsilon p_2 (\eta^2 - p^2)^{\frac{1}{2}}$ at the ends of the ray and the integral is taken along the ray. Through (9.15), the ellipticity adjustments δT can be computed when values of v and ϵ as functions of q are available. Provisional values of v derived ignoring ellipticity complications are sufficient for most purposes, the δT being small. Values of ϵ are obtained as in § 5.9.

In practice, the ellipticity adjustments are initially expressed in the form of tables which, for an assigned family of rays, give δT in terms of Δ, ϕ_F, and Z. Various simplifications can, however, be made (**B**, § 10.9.2).

9.3 Normal and abnormal seismic velocity variation

Observational evidence to be discussed later shows that, throughout much of the Earth's mantle, α and β slowly increase as the depth z increases, the rate of increase on the whole slowly decreasing. The same applies to α inside most of the core. It is convenient to call the variations of α and β normal where they conform to the predominant pattern of variation, and abnormal where they deviate from this pattern. This classification into normal and abnormal is not precise but serves provisionally for the purposes of the present chapter. Another classification is made in § 11.5.5 where a certain function of the velocity gradients is linked with degrees of departure of a region from uniform chemical composition and phase.

The variables ζ and ξ of § 9.1.5 provide measures of the rate of velocity variation. For P waves, the pattern of normal variation gives ζ as ranging from about $-2 \cdot 5$ near the top of the mantle to $-0 \cdot 3$ near the bottom, and from about $-0 \cdot 5$ near the top of the core to zero at the centre. For S waves in the mantle, the pattern is broadly similar to that for P. (Some further detail is given in § 9.6.2.) At any depth (except near the Earth's centre) where $\zeta = 0$ (or $\xi = 2$) the velocity gradient is zero and there is some degree of abnormality.

[*Refs. on p. 150*]

The abnormality is greater where $\zeta > 0$.

Abnormal velocity variations in particular regions of the Earth cause a variety of complications in the (T, Δ) curves. Features of the main abnormalities are outlined below. (For more complete detail, see B, Chapter 7.)

9.3.1 *Mild increase in velocity gradient*

Suppose that dv/dz is normal down to depth z_1, and then suddenly increases, being somewhat greater than normal for the range $z_1 < z < z_2$; let v be continuous throughout the range $z < z_2$. By (9.9), ζ is then normal for $z < z_1$ and somewhat less than normal for $z_1 < z < z_2$; and by (9.11) the upward curvatures κ of portions of rays in the range $z_1 < z < z_2$ are somewhat greater than in the range $z < z_1$. Correspondingly, the rate of increase of Δ with respect to decrease of p becomes reduced at a certain value of p and the downward curvature of the (T,Δ) curve becomes somewhat greater than normal for a range of values of p. (B, §7.3.4.)

9.3.2 *Strong increase in velocity gradient*

Now take circumstances as in §9.3.1 except that the increase in dv/dz at z_1 is stronger. When this increase is sufficiently strong, the increase in κ for portions of rays below z_1 causes $d\Delta/(-dp)$ to become negative for a range of values of p. The (T, Δ) curve then takes the form ACD of Fig. 9.4, with a cusp at C. The first branch AC corresponds to rays that are confined to the normal region above z_1, and the second branch CD to rays which penetrate below z_1. Assuming that normal behaviour of v is resumed below z_2, the (T,Δ) curve has a second cusp D and a third branch DE. Thus a marked increase in dv/dz at a particular value of z can cause triplication of the (T, Δ) curve. (B, §7.3.3.)

It has been shown (Bullen, 1960) that in general the cusp at D is associated with large amplitudes, but not the cusp at C.

9.3.3 *Discontinuous increase of velocity*

In the case where v increases discontinuously with z at z_1 and varies normally above and below $z = z_1$, the general form of Fig. 9.4 continues to apply. The branch AC is, as previously, associated with rays which lie entirely above the level z_1, the branch CD with rays

[*Refs. on p. 150*]

FIRST APPROXIMATION TO P AND S DISTRIBUTIONS [9.3

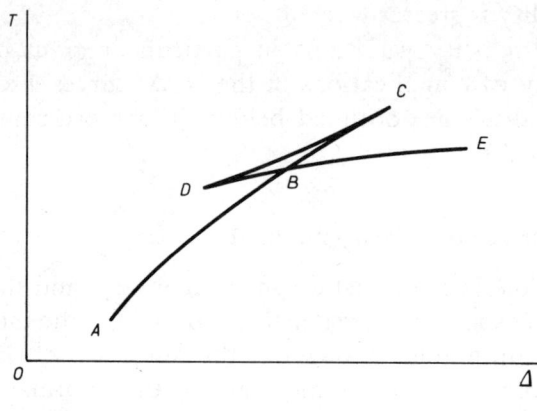

Fig. 9.4

reflected upward at the boundary $z = z_1$, and the branch DE with rays which have been refracted into the region below z_1. In general, the amplitudes are not abnormally large at either C or D. In practice, onsets corresponding to intermediate points on the branch CD may be unobservable because of low amplitudes. (B, § 7.3.5.)

9.3.4 *Mild negative velocity gradient*

In this subsection three cases are considered in which $0 \leqslant \zeta \leqslant 1$. By (9.8) and (9.9), the relations $2 \leqslant \xi \leqslant \infty$ and

$$0 \leqslant dv/dr \leqslant v/r \tag{9.16}$$

hold for these cases. The set of circumstances for each separate case will, for simplicity, be taken to hold throughout the model.

The first case is $\zeta = 0$, i.e. $dv/dz = dv/dr = 0$. The rays are then straight lines and $T = 2\eta_a \sin(\tfrac{1}{2}\Delta)$ when the focus is at the outside surface. The (T,Δ) curve is curved downward but less so than in the normal case for which dv/dz is positive. (B, § 7.5.2.)

The second case is $0 < \zeta < 1$. Then $dv/dr > 0$ and $dv/dz < 0$, with the further condition

$$dv/dr < v/r. \tag{9.17}$$

By (9.11) and (9.17), the downward curvature of any ray, though now positive, is everywhere less than r^{-1}. Thus a ray starting from the outside surface, though in general concave down at all points, reaches a lowest point and rises to the surface again. The (T,Δ) curve

[Refs. on p. 150]

is again curved downward, but the curvature is still less than in the case where v is constant. (B, §7.3.7.)

The third case is the extreme of (9.16) where $\zeta = 1$, i.e. $dv/dr = v/r$. Then (9.11) gives $\kappa = -r^{-1} \sin i$. Hence if any ray were to have a lowest point (other than the centre O), the downward curvature at that point would be r^{-1}. It follows that no ray (excepting the vertical ray through O) passing downward from any spherical surface of constant velocity could return to that surface. Rays starting from a point of the outside surface can therefore have another point on this surface only if they are wholly confined to the surface. Then $T = \eta_a \Delta$, and the (T, Δ) curve is a straight line. Otherwise, the rays spiral in to the centre.

9.3.5 Strong negative velocity gradient

Next let $\zeta > 1$ at $z = z_1$ (or $r = r_1$). Then the condition (9.17) is violated at z_1. For any ray which is horizontal at z_1, (9.11) now gives the downward curvature at z_1 as greater than r_1^{-1}. In this case, dv/dz is again negative and its magnitude (greater than in the third case of §9.3.4) is sufficient to preclude any ray from having its lowest point at the depth z_1. If the condition $\zeta > 1$ held throughout the model, all rays starting from the outside surface would spiral in to the centre.

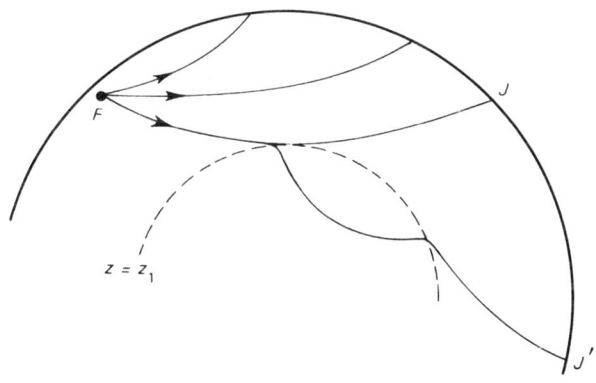

Fig. 9.5

Suppose that v behaves normally above the depth z_1 and that (9.17) is violated at z_1 and for a limited range of distance below.

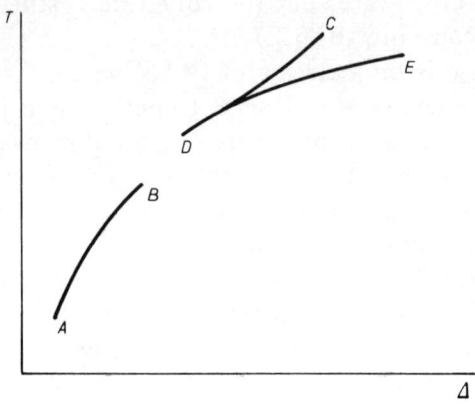

Fig. 9.6.

Consider rays leaving a focus F (above the level z_1) with continuously increasing steepness (decreasing p), and let FJ be the ray which just falls short of encountering the surface $z = z_1$ (Fig. 9.5). For the whole set of rays which do not penetrate below z_1, the (T, Δ) curve of course has the normal form AB of Fig. 9.6, the point B corresponding to the ray FJ. A ray just steeper at F than FJ will, however, as a consequence of the abnormality setting in at z_1, penetrate a finite distance below z_1 and emerge in general at a point, J' say, a finite distance beyond J. The (T, Δ) curve has a corresponding break from B to C, the values of p and therefore of $dT/d\Delta$ being equal at B and C. As p decreases further from the value at B and C, Δ and T in general decrease corresponding to the branch CD, and then increase corresponding to the branch DE. There is a cusp at D, usually associated with large amplitudes, and a 'shadow zone' between B and D. (**B**, §7.3.6.)

9.3.6 *Discontinuous decrease of velocity*

If v decreases discontinuously at some depth z_1, the essential results are again as depicted in Fig. 9.6. (**B**, §7.3.8.)

9.4 Bodily seismic phases

9.4.1 *Broad division of the Earth's interior*

It is necessary at this point to describe the nomenclature used for the principal phases of bodily seismic waves. For this purpose, reference

[*Refs. on p. 150*]

has to be made (in advance of detail to be given in later chapters) to the Earth's crust, the mantle extending to depth near 2900 km, and the central core consisting of the outer core of thickness about 2250 km and the inner core of radius 1220–1250 km. (See §10.2 for details.) The mantle-core and inner-core boundaries may here be treated as surfaces of mathematical discontinuity. Other divisions of the Earth's interior (including possible transition zones between the outer and inner core) — see §10.3 — will be ignored for the present. Throughout the mantle and crust (excluding the oceans), both P and S waves are observed to be transmitted. No S waves have been detected in the outer core and there is strong independent evidence (§10.2) that S waves are not transmissible there.

9.4.2 *Nomenclature for bodily seismic phases*

The record or 'signature' traced on a seismogram following an earthquake or explosion generally shows a series of phases, the onsets of which are associated with wave pulses that have travelled by various routes from the focus to the recording station. Many routes are possible because an incident P or S wave may generate refracted and reflected P and S waves at any internal surface of discontinuity, and reflected P and S waves at the outside surface of the Earth. A single record may therefore show many bodily-wave phases.

A bodily-wave phase is named according to the family of rays with which it is associated. The name thus depends on the boundaries which are encountered, the circumstances (refraction or reflexion) at the boundaries, and the P or S character between boundary encounters. The nomenclature below takes account of the boundaries separating the mantle, outer core, and inner core, and serves for broad purposes. Except when specifically mentioned the crust is ignored or regarded as part of the mantle.

Each of the following symbols (except c and i) corresponds to a segment occupying part or all of the length of a ray. The symbol P or S indicates a segment (or in simple cases a whole ray) lying inside the mantle, K a segment in P type inside the outer core, and I and J segments in P and S types inside the inner core. The symbols p and s indicate comparatively short segments which go from the source upward to the outside surface. The symbols c and i respectively indicate upward reflexions at the mantle-core boundary, N say, and at the boundary, L say, of the inner core.

[*Refs. on p. 150*]

FIRST APPROXIMATION TO P AND S DISTRIBUTIONS [9.4-

The phases P and S correspond to rays which lie wholly inside the mantle without suffering reflexion or change of type anywhere. Each of the phases PP, PS, SP, and SS involves two segments with a reflexion at the outside surface. Each of PPP, PPS, PSP, etc. involves three segments and two reflexions at the outside surface. The phase pSP differs from PSP only in that the first segment is short and goes upward from the focus. With phases such as PKP, SKP, and SKS, the first and third segments lie inside the mantle and the middle segment inside the outer core. With each of PKIKP, and PKJKP, there are five segments, the middle segment being inside the inner core. With PKKS, the two segments K in the outer core are separated by a downward reflexion at N. With phases such as PcP and PcS there are two segments, both in the mantle and separated by an upward reflexion at N. With PKiKP, there are four segments, the first and fourth being inside the mantle, the second and third inside the outer core, with upward reflexion at L.

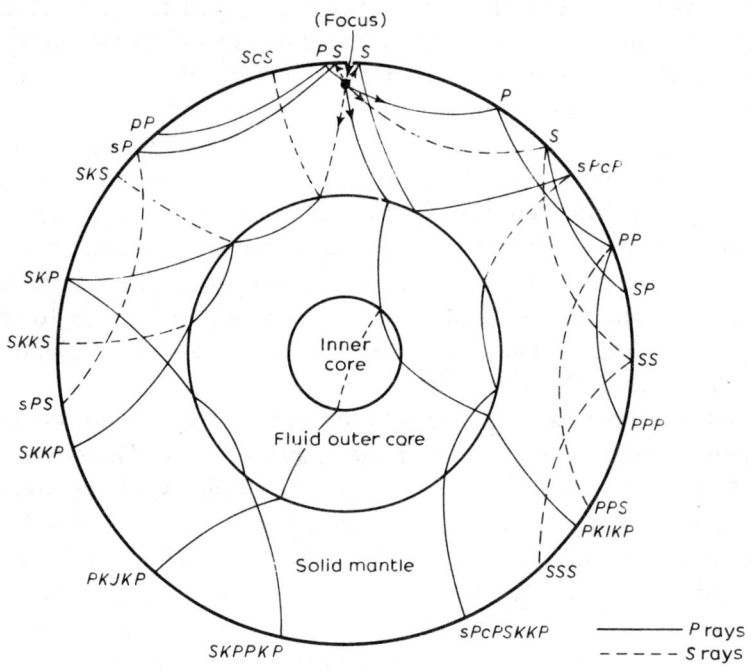

Fig. 9.7

[*Refs. on p. 150*]

The nomenclature for more complex cases is easily inferred. For example, the phase *sPcPSKKP* corresponds to a ray with seven segments: (i) from the focus upward to the outside surface in *S* type; (ii) reflexion down to *N* in *P* type; (iii) upward reflexion from *N* to the outside surface in *P* type; (iv) reflexion down to *N* in *S* type; (v) refraction into the outer core (in *P* type); (vi) downward reflexion at *N* back into the outer core; (vii) refraction upward through *N* and return through the mantle to the surface in *P* type. This and some other cases are illustrated in Fig. 9.7.

Symbolism for further phases will be introduced later as occassion arises.

9.5. Evolution of travel-time tables

A first task in applying seismic observational data to the study of the Earth's interior is to arrive at global travel-time tables for various bodily-wave phases for a suitably defined spherically symmetrical Earth model.

The evolution of reliable tables has been a long process starting from work of Oldham, Zöppritz, Gutenberg, Turner, and others (see §10.2) several decades ago. Essentially, the problem of constructing the tables is a least-squares problem interlocked with the determination of the source location and origin time of each event (when a natural earthquake) used for the purpose. In practice, account has to be taken not only of the complications discussed in §§9.2 and 9.3, but of such further matters as complicated source conditions, errors in reading seismograms, errors which do not follow the normal law, the erratic structure of the Earth's curst, and deviations of the Earth from spheroidal symmetry below the crust.

By 1940, tables had been evolved which gave values of T accurate within the order of one or two seconds for the main *P*, *PKP*, and *PKIKP* phases, and only a little less accurate for a number of other phases. Certain analyses made over the period 1932—39 of a large number of the best recorded natural earthquakes up to that time resulted in the J.B. tables (Jeffreys and Bullen, 1940, 1967). The tables included the travel times of *P* and *S* waves throughout the mantle and of *P* waves throughout the core. Curves corresponding to the J.B. tables for a shallow-focus earthquake are shown in Fig. 9.8. The tables apply to a spherically symmetrical Earth model X related to the oblate model Y as described in §9.2, and thus apply in a

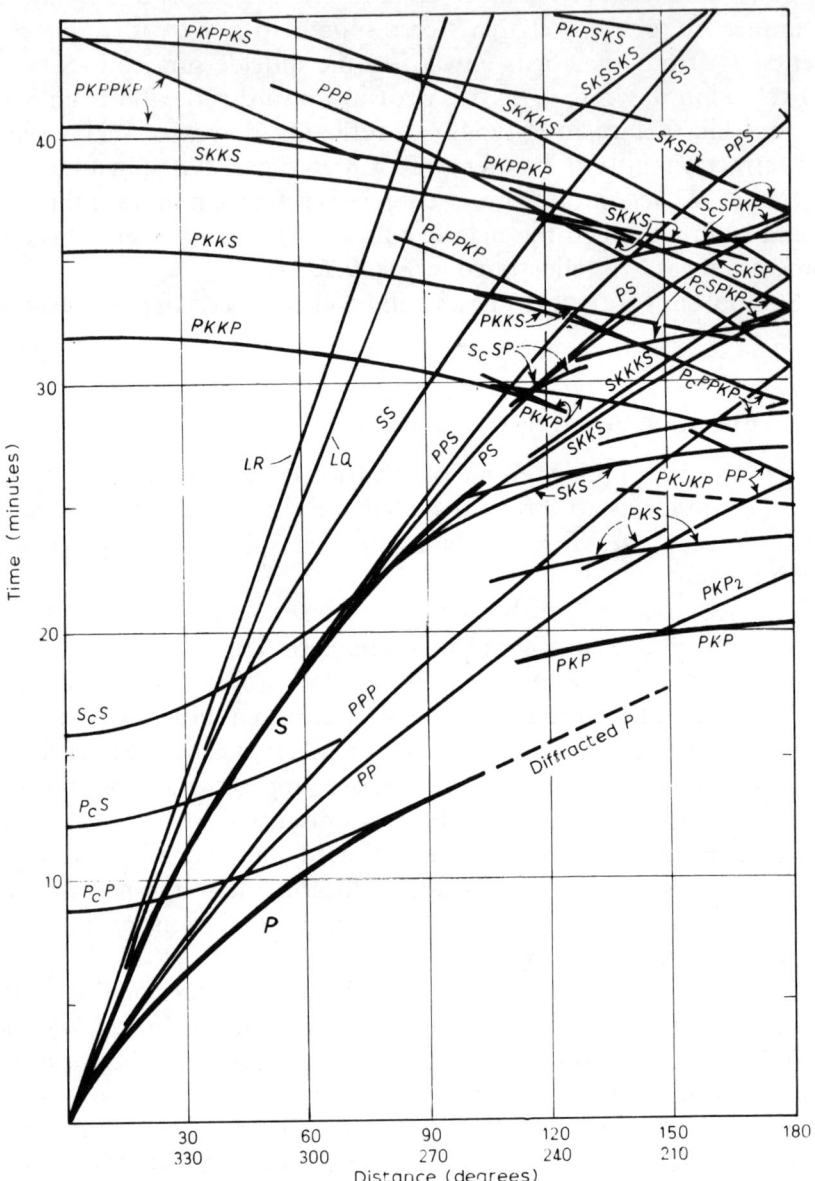

Fig. 9.8. *The Jeffreys-Bullen travel-time curves*

certain sense to 'average' global earthquakes (but see §§ 15.7 and 16.1.2).

Companion tables based on the theory of § 9.2 (Bullen, 1937–9)

[*Refs. on p.* 150]

gave the allowances δT for ellipticity. The allowances are applied in two ways. They are subtracted from observed onset times of phases during the process of deriving travel times for the spherically symmetrical Earth model X. They are added to the times in the tables finally arrived at for X to give travel times for the model Y.

Another widely based set of travel-time tables, independently derived by Gutenberg and Richter (1934–39), differed from the J.B. tables by amounts that were mainly very small compared with the overall changes that had been made from tables in general use around 1930. This degree of agreement confirmed the considerable reliability that had been attained by 1940. The application of statistical tests in the course of the work on the J.B. tables had indicated, moreover, that the stage had been reached where, on the existing seismic arrival-time data, no further significant improvement could be made to determinations of the average global bodily-wave travel times. As a consequence, the J.B. tables were adopted in preparing the International Seismological Summary for earthquakes occurring from 1937 on, and served as an unofficial standard for many years following.

Since 1940, great quantities of additional data bearing on the P and S travel times have been amassed. Analyses of this material (including further contributions from Jeffreys and Gutenberg, as well as others) have shown that the 1940 tables need some corrections. Apart from some limited exceptions, the corrections so far indicated may be regarded as second-order refinements. These refinements will be discussed in later chapters.

9.6 Derivation of P and S velocity distributions in the Earth

The close interconnection between (v,r) and (T,Δ), as seen from §9.1.7, shows that information on T and Δ can yield information on v (i.e. α and β) and hence on the density and other properties of the Earth's interior. The main procedure in deriving this information is outlined below.

9.6.1 *Formal method of derivation*

When the (T,Δ) relation is given for a family of rays in a spherically symmetric model Earth, (9.12) can, in normal conditions, be solved as an integral equation to give the corresponding (v, r) relation. The

[*Refs. on p. 150*]

procedure is nowadays often referred to as *inverting* the travel-time data.

The formal solution (B, §7.4) when it exists is

$$\int_0^{\Delta'} \cosh^{-1}(p/\eta') d\Delta = \pi \ln(r_a/r'), \qquad (9.18)$$

where η' is the value of η (or r/v) at an assigned level $r = r'$, and Δ' is the value of Δ for the ray whose lowest point is at this level. The solution (9.18) was essentially derived by Abel in his work on integral equations. Equivalent forms were evolved and adapted to seismological applications by Herglotz, Wiechert, and Bateman over the period 1907–10.

With the (T, Δ) relation given, p is by (9.6) a known function of Δ. Now assign Δ'. Then η' is known, being the value of $dT/d\Delta$ (or p) for $\Delta = \Delta'$. Through (9.18), the value of r' which corresponds to η' can thus be evaluated. Hence by (9.7) a pair of corresponding values of v and r is found. By repeating for different assigned Δ', a table giving v in terms of r can be derived.

The solution (9.18) avails for regions in which the variation of v with r is normal. An alternative procedure involving a process of successive approximation is also available (Bullen, 1960; see also B, §7.4.1, and Jeffreys, 1966).

Abnormal velocity variation may cause procedures based on (9.18) to fail. When the condition (9.17) is violated, there is failure in principle because of trouble connected with the divergent integrand in (9.12). There may also be failure in practice even where (9.17) is not violated. For example, when the (T, Δ) curve is triplicated over a range of values of Δ (Fig. 9.4), so that the curve has a 'loop', evaluation of the left side of (9.18) involves integration round the loop for a range of values of Δ'; and where such loops occur, there are usually difficulties on the observational side in determining their complete shapes to desirable precision. Most of the abnormalities referred to in subsections of §9.3 appear to occur inside the Earth. Fortunately, however, they are significant only inside a limited number of fairly narrow ranges of depth and (9.18) has therefore proved to be serviceable for much of the Earth's interior.

In treating the abnormal regions, various supplementary considerations are brought to bear. For example, direct evidence on $dT/d\Delta$ (sometimes called the *slowness*) provided by seismic array stations (see §12.1) assists in determining $d\alpha/dz$ and $d\beta/dz$ over some difficult

[Refs. on p. 150]

ranges of depth. Supplementary information is also supplied by analyses of seismic surface waves (Chapter 13), free Earth oscillations (Chapter 14), and bodily-wave amplitudes (§ 12.1.5).

Where procedures based directly on (9.18) fail to determine the velocity variations, they may still be adapted to supply useful bounds to the possible velocities. Inside these bounds, it is often appropriate to set down parametrically simple conventional velocity distributions; an example is given in the second paragraph of §9.6.2.

9.6.2 Numerical results

Numerical estimates of α and β at depth in the Earth began to appear when the method of §9.6.1 become available. Early contributors include Gutenberg and his Göttingen colleagues (who published frequent papers on the subject over the whole period 1907–30), Knott (1919), Witte (1932), and Wadati (1933–4). The early estimates were mostly based on data from small numbers of earthquakes and the uncertainties were high. Around 1930, Jeffreys and Gutenberg perceived the need for bringing more comprehensive bodies of data to bear, and by 1939 Jeffreys had produced the first statistically well-based distributions of α and $\beta - \alpha$ throughout the whole interior of the Earth, and β throughout the mantle. Over the period 1939–58, Gutenberg also published several distributions based on large quantities of data.

The 1939 velocity distributions of Jeffreys were derived in conjunction with the J.B. tables, the method of §9.6.1 being applied considerably but not exclusively. For ranges of depth where normal velocity variations were indicated, various smoothing devices were applied to remove superfluous parameters. In respect of certain limited ranges of depths for which complexities or uncertainties in the observational travel-time data forbade a close determination of the velocities, conventional velocity representations containing minimum numbers of parameters were taken inside the limits of the uncertainties. (For example, α was conventionally taken by Jeffreys as proportional to r between 4980 and 5120 km depth.) The final J.B. travel times incorporated the smoothed and the conventional velocities. Thus the procedures for determining the velocities and the travel-time tables were interlocked. [It needs to be noted — see Haddon and Bullen (1969) — that the 1939 velocities are not entirely compatible with the J.B. tables. The discrepancies are insignificant

[Refs. on p. 150]

except in contexts involving fine numerical computations.]

The 1939 velocity distributions included a conventional Earth's crust, consisting of a 'granitic' layer 15 km thick in which $\alpha, \beta = 5\cdot57, 3\cdot36$ km/s, and an 'intermediate' layer 18 km thick in which $\alpha, \beta = 6\cdot50, 3\cdot74$ km/s.

Table 9.1 gives values of the Jeffreys velocities at some representative depths; the upper part of the table relates to the mantle, the lower part to the core. (For fuller details, see B, § 13.1.4.)

TABLE 9.1
P and S velocities (Jeffreys, 1939)

Depth (km)	33	400	1000	2700	2898
α (km/s)	7·75	8·93	11·42	13·62	13·65
β (km/s)	4·35	4·94	6·36	7·26	7·30
Depth (km)	2898	4500	4980	5120	6371
α (km/s)	8·10	9·97	10·44	9·40, 11·16	11·31

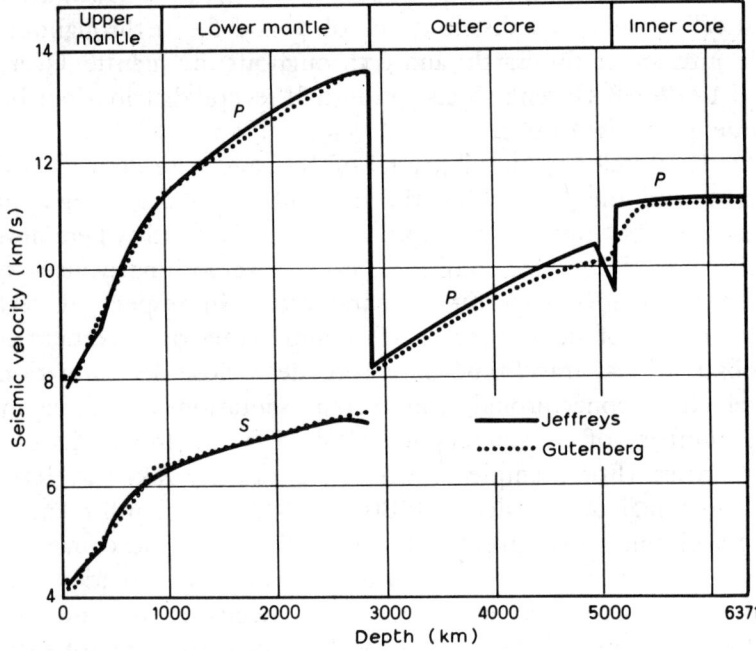

Fig. 9.9. *P and S velocity distributions of Jeffreys and Gutenberg*

Velocity distributions derived by Gutenberg and Richter (1939) usefully confirmed the general reliability of the 1939 Jeffreys distributions. Fig. 9.9 shows a comparison of the Jeffreys distributions with a later set due to Gutenberg (1951, 1958). The comparison indicates the general order of uncertainty in the estimated α and β at various depths. The differences between the Jeffreys and Gutenberg curves are small except in the outer part of the mantle and in the transition zone between the outer and inner core. Later work has shown that these are the regions where the needed corrections are largest.

In regions where the velocities appeared (on the Jeffreys distributions) to vary normally, the average values of ζ are approximately: -2.3 for P waves and -2.1 for S waves in the outermost 400 km of the mantle; -0.5 for P and -0.35 for S between 1000 and 2700 km; -0.3 for P between 2900 and 4500 km (see also §§ 9.3 and 11.5.5). These values provide a tentative basis for defining normal and abnormal regions.

An interesting feature, to be discussed later in some detail (§§ 10.3.4, 11.2.2, and 12.1.4) is that the distributions of Jeffreys and Gutenberg both show rapid declines in the P and S velocity gradients inside the lowest 200 km of the mantle. Work of Dahm (1936) and others had earlier pointed to some abnormality near the bottom of the mantle.

By (8.6,7), information on α and β at any point carries with it information on the physical quantities k/ρ and μ/ρ. Thus by 1940, the important stage had been reached where values of k/ρ and μ/ρ, reliable within well-determined bounds of uncertainty, had become available throughout the Earth's mantle. Independent evidence (see § 10.2) indicates that, in the outer core, μ can be neglected and β therefore taken equal to zero. With α known, similarly reliable values of k/ρ and μ/ρ thus became available throughout the outer core as well. [For numerical values of k/ρ and μ/ρ, derived from the J.B. tables, see Bullen (1956), p. 105.]

The quality of the results that became available between 1935 and 1940 (see Jeffreys and Bullen, 1935, 1940), in conjunction with other evidence, made it possible to determine reliable details on the Earth's internal density distribution for the first time, and within assessable bounds of uncertainty. The procedures used are discussed in Chapters 10 and 11. Second approximations incorporating amendments to the 1940 travel-time tables and the corresponding values of

α, β, k/ρ, and μ/ρ are considered in Chapter 12 and later chapters.

REFERENCES

Bateman, H. (1910). The solution of the integral equation which connects the the velocity of propagation of an earthquake wave in the interior of the Earth with the times which the disturbance takes to travel to different stations on the Earth's surface. *Phil. Mag.* (6), **19**, 576–587.

Bullen, K.E. (1937–39). The ellipticity correction to travel-times of P and S earthquake waves. *Mon. Not. R. Astr. Soc., Geophys. Suppl.*, **4**, 143–57, 317–31, 332–5, 469–71; and later papers on ellipticity corrections.

Bullen, K.E. (1956). Seismic wave transmission. *Encyclopedia of Physics*, **47**, 75–118. Berlin: Springer-Verlag.

Bullen, K.E. (1960). A new method of deriving seismic velocity distributions from travel-time data. *Geophys. J., R. Astr. Soc.*, **3**, 258–269.

Dahm, C.G. (1936). Velocities of P and S waves calculated from the observed travel times of the Long Beach earthquake. *Bull. Seismol. Soc. Amer.*, **26**, 159–171.

Gutenberg, B. (1914). Über Erdbebenwellen. VIIA. Beobachtungen an Registrierungen von Fernbeben in Göttingen und Folgerungen über die Konstitution des Erdkörpers. *Nachr. Ges. Wiss. Göttingen, Math.-Phys. Klasse*, pp. 1–52 and 125–176.

Gutenberg, B. (1951). PKKP, P'P', and the Earth's core. *Trans. Amer. Geophys. Un.*, **32**, 373–390.

Gutenberg, B. (1958). Wave velocities in the Earth's core. *Bull. Seismol. Soc. Amer.*, **48**, 301–314.

Gutenberg, B. and Richter, C.F. (1934–39). On seismic waves. *Beitr. Geophys.*, **43**, 56–133; **45**, 280–360; **47**, 73–131; **54**, 94–136.

Haddon, R.A. and Bullen, K.E. (1969). An Earth model incorporating free Earth oscillation data. *Phys. Earth Planet. Interiors.*, **2**, 35–49.

Herglotz, G. (1907). Über das Benndorfsche Problem der Fortpflanzungsgeschwindigkeit der Erdbebenstrahlen. *Phys. Z.*, **8**, 145–147.

Jeffreys, H. (1939a). The times of P, S and SKS and the velocities of P and S. *Mon. Not. R. Astr. Soc., Geophys. Suppl.*, **4**, 498–533.

Jeffreys, H. (1939b). The times of the core waves. *Mon. Not. R. Astr. Soc., Geophys. Suppl.*, **4**, 548–61, 594–615.

Jeffreys, Sir H. (1966). Estimation of small changes in a function. *Geophys. J., R. Astr. Soc.*, **12**, 113.

Jeffreys, H. and Bullen, K.E. (1935). Times of transmission of earthquake waves *Publ. Bur. Centr. Séismol. Internat.* A, **11**, 1–202.

Jeffreys, H. and Bullen, K.E. (1940, 1967). Seismological tables. Brit. Ass., Gray-Milne Trust, 50 pp.

Knott, C.G. (1909). The propagation of earthquake waves through the Earth, and connected problems. *Proc. R. Soc. Edinb.*, **39**, 157–208.

REFERENCES

Wadati, K. and others (1933—34). On the travel time of earthquake waves. *Geophys. Mag., Tokyo*, 7, 87—153, 269—90; 8, 187—94.

Wiechert, E. (1907). Über Erdbebenwellen. I. Theoretisches über die Ausbreitung der Erdbebenwellen. *Nachr. Ges. Wiss. Gottingen, Maths.-Phys. Klasse*, pp. 415—529.

Witte, H. (1932). Beiträge zur Berechnung der Geschwindigkeit der Raumwellen im Erdinnern. *Nachr. Ges. Wiss. Göttingen, Math.-Phys. Klasse*, pp. 199—241.

Zöppritz, K. (1907). Über Erdbebenwellen. II. Laufzeitkurven. *Nachr. Ges. Wiss. Göttingen, Math.-Phys. Klasse*, pp. 529—549.

CHAPTER 10

Earth models of type A

Knowledge of values of the Earth's radius a, mass M, and moment of inertia I provides two criteria (§6.6) which must be satisfied by any acceptable representation of the Earth's internal density (ρ) distribution. As seen in Chapter 6, these criteria can be satisfied by widely different density distributions. When seismology began to provide values of the P and S velocities α and β for specific values of z, it became possible to narrow the allowable density distributions. An early application of seismic data to the density problem has been described in §6.9.2. The present chapter will be concerned with developments over the period 1936–42, leading to a close first approximation to the Earth's density distribution.

Earth models will continue to be taken as spherically symmetrical except where otherwise stated, and defined to be related to corresponding spheroidal models as in §9.2. Thus the main physical properties will for the present usually be treated as functions of the distance r from the centre or the depth z below the surface.

By 1935, the evolution of seismic travel-time tables had reached the point where it had become desirable to evaluate and apply corrections for the ellipticities ϵ of internal surfaces of constant density. As seen in §§5.9 and 9.2, the estimation of these corrections requires knowledge of ρ as a function of r. Since the values of ρ available in 1935 were too unreliable for the purpose, specific attention was then given to the problem of deriving a reliable density distribution.

By the methods described below, a much improved density distribution, adequate for the ellipticity problem, was derived (Bullen, 1936). This distribution was amended in some of the detail (Bullen, 1940, 1942) after the improved travel times of 1940 (§9.5) became available. The results were incorporated in a series of Earth models,

[Refs. on p. 180]

later referred to as the A-type models. In addition to the density ρ, these models included distributions of the Earth's incompressibility k, rigidity μ, pressure p, gravitational intensity g, and of several derived variables — Young's modulus E, Poisson's ratio σ, etc. — and so supplied a useful first approximation to the main internal physical structure of the Earth.

The present chapter is largely concerned with the development of the original family of A-type models. Two of the models, Models A and A', came to be widely used till the early 1960s. Some details will also be included of a further model, A'', which differs from A' only in that the revised value (§5.7) is taken for the Earth's moment of inertia. A related series of Earth models, the B-type models, is discussed in Chapter 11.

10.1 Introductory theory of density variation in the Earth

10.1.1 *General expression for density gradient*

Strictly, the density at a point of the Earth's interior depends on the stress, temperature (τ), chemical composition and phase at the point. The stress will here be represented in terms of the pressure p, as defined in §7.8.1. The temperature dependence is for some purposes most suitably represented in terms of the entropy s. The composition and phase will be represented in terms of indefinite sets of parameters n_i and q_i, respectively. The parameters q_i depend on p, τ, and the n_i, but because there is in general a time-lag in changes of q_i when p and τ change, it is appropriate to treat the q_i as independent variables in considering the short-lived changes which accompany seismic wave transmission.

Thus inside a region of the Earth where ρ, n_i, and q_i can be treated as varying continuously, we take $\rho = \rho(p, s, n_i, q_i)$, so that

$$\frac{d\rho}{dz} = \frac{\partial \rho}{\partial p}\frac{dp}{dz} + \frac{\partial \rho}{\partial s}\frac{ds}{dz} + \sum \frac{\partial \rho}{\partial n_i}\frac{dn_i}{dz} + \sum \frac{\partial \rho}{\partial q_i}\frac{dq_i}{dz}$$
$$= P + S + N + Q, \quad \text{say.} \tag{10.1}$$

Immediate application of (10.1) to determine density gradients inside the Earth would require prior knowledge of the variation of all of p, s, n_i, and q_i with z. Since this full knowledge is far from adequately available, it is necessary first to estimate the relative

[Refs. on p. 180]

EARTH MODELS OF TYPE A [10.1]

importances of the terms P, S, N, and Q in particular regions of the Earth and carry out trial calculations in which certain of the terms are neglected. The trials result in provisional models based on the more readily available observational data. The provisional models are tested against further evidence and amended by successive approximation.

10.1.2 Contribution from changes in pressure

In the early stages of determining the Earth's density distribution, there was little definite information on chemical composition and phase except close to the surface. Hence it was appropriate to start by ignoring N and Q in (10.1) and thus take $\rho = \rho(p, s)$; then, using (7.15),

$$P = \left(\frac{\partial \rho}{\partial p}\right)_s \frac{dp}{dz} = \frac{\rho}{k} \frac{dp}{dz}, \qquad (10.2)$$

where k is the adiabatic, or isentropic, incompressibility.

Introducing

$$\phi = k/\rho = \alpha^2 - 4\beta^2/3 \qquad (10.3)$$

[see (8.6,7)], and assuming the hydrostatic and gravitational relations

$$dp/dz = g\rho, \qquad g = Gm/r^2, \qquad (10.4, 5)$$

where G is the constant of gravitation and m is the mass inside the sphere of radius r (cf. §§ 6.4 and 6.4.1), we obtain

$$P = g\rho/\phi = Gm\rho/(r^2 \phi). \qquad (10.6)$$

When, further, the *super-adiabatic* temperature gradient ϑ, i.e. the excess of $d\tau/dz$ over the adiabatic gradient, is ignored, $d\rho/dz$ becomes equal to P and (10.1) reduces to the 'Williamson-Adams equation'

$$d\rho/dz = g\rho/\phi = Gm\rho/(r^2 \phi). \qquad (10.7)$$

It was through the equation (10.7), along with (10.3), that Williamson and Adams (§6.9.2) first brought seismic data to bear to estimate $d\rho/dz$ in parts of the Earth.

10.1.3 Contribution from changes in temperature

At depth z, let γ and c denote the coefficient of cubical expansion and the specific heat, respectively, at constant pressure. The definitions of γ and ϑ give

$$\gamma = -\rho^{-1}(\partial \rho/\partial \tau)_p, \tag{10.8}$$

$$\vartheta = \frac{d\tau}{dz} - \left(\frac{\partial \tau}{\partial p}\right)_s \frac{dp}{dz} = \left(\frac{\partial \tau}{\partial s}\right)_p \frac{ds}{dz} \tag{10.9}$$

$$= -(\gamma\rho)^{-1}\left(\frac{\partial \rho}{\partial s}\right)_p \frac{ds}{dz}. \tag{10.10}$$

Hence, by (10.1) and (10.3),

$$S = -\gamma\rho\vartheta = -\lambda c\rho\vartheta/\phi, \tag{10.11}$$

where λ is Grüneisen's ratio, defined by

$$\lambda c\rho = \gamma k. \tag{10.12}$$

By (10.6) and (10.11),

$$S/P = -\gamma\phi\vartheta/g = -\lambda c\vartheta/g, \tag{10.13}$$

a formula due to Birch (1952) — see also Bullen (1954).

Detail in §10.8 will show that S/P is likely to be substantial inside part of the upper mantle, but sufficiently small below 1000 km to be neglected to a good first approximation. Thus for most of the Earth, the term P strongly predominates over S in the contribution to $d\rho/dz$.

10.1.4 *Contribution from changes in chemical composition and phase*

It has already been mentioned that, whereas the composition parameters n_i are fully independent variables, the phase parameters q_i strictly depend on p, τ, and the n_i. For this reason, discrimination between the terms N and Q of (10.1) can be important in some contexts on density variation, for example, in comparisons of changes of density with pressure inside different planets in the equilibrium state (Bullen, 1966), and in the construction of Earth models designed to test specific geochemical assumptions. Direct chemical evidence on distinctions between compositional and phase changes inside the Earth is, however, not sufficient to contribute usefully to the main procedures of deriving the Earth's density distribution (although some geochemical investigations do have the potentiality of contributing small refinements to density distributions otherwise derived — see e.g. §15.2.3.)

[Refs. on p. 180]

Hence, except where otherwise stated, the n_i and q_i will be treated as a single group. It will be convenient to describe a region of the Earth as 'homogeneous' if N and Q are both negligible (i.e. if there is no evidence of changes in composition or phase); 'inhomogeneous' if there is evidence that one or both of N and Q is significant.

Throughout a sizeable part of the interior of the Earth, it transpires that $(N + Q)$ may be neglected, compared with P, to a useful first approximation. Inside regions where inhomogeneity is marked, forms such as $\rho = c_1 + c_2 r + c_3 r^2$ are usually serviceable because the ranges of depth involved are usually fairly small. Since ρ is connected with α and β by simple algebraic formulae (8.6,7), ρ will normally be expected to be continuous or discontinuous where α and β are continuous or discontinuous; and similarly with the gradients $d\rho/dz$, $d\alpha/dz$, and $d\beta/dz$. This property assists in the evaluation of the coefficients c_i. (See e.g. the procedure described in §10.7.1 for the region C.)

The discussion in §§10.1.2–10.1.4 makes it appropriate in general to place the first emphasis on the term P of (10.1) in estimating $d\rho/dz$, and so substantiates the important role played by the formula (10.7) in the early evolution of a reliable density distribution.

At the same time, it should be noted that the degree of inhomogeneity for some ranges of depth may be so great as to render (10.7) useless even as a crude approximation. As an extreme example, the 1939 Jeffreys P velocity distribution would (see §11.6.6) entail, inside the region F (§10.3.6) of the core, a density gradient of order 30 times that given by (10.7).

Below 1000 km depth, theory on compressibility to be given in Chapter 11 provides an effective method of estimating both $d\rho/dz$ and the degree of inhomogeneity in some inhomogeneous regions.

10.2 Historical background on the Earth's internal layering

It was known last century (§6.7.2) that the Earth's central density ρ_0 must exceed 7 g/cm^3 and most model density distributions considered at the time gave values of ρ_0 between 10 and 12 g/cm^3. The comparatively high central density thus suggested, along with evidence from meteorities, led to the speculation, voiced by Wiechert in 1896, that the Earth consists of an outer shell ('*mantel*') surrounding a denser metallic core. In 1906, Oldham supplied positive

[*Refs. on p. 180*]

seismological evidence of the presence of a central core. He found a substantial delay in the arrival of P waves at angular distances beyond 120° from an earthquake focus, and inferred that the Earth contains a central region characterized by an average P velocity appreciably less than that in the surrounding shell. Oldham made a rough estimate of 1600 km for the radius of the core. Later the shell came to be called the *mantle*. (The term 'shell' had been used as early as 1688 by Halley for the solid outer part of the Earth.)

Around 1906, it was also found that the mantle everywhere transmits P and S waves and so is solid in the sense defined in §7.7. No S waves were observed below the mantle and it was surmised that most of the core is molten.

By 1907, Zöppritz had evolved workable travel-time tables for a few seismic phases. By 1914, Gutenberg had extended earlier work of Oldham, Wiechert, and Zöppritz to arrive at an estimate of about 2900 km for the depth of the mantle-core boundary; having regard to the observational evidence used, Gutenberg's estimate was uncertain by 50 to 100 km. Jeffreys obtained 2898 ± 3 km using data available by 1939. More recent estimates will be given later.

Andrija Mohorovičić (1909) found evidence in Croatia of a sharp increase in the P velocity at a depth which he placed as 54 km below the Earth's surface. A great deal of later work by others showed such an increase to be world-wide, and the boundary where it occurs came to be called the *Mohorovičić discontinuity*. The depth of this discontinuity is in general about 35 km below the surface in continental shield areas, reaches 70 km under some mountain ranges, and can be as little as 5 km or so below the floors of deep oceans. The region of the Earth above the Mohorovičić discontinuity is now called the *crust* (sometimes *crustal layers*, or *upper layers*).

In 1936, Inga Lehmann interpreted certain seismological evidence as showing that the core consists of at least two distinct regions, the outer core and the inner core. The reduction in the velocity α of P waves on passing from the mantle into the core causes descending P rays to be bent abruptly downward, resulting in a 'shadow zone' between angular distances of about 100° and 142° for P waves recorded at the Earth's surface. The zone is not a complete shadow, however, and Lehmann attributed certain waves recorded between 110° and 142° to the presence, deep down in the core, of a sharp increase in α which causes the descending P rays to be bent strongly upward.

[*Refs. on p. 180*]

In their detailed travel-time studies (§9.5), Gutenberg and Ritcher (1938) and Jeffreys and Bullen (1940) confirmed Miss Lehmann's interpretation, and Jeffreys (1939a) showed that the principal competing hypothesis, that the waves in question are all due to diffraction, is untenable. The interpretation was further supported in work of Bullen and Burke-Gaffney (1958) on nuclear explosions. Jeffreys and Gutenberg showed that the radius of the inner core lies between 1200 and 1250 km (see §10.3.7).

In an extension of Kelvin's work (§6.8.2), Jeffreys (1926) showed that the mean rigidity μ_1 of the Earth's core is significantly less than the mean rigidity μ_2 of the mantle, and is possibly zero, thereby strengthening the surmise that most of the Earth's core is in an essentially fluid state. In arriving at this result, Jeffreys combined evidence from tidal yielding on the mean rigidity of the whole Earth with a provisional estimate of $1 \cdot 7 \times 10^{11}$ N/m² for μ_2. (The estimate of μ_2 was obtained assuming the mean values of ρ and β in the mantle to be 4·27 g/cm³ and 6·30 km/s.) When, later, the calculations for Earth models of type A had yielded detailed values of the rigidity throughout the mantle, it became possible to set bounds to μ_1. Takeuchi (1950) and Molodenski (1953) showed in this way that the rigidity of the outer core does not exceed a small fraction of the value in the lower mantle. (Takeuchi gave $\mu_1 < 10^9$; Molodenski gave $\mu_1 < 5 \times 10^{10}$ N/m².) The calculations fell short, however, of providing any useful estimate of the rigidity in the inner core. On other grounds (Bullen, 1946 — see §11.7), it became strongly probable that the inner core has significant rigidity.

Thus in broad outline the interior of the Earth consists of: a thin crust of variable and relatively complicated structure; a less irregular solid mantle extending from the crust to a depth of about 2900 km; a fluid outer core possibly 2200 km thick; and a probably solid inner core of radius about 1200 km. We now proceed to further detail.

10.3 The regions A, B, C, D, E, F, and G

In the course of constructing the Earth models of type A, the nomenclature A, B, C, D, E, F, G (Bullen, 1940–42) was introduced for seven concentric regions of the Earth. The nomenclature came to be widely used and, with certain modifications, continues to serve as a basis for describing the Earth's internal layering.

In general (but not quite exclusively), the boundaries between layers are taken where there are fairly sharp changes in α and β or their gradients. Several of the boundaries have been treated as surfaces of mathematical discontinuity in Earth model representations. The suitability of a representation in terms of a mathematical discontinuity depends on the context, in this case on the rate of change of physical properties in the neighbourhood of the boundary in relation to the lengths of the P and S waves being considered.

The original delineation of the layers A to G is intimately connected with the J.B. travel-time tables (§ 9.5), the corresponding values of α and β derived by Jeffreys in 1939, and the writer's early work on density. Details are given in the following subsections along with a preliminary indication of likely modifications in the light of later evidence.

10.3.1 *Region A*

The structure of the Earth's crust is not only extremely complicated but is markedly different in different geographical areas. But because the crust occupies only a very small fraction of the Earth's volume, it is generally sufficient when investigating the large-scale structure of the Earth's interior to treat the crust as a single unit, to be referred to as the region A, and to represent it in terms of simple models with few parameters.

The conventional representation of the crust used in the A-model calculations, to be given in § 10.4.1, was based mainly on studies in continental areas, the crustal thickness being taken as 33 km (see also § 9.6.2). This representation is substantially departed from in other areas, especially oceanic areas (see § 10.2), but the departures are usually 'compensated' (§§ 2.1.1 and 2.3) to some extent by excesses or deficiencies in the densities just below. Thus taking a spherically symmetrical Earth model with a continental-type crust for the region A does not serve too badly in the process of determining a first approximation to the internal density distribution of the Earth as a whole. Amended crustal representations will, however, be later introduced (see §§ 13.5 and 14.7).

The Mohorovičić discontinuity which separates the region A from the sub-crust below is not sufficiently sharp to have caused, except in a few limited areas, undisputed observations of bodily waves reflected from it. The favoured view is that in the vicinity of this discontinuity there are changes in α from about 7 to 8 km/s, and analogous changes in β, which are likely to be spread over a range of one or more kilometres depth.

10.3.2 Region B

Work of Byerly (1926), Jeffreys and Bullen (1933, 1935), and Lehmann (1934) indicated sharp downward bends in the P and S travel-time curves near $\Delta = 20°$, a phenomenon referred to as the '20° discontinuity'. The expected form of the (T, Δ) curve in the vicinity of such a bend is shown in Fig. 9.4, which has a loop BCD and triplication over a range of values Δ around the point B where the bend occurs. The cusp at C corresponds to a sharp change in velocity or velocity gradient at some particular depth (§§ 9.3.2 and 9.3.3). Points on the loop correspond to arrivals on seismograms later than the first and, as indicated in § 9.6.1, are not usually determined in practice to good precision. The same applies to the part of the velocity distribution associated with the loop. In the particular case of the 20° discontinuity, it was found difficult to determine even whether the corresponding discontinuities in velocity were 'first-order', i.e. involving sudden increases in α and β, or 'second-order', involving sudden increases in $d\alpha/dz$ and $d\beta/dz$. Jeffreys initially assumed first-order discontinuities but later (1939b) inferred that the balance of evidence slightly favoured second-order discontinuities. He provisionally placed the sudden changes of velocity gradient at a depth of $0·06R$ below the crust, where R here denotes the mean sub-crustal radius of the Earth; thus $0·06R \approx 380$ km.

The region B was originally delineated as the 380-km-thick region extending from the bottom of the crust to where these second-order discontinuities occurred, and the 1939 P and S velocity distributions were taken as steady and normal throughout B. This set of circumstances holds also in the original Earth models of type A.

As a result of amendments subsequently needed to the 1939 picture, neither the P nor S velocity variation can now be assumed to be quite normal anywhere inside B. The major underlying causes, which are fairly special to B, are lateral variations of property and abnormal temperature gradients (see § 10.8).

[Refs. on p. 180]

Part of the later evidence came from critical studies of the 20° discontinuity itself, the existence of which became intermittently a matter of some controversy after 1940. Some of the later investigations, including those of Jeffreys (1954, 1958), Gutenberg (1959), and Lehmann (1959), indicated that in certain geographical areas the first sharp bend in the (T, Δ) curve for P waves below the crust occurs at values of Δ significantly less than 20°, e.g. 15–16° in Europe. Beneath these particular areas, the corresponding sharp increases in α or $d\alpha/dz$ appeared to occur at depths significantly less than 400 km. At the same time, it seems fairly well established that the bend is in fact near 20° for some other areas, e.g. Japan. This geographical variability was further evidence that lateral variations extend considerably below the crust.

In addition to the difficulty of diagnozing the velocity changes in the neighbourhood of the depth z_C which corresponds to the cusp C, the presence of the loop makes it difficult to determine the velocity variation for some distance below z_C. In particular, the presence of one or more additional sharp changes of velocity is not precluded. Other evidence, to be presented in more detail in §14.7, indicates that the region B needs to be subdivided into at least two sub-regions B' and B'', with a possibly modified depth of the lower boundary of B.

A related question, first raised by Gutenberg in 1926, is the question of the existence of a 'low velocity layer' inside the region B. This question is discussed in §12.1.1 and later.

Notwithstanding the complications, it will be seen in later chapters that the amendments needed in the region B to the 1939 values of α and β, considered as laterally averaged values, are likely to be comparatively small. It is therefore satisfactory to use the 1939 velocities in seeking a good first approximation to the density distribution.

On the other hand, the amendments needed to $d\alpha/dz$ and $d\beta/dz$ are likely to be quite substantial in some parts of B, and particular caution is needed in drawing inferences which depend on assessments of these velocity gradients. The uncertainties in the velocity fluctuations incidentally complicate the problem of deciding on the most suitable forms of parametric representation to take for the velocity distributions in B. A number of (differing) recent attempts to replace the 1939 velocities in B have suffered from parametric profligacy. A further point is that, in spite of the fluctuations

[Refs. on p. 180]

inside B, the mean values of $d\alpha/dz$ and $d\beta/dz$ over the whole of B are not much less than normal. In respect of $d\alpha/dz$ and $d\beta/dz$, the region B is noticeably different from C.

10.3.3 *Region C*

The original region C extended from the lower boundary of B to a depth of $0.15R$ below the crust, or 984 km (sometimes taken conventionally as 1000 km) below the surface. Estimates of the depth of the lower boundary of C are not precise (see below) and are possibly reducible to 900 km or a little less. The region C is characterized by velocity gradients greater than normal but diminishing towards normal as the region D is approached. Evidence requiring subdivision of C into C' and C'' will be given in §12.1.2.

Application of the theory of §10.1 showed that appreciable inhomogeneity must occur over some range of depth between crust and core (Bullen, 1936), and the inhomogeneity was associated (Bullen, 1940) with the region C because of the abnormal velocity gradients. Details of the derivation of and supporting evidence on this point are given in §§10.5 and 11.6.3.

The regions B and C together are often referred to as the *upper mantle*. (Some investigators have referred to the outermost 400 km alone as the upper mantle. In order to avoid ambiguity, considerable reference will be made in the present book to specific regions such as B, C, etc.)

10.3.4 *Region D*

The boundary between the regions C and D is not sharply defined, being exceptional in that α, β, $d\alpha/dz$, and $d\beta/dz$ are all usually treated as continuous there. The boundary is intended to mark the depth at which $d\alpha/dz$ and $d\beta/dz$ have reverted to approximately normal behaviour, but precise physical significance cannot be attached to the depth conventionally taken.

The region D, often referred to as the *lower mantle*, extends down to the mantle-core boundary N. Throughout it down to about 2700 km depth, the 1939 velocity gradients remain fairly steady and normal. This was taken as evidence of approximate homogeneity by Bullen (1936, 1949) and Birch (1952) – see §§11.2.2, 11.5.5, and 11.6.4.

[*Refs. on p.* 180]

Between 2700 and 2900 km, the velocity gradients as determined by Jeffreys and Gutenberg fall to the order of zero and could be negative. A scrutiny of physical implications of this feature led to a subdivision (Bullen, 1949, 1950) of the region D into D' ($1000 < z < 2700$ km) and D'' ($2700 < z < 2900$ km) – see §§10.8.4 and 11.2.2.

Several lines of evidence have lately indicated that the depth of N may be somewhat less than the value 2898 ± 3 km (§10.2) obtained by Jeffreys in 1939 (see §§12.4, 14.6.2, and 16.4).

10.3.5 *Region E*

The terms 'outer core proper' and 'inner core proper' will refer to the two main parts of the Earth's core extending from the mantle down and from the centre up, excluding any possible transition regions lying between. The term 'lower core' will refer to the part of the Earth below the outer core proper; i.e. it will include the inner core proper and any transition regions.

The Jeffreys 1939 values of $d\alpha/dz$ are fairly steady and normal between N and a distance $0.40R'$ from the Earth's centre, where R' is the mean core-radius (here taken as 3473 km), i.e. between depths of about 2900 and 4980 km. The region E, or outer core proper, was thus originally delineated as lying between these depths.

The work of Jeffreys, Takeuchi, and others referred to in §10.2 enables the region E to be treated as fluid and the S velocity in E to be taken as effectively zero.

The P and S seismic velocities can therefore be assumed known to good approximation from the surface of the Earth to the base of the region E.

Most of E can also (Bullen, 1949; Birch, 1952) be reasonably treated as approximately homogeneous – but see also §§11.2.3 and 12.2. On the dynamo theory of the Earth's magnetic field, the region E is assumed to be the seat of convection currents which contribute the main part of the field.

10.3.6 *Region F*

The Jeffreys 1939 P velocity distribution (see Fig. 9.9) included a transition layer, later called the region F, extending from $r = 0.40R'$ to $r = 0.36R'$, i.e. from 4980 to 5120 km depth. Jeffreys could not avoid a negative P velocity gradient in any simple way in this part of the Earth, and he conventionally postulated the simplest velocity law compatible with the then available seismic travel-time data; the law gave $d\alpha/dz \approx -0.0075$ s^{-1} throughout F. At the bottom of F, Jeffreys gave a discontinuous jump in α, from 9.40 to 11.16 km/s.

Gutenberg (1951), on the other hand, gave α (but not $d\alpha/dz$) as continuous throughout the whole core, but with large positive gradients inside a range of depth approximating to the region F, as also shown in Fig. 9.9. Later work (§§ 11.6.6 and 12.3.2) indicated that both the Jeffreys and the Gutenberg velocity distributions here need substantial amendments. Among other things, it no longer seems likely that $d\alpha/dz$ is strongly negative or strongly positive anywhere inside internal regions of the core. But the crucial finding of Lehmann, namely that there is a sharp jump in α at the boundary of the inner core, remains well established; this finding carries far-reaching implications.

10.3.7 *Region G*

For the innermost region G, the inner core proper, Jeffreys and Gutenberg gave abnormally small values of $d\alpha/dz$. A similar result has been indicated in most recent work, though the possibility that $d\alpha/dz$ may be nearly normal inside G cannot be quite ruled out. In 1939, Jeffreys gave the radius of the inner core tentatively as $0.36R'$, or 1250 km, later (1942) suggesting a reduction towards 1200 km. Gutenberg's and most recent estimates also lie between these values (see §§ 12.3.2 and 16.4).

Evidence from theory on compressibility that the inner core is solid (§ 11.7) led to estimates of β in the inner core (see Table 12.4, § 12.7.2). Later independent evidence confirmed the order of magnitude of the results on β and added some precision (see §§ 14.9 and 16.4).

10.3.8 *Summary*

Table 10.1 shows the depths of the boundaries of the seven regions as delineated by the writer in 1942, and summarizes features of the 1939 velocity variations. Detail on modifications required to the original delineation will be given in later chapters.

[*Refs. on p. 180*]

TABLE 10.1
Internal regions of the Earth as delineated in 1942

Region	Approximate range of depth (km)	Characteristics of velocity gradients
A	0–33	Irregular (crustal layers)
B	33–410	Normal (P and S)
C	410–1000	Greater than normal (P and S)
D	1000–2900	Normal (P and S) except near bottom
E	2900–4980	Normal (P only)
F	4980–5120	Negative (P only)
G	5120–6370	Very small but not negative (P only)

10.4 Density near the Earth's surface

10.4.1 *The crustal layers*

As already mentioned (§10.3.1) the uncertainties about the crustal layers have only slight effects on calculations for the deeper interior. Along with the 1939 conventional representation (§9.6.2) in terms of uniform granitic and intermediate layers of thicknesses 15 and 18 km, the densities of the layers were taken as 2·65 and 2·87 g/cm³, respectively, in the A-type models. A summary of evidence relating to this conventional continental-type crust, which served as a model for many years, and an indication of its degree of suitability in the light of recent evidence are given in *B* (§13.7) and in papers of Birch (1955, 1958). In several recent model calculations, a single-layer crust of constant density equal to 2·84 g/cm³ has been conventionally used [see e.g. Worzel and Shurbet (1955), Haddon and Bullen (1969); see also §§13.5 and 16.4].

10.4.2 *Correlation between seismic velocities and densities of rocks*

Useful evidence on the densities ρ of subcrustal materials has come from a long-known correlation between ϕ and ρ for rocks in general whose densities exceed 2·5 g/cm³. An empirical relation

$$\alpha \approx \{3 \cdot 31\rho - f(w)\} \pm 0 \cdot 28 \text{ km/s}, \qquad (10.14)$$

derived by Birch (1961) from his laboratory experiments on silicates in the density range 2·6–5·0 g/cm³, goes some distance towards

[*Refs. on p. 180*]

formulating this correlation (see also §15.1.4). In (10.14), α is in km/s, ρ is in g/cm^3, w is the mean atomic weight, and $f(w) = 2\cdot55$ and $5\cdot7$ units for $w = 21$ and 25 respectively. For a selection of rocks including granites, igneous rocks, dunites, peridotites, and eclogites, Birch gave values of w ranging between $20\cdot9$ and $22\cdot1$. He found that the serviceability of (10.4) is heightened by its approximate independence of pressure and of any phase changes that may have been caused by pressure. He also found that temperature is the most serious source of uncertainty in applying (10.14).

The formulation of (10.14) would have been better in terms of ϕ, ρ, and w than α, ρ, and w, since ϕ [equation (10.3)] is independent of μ, and temperature effects are likely to affect k/ρ much less than μ/ρ. Although this improvement is desirable in theoretical discussions, the numerical effects are not very large. Thus (10.14) as it stands has proved to be quite useful for various applications to the Earth's interior. Further reference to (10.14) and alternative equations is made in §§15.1.4 and 15.2.1.

10.4.3 *Density just below the crust*

Provided that departures from the conditions stated in §10.1.2 are not too great, the heavily simplified relation

$$d\rho/dz = Gm\rho/(r^2 \phi) \qquad (10.7)$$

can be used to assess $d\rho/dz$ to a useful first approximation inside some regions of the Earth. Sufficient numerical information on the quantities on the right side of (10.7) must be available. The value of G is of course known to more than sufficient accuracy. Values of ϕ are supplied principally from seismic data, using (10.3). Starting from the Earth's surface $r = a$, at which $m = M$ (M and a being the Earth's known mass and mean radius), and integrating downward using $dm = 4\pi r^2 \rho dr$ simultaneously with (10.7), values of m can be derived in terms of r. (The values of m must of course also fit the condition $m = 0$ at $r = 0$; values of m for $r > 0$ are sometimes derived by postulating a value for the central density ρ_0, and integrating upward from $r = 0$.) The remaining prerequisite is the provision of a value of ρ at some one point inside each region to which (10.7) is applied.

The subscript b will denote values of variables at the top of the region B, i.e. just below the crust. In integrating (10.7), it has been

[Refs. on p. 180]

usual to postulate a value for the density ρ_b. The selection of this value has occupied a prominent place in the historical development of knowledge of the Earth's density distribution. In 1923, Williamson and Adams (§6.9.2) assumed $\rho_b \approx 3\cdot 3$ g/cm^3.

Most seismological determinations of $(k/\rho)_b$, i.e. ϕ_b, lie between 35 and 38 km^{2}s^{-2}. This range of values agrees with laboratory determinations for only a few (comparatively rare) known rocks, chiefly dunite, peridotite, pyroxenite, and eclogite, which, at the pressures and temperatures of interest, have densities mostly near $3\cdot 3 - 3\cdot 4$ g/cm^3. In view, moreover, of the fairly close correlation between ρ and ϕ for rocks whose densities exceed $2\cdot 5$ g/cm^3, even should the sub-crust consist of material other than the rocks mentioned, it becomes strongly probable that $\rho_b \approx 3\cdot 3 - 3\cdot 4$ g/cm^3.

A minority of investigators have suggested that the region B may consist of a phase modification, brought about by pressure, of the material of the lower crust. On Birch's evidence that (10.14) is largely independent of phase changes, this possibility would not affect the estimate of ρ_b significantly.

Evidence from several additional sources (see B §13.4.3) also supports this value for ρ_b. For example, on most theories of the origin of the Earth and Moon, it is thought rather unlikely that ρ_b could exceed the Moon's mean density of 3.34 g/cm^3. At the same time, the currently available evidence does not appear to permit a value of ρ_b significantly less than $3\cdot 3$ g/cm^3. An extreme upper limit of $3\cdot 6$ g/cm^3 has been suggested (see Birch, 1961).

Thus the long-standing conclusion of Williamson and Adams that $\rho_b \approx 3\cdot 3$ g/cm^3 continues to be reasonably in line with modern evidence. [See also Jeffreys (1970), §3.06.] It is very likely that the immediate sub-crust is composed of rocks of the type of dunite and peridotite, consisting of magnesium-iron olivines and pyroxenes. An eclogite (pyroxene-garnet) composition has been favoured by some investigators. See also Press (1969) and §15.3.3.

10.5 Early evidence on inhomogeneity inside the upper mantle

In preliminary studies (Bullen, 1936, 1937) of the density distribution in the Earth's mantle, the relation (10.7) was first applied exploratively throughout the regions B, C, and D, using the then available data on α and β. The crustal layers were taken approximately as in §10.4.1, ρ_b was taken equal to $3\cdot 32$ g/cm^3, and ρ was taken

[*Refs. on p. 180*]

continuous between crust and core. On these postulates, a trial density distribution was formally computed for the mantle and the corresponding value of the moment of inertia I' of the core derived by subtraction from the value for the whole Earth.

The result was $I' = yM'R'^2$, where M' and R' are the mass and radius of the core, and $y = 0.57$. Since 0.57 exceeds the value $y = 0.40$ which applies to a sphere of constant density, the result would entail a substantial overall diminution of density with depth in the core and so was quite untenable.

Raising the assumed value of ρ_b to nearly 3.8 g/cm^3 and still using (10.7) would enable y to be reduced to 0.40. In view of the evidence on ρ_b (§ 10.4.3), this way out of the impasse was rejected as very improbable, leaving the strong conclusion that there is a region of significant inhomogeneity somewhere between crust and core. [The effect of temperature was ignored as being fairly minor in the preliminary calculations. Application of (10.13) later showed that allowance for temperature if significant would increase, not decrease, the calculated value of y. See also § 10.8.3.]

In the absence of discriminatory evidence, the inhomogeneity thus established could either be concentrated at a single discontinuity inside the mantle or be spread over a finite range of depth. In the 1936 calculations, a single discontinuity was taken at $z = 350$ km (changed in 1937 to 480 km), in line with the information then available on α and β. When subsequent amendments to the bodily-wave travel times favoured approximately continuous variation of α and β throughout the upper mantle, the inhomogeneity was associated with the region C and so taken as being spread between about 400 and 1000 km depth (Bullen, 1940).

This finding of an extended region of inhomogeneity inside the mantle meant that (10.7) could no longer be used throughout the whole mantle, and that one or both of the terms N and Q is significant inside C. The failure of (10.7) over a range of depth forced a revision of the initial trial procedure. The revised procedure is given in § 10.7.

10.6 Minimum central density

Although, prior to 1936, many Earth models had central densities ρ_0 of order up to 12 g/cm^3, the established lower bound to ρ_0 was only between 7 and 8 g/cm^3 (see § 6.7.2). The 1936 calculations

[Refs. on p. 180]

raised this lower bound to about 12·3 g/cm³, but an uncertainty of order 0·2 g/cm³ persists, principally because of remaining uncertainties in the distributions of α and β. The revision of the estimated moment of inertia coefficient for the Earth from 0·3335 to 0·3308 also had a small effect on the minimum ρ_0.

10.7 Earth models of type A

A forerunner of the A-type models was produced in 1936. Improvements in the determination of α and β between 1936 and 1940 made it desirable to revise and extend the detail of the 1936 model. The equivalent of calculations (Bullen, 1940, 1942) which adhered to the main procedures used for the 1936 model, and which resulted in the original family of A-type models, will now be described.

10.7.1 *Procedures in construction*

With the original A-type models, layering as in §10.3 was assumed and the Jeffreys P and S velocity values (§9.6.2) were taken, along with the conventional crust (region A) described in §§10.3.1 and 10.4.1. The S velocity was taken as zero throughout the region E (§10.3.5). In line with the evidence in §10.5 on homogeneity, (10.7) was assumed in the regions B, D, and E, but not in C, F, or G. Temperature corrections to (10.7) were ignored at this stage.

In the light of details in §10.4.3, ρ_b was taken as 3·32 g/cm³ (the value for an average dunite at the pressure involved). A definite density distribution was then derived for the region B, using (10.7).

For the region C, the form $\rho = c_1 + c_2 r + c_3 r^2$ was assumed. Two further parameters, c_4 and c_5, were introduced to represent unknown densities at the tops of the regions D and E, respectively. Application of (10.7) in D and E then gave the density distribution throughout C, D, and E in terms of the five parameters $c_1, \ldots, c_5$. In accordance with the principle referred to in §10.1.4 on matching discontinuities in ρ, α, and β, ρ was assumed continuous at the B–C and C–D boundaries and $d\rho/dz$ at the C–D boundary, corresponding to the seismically indicated behaviour of α and β at these boundaries. These conditions gave three equations connecting the c_i. Thus the density distribution all the way down to the bottom of E was yielded in terms of just two independent unassigned parameters.

[*Refs. on p. 180*]

Knowledge of the Earth's mass M and moment of inertia I supplies two further equations of condition on parameters involved in the expression of the density distribution for the whole Earth. Thus a formally determinate solution, giving a definite Earth density model, could be arrived at if, but only if, no additional parameters were introduced in representing the density variation below E.

The procedures below the region E for the various models were in effect as follows. First, a value of the central density ρ_0 was assigned. A Roche-type distribution (§6.5.1) was then assumed for the region G, with $\rho^{-1} d\rho/dz$ taken (approximately) as expected for a region of uniform composition. (A precise theory was not required since the total variation of ρ inside G, as thus determined, did not exceed 0.5 g/cm^3 in any of the models.) At the E–F boundary, ρ was taken continuous in all the models. For the model with minimum ρ_0, the Roche distribution was extended continuously upward from G through the transition region F. For each of the other A-type models, ρ was taken to increase approximately linearly with z inside F, giving a total density increase Δ_1, say, inside F, and to jump discontinuously, by Δ_2 say, at the F–G boundary (corresponding to the jump in α at this boundary). In order to arrive at definite models, the arbitrary condition $\Delta_1 = \Delta_2$ was imposed.

The value assigned to ρ_0 noticeably affects the density distribution above the lower core since the overall model has to fit the values of M and I. But once ρ_0 is assigned, the fine detail of the prescription in the lower core affects the outside density distribution only very slightly, since F and G together occupy only about 1 per cent of the Earth's volume.

The original A-type models all had identical density distributions inside the regions A and B, but differed at all levels below B because of the different ρ_0. For models with $\rho_0 \leqslant 18 \text{ g/cm}^3$, the differences from model to model were found, however, to be fairly slight down to the bottom of the region E. Thus, whereas before 1936 the estimated density values inside the Earth were quite unreliable below about 300 km depth [the Williamson-Adams model (§6.9.1) is fairly similar to the A-type models above this depth], by 1942 the density distribution of the whole Earth had become fairly well determined to nearly 5000 km depth, i.e. throughout 99 per cent of the Earth's volume.

[Refs. on p. 180]

In practice, the original calculations were rather more complicated than the above outline indicates, the non-linearity of (10.7) and the interplay between the parameters c_i and ρ_0 making it initially a task of some difficulty to reach a useful first approximation. (Electronic computers were not available at this epoch.) However, once an approximate solution was reached, it was quite easy to proceed by successive approximation. During the following twenty years, a number of other attempts, some with the aid of modern computers, were made to derive density distributions, using essentially the same principles as those described above (see e.g. Bullard, 1957). The results gave only inconsequential deviations from the 1942 results. (In other calculations, specific deviations from the A-type models were sometimes deliberately postulated with a view to testing the consequences.)

Two of the original A-type models, called A and A' (Bullen, 1950, 1952), were worked out in considerable detail. Model A' had the minimum value of ρ_0, taken as 12·3 g/cm³, with ρ smoothly extrapolated from E to the Earth's centre. Model A had $\rho_0 = 17\cdot3$ g/cm³. The differences in ρ for this pair of models reach a maximum of 0·04 g/cm³ in the mantle, and 0·4 g/cm³ in the outer core.

10.7.2 *Numerical detail for an A-type model*

The Earth models considered in this book will generally be regarded as specified when numerical values of the set of variables (ρ, k, μ) are specified for all z. This is equivalent to specifying (ρ, α, β), or ρ and another pair of elasticity parameters such as (λ, μ) and (E, σ), where λ and μ are the Lamé parameters, and E and σ are Young's modulus and Poisson's ratio. When a distribution of ρ is arrived at for any model and values of α and β are available, the distributions of k and μ are immediately derived using (8.6,7). The parameters λ, E, and σ are then derived using (7.8,19,20). It may be noted that σ can be determined from α and β alone through the relation

$$2\sigma = (\alpha^2 - 2\beta^2)/(\alpha^2 - \beta^2), \quad (10.15)$$

which is readily deducible from (7.20) and (8.6,7). Values of p and g can be determined using (10.4) and (10.5).

Over a long period starting from 1942, Model A was used as the basis of a variety of calculations on the physics of the interior of the Earth. Investigations during 1961–7 (see §12.5) pointed, however, to a value of ρ_0 nearer the minimum, thus favouring Model A' rather than Model A. As mentioned, the differences between the two models are small outside the lower core.

When the original Model A and A' calculations were carried out, the current estimate of I was $0.3335 Ma^2$. A model whose density distribution agrees with the newer estimate $0.3308 Ma^2$, whose construction follows the procedures in §10.7.1 in all other respects (except that a single-layer crust is taken), and which has ρ_0 a minimum, has been called Model A'' (Bullen and Haddon, 1967). This model was constructed in 1965 as a preliminary step towards incorporating free Earth oscillation data (§14.7.1).

Numerical values of properties of Model A'' are shown in Table 10.2. The values in brackets for crustal properties are conventional. Values of properties dependent on observational evidence on β and μ are included only to the base of the region E. Values below E were later estimated through compressibility theory (§11.7) and further evidence (§§12.6 and 14.9).

The initial calculations on A-type models brought to light the interesting result that g lies within 1 per cent of 9.9 m/s^2 at all depths down to 2000 km, and within 2 per cent of 10.0 m/s^2 down to 2400 km. Hence, to good accuracy, g may be for many purposes treated as constant throughout most of the mantle.

In connection with comparisons to be made later, it is convenient to note also that the distributions of k and p for Model A'' give k as nearly continuous at the mantle-core boundary, and yield $dk/dp \approx 3.0$ and 2.8 at $z = 2700$ km and at the top of the core, respectively.

10.8 Corrections for temperature and inhomogeneity

Because of the limited evidence available at the time, no allowance was made for super-adiabatic temperature gradients ϑ in the original A-type models. Knowledge of the distributions of the temperature τ and of ϑ in the Earth still remains much less precise than knowledge of the distributions of α, β, ρ, k, μ, p, and g, but is now sufficient to permit useful assessments of the term S of §10.1.3 in some parts of the Earth. In §§10.8.1–10.8.3, numerical detail on S is discussed and applied to estimating the orders of magnitude of temperature corrections needed in the A-type models. In §10.8.4,

TABLE 10.2
Properties of Model A''

Region	Depth (km)	ρ (g/cm³)	p	k (10^{11} N/m²)	μ	g (m/s²)
A	0	(2.84)	0.000	(0.65)	(0.36)	9.822
B	33	3.32	0.009	1.15	0.63	9.846
C	413	3.64	0.141	1.73	0.90	9.960
D	984	4.55	0.379	3.49	1.83	9.966
	2000	5.11	0.87	5.10	2.45	10.01
	2898	5.56	1.36	6.39	2.97	10.73
	2898	9.98	1.36	6.55	0	10.73
E	4000	11.42	2.47	10.33	0	7.87
F	4980	12.17	3.20	13.26	0	4.78
G	5120	12.25	3.28			4.31
	6371	12.51	3.61			0

Region	Depth (km)	α (km/s)	β	E (10^{11} N/m²)	σ
A	0				
	33	(6.30)	(3.55)	(0.91)	(0.267)
B	413	7.75	4.35	1.60	0.270
C	984	8.97	4.96	2.29	0.280
D	2000	11.42	6.35	4.68	0.276
	2898	12.79	6.92	6.33	0.293
	2898	13.64	7.30	7.71	0.299
	2898	8.10	0	0	0.5
E	4000	9.51	0	0	0.5
F	4980	10.44	0	0	0.5
	5120	9.40			
	5120	11.16			
G	6371	11.31			

effects of departures from homogeneity in particular regions are discussed.

10.8.1 *Estimates of super-adiabatic gradients*

The temperature gradient $d\tau/dz$ at a point of the Earth is proportional to the ratio of the rate F of heat outflow per unit area to the thermal conductivity κ. Over the Earth's surface, measurements of F cluster around 0·011 and more than 80 per cent of the measurements give F less than 0·02 cal-m^{-2}s^{-1}. Thus F is roughly constant within a factor of order 2 over much of the Earth's surface. On the other hand, κ and therefore $d\tau/dz$ are widely variable in different geographical areas. A rough average figure for $d\tau/dz$ near the surface is 25°C/km, but the value falls rapidly with depth, being probably not greater than about 6 deg/km immediately below the crust. [There have been suggestions (Lubimova, 1969) that $d\tau/dz$ may increase for some tens of kilometres just below the crust, before starting to fall again. See also Birch (1969a, 1969b).] At $z = 100$ km, $\tau \approx 1000°$C.

The fact that the whole mantle transmits S waves and so is not molten enables upper limits to be set to temperatures throughout it. For the bottom of the mantle, Uffen (1952) gave an upper limit of about 5000°C, and the values given by most investigators [see e.g. Verhoogen (1956), Jacobs (1956), Gilvarry (1957), Lubimova (1958)] lie between about 2000°C and 7000°C; the preferred estimate is 3000°C to 4000°C. Below 100 km depth, the average value of $d\tau/dz$ in the mantle is, on these figures, likely to be about 1 deg/km. Adaptation of a rough calculation of Birch (1952) would give about 0·2 deg/km for the average adiabatic temperature gradient in the mantle. Hence the average value of ϑ between 100 and 2900 km depth is likely to be a little less than 1 deg/km.

It is well established that ϑ exceeds the average mantle value inside the region B; through an appreciable part of B, ϑ may be of the order of 6 deg/km. The average mantle value is also probably exceeded inside the region C. Inside D, ϑ must consequently be rather less than 1 deg/km; the most probable average value in D has been assessed as about 0·5, with an uncertainty of at least 0·3 deg/km.

Inside the core, $d\tau/dz$ is likely to be appreciably smaller than in the mantle because of the fluidity of the outer core, and the temperature rise inside the core is unlikely to be much more than 500°C. Birch (1972) stated that estimates of 4000°K and 5000°K for the temperatures at the top and bottom of the core "may have some standing as upper limits". Inside the outer core, Higgins and Kennedy (1971) suggested that ϑ may actually be negative, with magnitude 0·3–0·4 deg/km; the suggestion is at present controversial.

[Refs. on p. 180]

Below the outer core, the effects on $d\rho/dz$ of any deviations of ϑ from zero may be neglected as trivial compared with the effects of other uncertainties.

10.8.2 Temperature corrections to density gradients

Factors in the expression (10.13) for S/P were examined numerically by Birch (1952). For the mantle, using theory of solids and laboratory evidence, he estimated that $\lambda \approx 1-2$, and took $g = 10$ m/s^2 and $c \approx 1\cdot2 \times 10^3$ m^{2}s^{-2} deg^{-1}. These data give $\lambda cg^{-1} \approx 0\cdot1-0\cdot2$ km-deg^{-1}.

Inside part at least of the region B, where ϑ is large, it then follows by (10.13) that the magnitude of the temperature correction term S may approach or even exceed P. A reasonable tentative estimate of the average value of S/P inside B is about $-0\cdot5$. Thus values of $d\rho/dz$ derived through (10.7) for the region B, as in the A-type models, are likely to need considerable reduction inside part of D. Independent evidence (see e.g. § 14.7.2) supports this conclusion.

For the region C, estimates of the term S are not required on the Model A procedures, the use of (10.7) having been avoided in determining $d\rho/dz$ in C.

For the region D, Birch assumed $\lambda = 1\cdot2$, $0\cdot8$ and obtained $\lambda cg^{-1} \approx (0\cdot14, 0\cdot09)$ km-deg^{-1} near the top and bottom, respectively. Taking $\vartheta = 0\cdot5$ deg/km would by (10.13) then give S/P as ranging from about $-0\cdot07$ to $-0\cdot045$ inside D.

For the outer core, Birch assumed $\lambda = 1$ and $c = 0\cdot6 \times 10^3$ m^{2}s^{-2} deg^{-1}, and took $g = (10, 7)$ m/s^2 at $z = 2900, 4600$ km. These data yield $S/P = -(0\cdot06, 0\cdot09)\vartheta$ at the two levels, where ϑ is in deg/km. The results of Higgins and Kennedy would then give $S/P \approx +0\cdot02-0\cdot03$; this correction is small and, being doubtful, can be ignored pending the emergence of further evidence on ϑ in the core.

10.8.3 Temperature corrections in A-type density distributions

A uniform reduction by a fraction x, say, in $d\rho/dz$ throughout the mantle would, in the absence of other changes, reduce the estimated density near the base of the mantle by about $2\cdot3x$ g/cm^3. Several investigators inferred from this result that quite large temperature corrections to ρ were needed in the deeper interiors of A-type

[Refs. on p. 180]

models. This, however, is not the case because the model A procedures have a built-in device (involving the procedure for the region C) which much reduces the corrections required below B (Bullen, 1956, 1968). Table 10.3 gives the corrections $\Delta\rho$ to ρ needed in A-type models corresponding to assigned reductions in $d\rho/dz$ in the regions B and D. The assigned reductions are constant throughout the regions to which they apply and so may be considered as averages. In preparing the table, the value $3 \cdot 32$ g/cm^3 for ρ_b has been kept unvaried.

TABLE 10.3

Corrections $\Delta\rho$ (in g/cm^3) corresponding to assigned proportionate reductions in $d\rho/dz$ in the regions B and D

Proportionate reduction in $d\rho/dz$		Increase ($\Delta\rho$) in ρ (in g/cm^3)				
B	D	Top of region B	B–C boundary	C–D boundary	Base of mantle	Throughout core
1·0	0·0	0	−0·32	+0·15	+0·15	−0·21
0·0	1·0	0	0	+0·19	−0·84	0·0
0·1	0·1	0	−0·03	+0·03	−0·07	−0·02
0·5	0·1	0	−0·16	+0·09	−0·01	−0·10
1·0	0·5	0	−0·32	+0·24	−0·27	−0·21

The values of $\Delta\rho$ in Table 10.3 are sufficiently small for lines of the table to be treated as linearly superposable to good approximation. This can be verified by deriving lines 3, 4, and 5 by linear combinations from lines 1 and 2. Values of $\Delta\rho$ for other postulated reductions in $d\rho/dz$ can likewise be simply derived.

From line 3, it is seen that assigned reductions of 10 per cent in $d\rho/dz$ inside both B and D would entail a maximum correction $|\Delta\rho|$ of only $0 \cdot 07$ g/cm^3, less than one-third of the correction $0 \cdot 23$ g/cm^3 formally required if $d\rho/dz$ were reduced by 10 per cent throughout the mantle. Reduction of $d\rho/dz$ to zero inside B (lines 1 and 5) would reduce ρ by $0 \cdot 32$ g/cm^3 at the B–C boundary; below the B–C boundary, $|\Delta\rho|$ would be everywhere less than $0 \cdot 32$ g/cm^3 for any simultaneous reduction of $d\rho/dz$ inside D up to 50 per cent. The case (line 4) where $d\rho/dz$ is reduced by 50 per cent inside B and 10 per cent inside D roughly fits the recent evidence discussed in §10.8.2, and gives $|\Delta\rho| > 0\cdot1$ g/cm^3 inside only a small part of B and nowhere inside D.

On the data in §10.8.2, any modifications to Table 10.2 to allow for temperature corrections in the core are likely to be slight. Thus, throughout the core, the temperature corrections $\Delta\rho$ in the A-type models are also likely to be quite small.

For the region B, where the temperature effects are expected to be greatest, $d\rho/dz$ can now be estimated (§14.7) without recourse to (10.7). Although the precision is still not high, the result is probably superior to the net result of using (10.7) and attempting to apply temperature and other corrections.

10.8.4 *Corrections for inhomogeneity*

The corrections needed to (10.7) for departures from homogeneity were, in the early stages of the Earth's density determination, even more uncertain than the temperature corrections. So it was appropriate to start by treating each region of the Earth as homogeneous and then test the consequences against evidence from other sources. For example, as seen in §10.5, the assumption of a wholly homogeneous mantle led to a test from which the region C was inferred to be inhomogeneous.

Normalcy (§9.3) of the seismic velocity gradients was suggested (Bullen, 1940) as a broad criterion of homogeneity and the criterion was subsequently (see §11.5.5) shown to be generally reliable. Application of the criterion confirms that the regions D' and E of the Earth are homogeneous to a good first approximation. Although C is inhomogeneous, its density distribution was satisfactorily estimated without recourse to (10.1) or (10.7). Since the regions C, D', and E together occupy nearly 80 per cent of the Earth's volume, it is therefore to be expected that corrections for inhomogeneity will not greatly affect the overall density distributions of A-type models.

In general, the presence of inhomogencity in any region increases $d\rho/dz$; hence inhomogeneity corrections to $d\rho/dz$ act to offset temperature corrections (except possibly in the core, should the super-adiabatic gradient prove to be negative). Table 10.3 can be applied to assessing the effects on the overall density distributions not only of super-adiabatic temperature gradients, but also of inhomogeneities in particular regions.

[*Refs. on p. 180*]

Some comments will now be made on questions of inhomogeneity in a few specific regions.

The model A calculations treated the region B, as well as D and E, as homogeneous because of the normalcy of the Jeffreys 1939 velocity gradients inside B. However, the later evidence (§10.3.2) on departures from normalcy now forbids the unqualified use of (10.7) anywhere in the upper mantle. Both temperature and inhomogeneity may contribute to these departures and it is not yet possible inside B to separate out finely the contributions from the terms S and $(N + Q)$ to $d\rho/dz$. Independent evidence, including the matching of the results of laboratory experiments on rocks against the observed values of α and β, makes it probable, however, that in B the inhomogeneity corrections are small compared with the temperature corrections, already indicated as fairly large (§10.8.2). Any inhomogeneities inside B are therefore likely to have the effect merely of slightly reducing errors due to neglect of superadiabatic temperature gradients. (See also §15.3.3.)

For the region C, Birch (1952) has discussed the separate contributions of S, N, and Q, and Clark and Ringwood (1967) of N and Q, in considerable detail.

The possibility that D'' is inhomogeneous has to be taken into account because of the likelihood that the velocity gradients are abnormally small (§10.3.4). It will be shown in §11.2.2 that the maximum likely effect of inhomogeneity inside D'' would be to raise by 0.2 g/cm^3 the density increase through D'' as calculated on the Model A procedures. The region D'' being small, the effects on density values outside D'' are negligible. Thus it is appropriate to treat the whole of D as homogeneous in the first approximation to the Earth's density distribution.

Although there may be some mild inhomogeneity inside the region E (see §12.2.1), it remains appropriate to treat E also as homogeneous in the first approximation. Below E, it is difficult to assess departures from homogeneity directly from seismic data. For this reason, the original A-type models allowed for a variety of density distributions in the lower core, and Table 10.2 gives only minimum values of ρ below 4980 km depth. Seismological and other evidence on density and homogeneity in the lower core will be examined more closely in Chapter 12.

10.9 Critique of A-type models

The chief assumptions made in constructing the A-type models were (a) the value assumed for ρ_b; (b) the values assumed for α and β and hence ϕ; (c) the appropriateness of (10.7) in the regions B, D, and E. Uncertainties below E were allowed for by providing a family of A-type models, of which the models A, A', and A'' are representative.

The degree of reliability of (a) has been discussed in §10.4.3. Changing the assumed value of ρ_b within the limits indicated in §10.4.3 would change the estimates of ρ and $d\rho/dz$ inside B and C, but would not greatly affect the estimates below C. (After prodigious numerical testing, some modern investigators appear to have surprised themselves on finding how narrow are the limits within which ρ is determined near the top of the region D.) If ρ_b were taken as high as 3.8 g/cm^3, an A-type model could be constructed compatibly with the whole mantle being homogeneous between crust to core.

A provisional discussion of the reliability of (b) has been given in §§9.6.2 and 10.3.1–10.3.7, the essential position being as follows. For the greater part of the regions D and E, the 1939 seismic data on α, β, and ϕ are generally reliable (comparisons with recent data are given in §§12.1.3–12.2, and Tables 12.2 and 12.3). There still remain appreciable uncertainties in determinations of α and β in the region B and the upper part of C. As stated, the uncertainties affect principally the velocity gradients rather than α and β themselves; $d\alpha/dz$, $d\beta/dz$, and $d\phi/dz$ are not finely determined until depths exceeding 650 km are reached. But ϕ itself is sufficiently well known to yield through (10.7) closely reliable values of the contribution to $d\rho/dz$ arising from adiabatic compression throughout the mantle and the outer core proper. Amendments required to ϕ are more substantial in the lower core and carry implications specially considered in §§12.3 and 12.7.1. These amendments do not require revision of the more essential Model A procedures, but markedly affect the numerical assessments of ρ, k, and μ in the lower core.

The reliability of the assumption (c) is indicated in the discussions in §§10.8–10.8.4. Inside B, corrections to ρ due to departures from (10.7) are likely to range from zero to $-\theta$, where $\theta \leqslant 0.3$ g/cm^3. Inside D', (10.7) is likely to be reliable within 5 to 10 per cent. Inside D'', increases in ρ up to 0.2 g/cm^3 are possibly

[Refs. on p. 180]

required. Inside E, (10.7) is likely to be reliable within 10 per cent, but the depth to which E reaches may be greater than that taken in the A-type models (see §§15.6.3, 16.4). Corrections made to ρ in the mantle in general entail additional corrections in the core; such additional corrections can be derived with the help of Table 10.3.

Estimates of the core densities are noticeably affected at all depths by the large uncertainties near the Earth's centre. For example, an increase of 5 g/cm³ in the postulated central density ρ_0 would be accompanied (see B, §13.4.3) by corresponding sizeable increases in ρ throughout most of the lower core and by decreases up to 0·4 g/cm³ outside the inner core proper. (The effects are minor outside the core.)

A broad summary is that the densities in Table 10.2 are likely to be reliable within about 0·3 g/cm³ inside the mantle and about 0·5 g/cm³ inside the outer core. The central density is at least 12 g/cm³, but upper bounds to ρ inside the lower core cannot be usefully assessed without recourse to further evidence given in later chapters. It will be seen incidentally that the further evidence bears out the foregoing assessments of the reliability of Table 10.2 in respect of the mantle and outer core.

The uncertainties in the observationally determined α and β are necessarily rather smaller than the uncertainties in ρ at most depths. By (8.6,7), k and μ are therefore determined in general to about the same reliability as ρ at most depths, and the same applies to derived quantities such as Young's modulus E and the Lamé elasticity parameters. (There are some ranges of depth, including B and D'', inside which the variation of μ is rather more uncertain than that of k.) Poisson's ratio σ is somewhat more reliably determined than k and μ since, by (10.15), σ is derivable from α and β. The reliabilities with which p and g are determined are indicated by comparing results for different models, e.g. Model A'' and the models in §16.4.

REFERENCES

Birch, F. (1952). Elasticity and constitution of the Earth's interior. *J. Geophys. Res.*, **57**, 227–286.

Birch, F. (1955). Physics of the crust. *Geol. Soc. Amer. Spec. Paper*, **62**, 101–118.

Birch, F. (1958). Interpretation of the seismic structure of the crust in the light of experimental studies of wave velocities in rocks. In *Contributions in Geophysics*, London. Pergamon, pp. 158–170.

REFERENCES

Birch, F. (1961). Composition of the Earth's mantle. *Geophys. J. R. Astr. Soc.*, 4, 295–311.

Birch, F. (1969a). Density and composition of the upper mantle: first approximation as an olivine layer. In *The Earth's Crust and Upper Mantle* (ed. P.J. Hart), American Geophys. Union, Washington, pp. 18–36.

Birch, F. (1969b). Interpretations of the low-velocity zone. *Phys. Earth Planet. Interiors*, 3, 178–181.

Birch, F. (1972). The melting relations of iron, and temperatures in the Earth's core. *Geophys. J. R. Astr. Soc.*, 29, 373–387.

Bullard, E.C. (1957). The density within the Earth. *Verh. geol. mijnb. Genoot. Ned. Kolon.*, 18, 23–41.

Bullen, K.E. (1936). The variation of density and the ellipticities of strata of equal density within the Earth. *Mon. Not. R. Astr. Soc., Geophys. Suppl.*, 3, 395–401.

Bullen, K.E. (1937). Note on the density and pressure in the Earth. *Trans. R. Soc. N.Z.*, 67, 122–124.

Bullen, K.E. (1940). The problem of the Earth's density variation. *Bull. Seismol. Soc. Amer.*, 30, 235–250.

Bullen, K.E. (1942). The density variation of the Earth's central core. *Bull. Seismol. Soc. Amer.*, 32, 19–29.

Bullen, K.E. (1946). A hypothesis on compressibility at pressures of the order of a million atmospheres. *Nature, Lond.*, 157, 405.

Bullen, K.E. (1949). Compressibility-pressure hypothesis and the Earth's interior. *Mon. Not. R. Astr. Soc., Geophys. Suppl.*, 5, 355–368.

Bullen, K.E. (1950). An Earth model based on a compressibility-pressure hypothesis. *Mon. Not. R. Astr. Soc., Geophys. Suppl.*, 6, 50–59.

Bullen, K.E. (1952). On density and compressibility at pressures up to thirty atmospheres. *Mon. Not. R. Astr. Soc., Geophys. Suppl.*, 6, 383–401.

Bullen, K.E. (1954). On the homogeneity, or otherwise, of the Earth's upper mantle. *Trans. Amer. Geophys. Un.*, 35, 838–841.

Bullen, K.E. (1956). The influence of temperature gradient and variation of composition in the mantle on the computation of density values in Earth model A. *Mon. Not. R. Astr. Soc., Geophys. Suppl.*, 7, 214–217.

Bullen, K.E. (1966). On the constitution of Mars; third paper. *Mon. Not. R. Astr. Soc.*, 133, 229–238.

Bullen, K.E. (1967). Basic evidence for Earth divisions. In *The Earth's Mantle* (ed. T.F. Gaskell), Academic Press, London, pp. 11–39.

Bullen, K.E. (1968). Effects of varying density gradients in the Earth's mantle. *Bull. Internat. Inst. Seism. Earthq. Eng.* (Tokyo), 5, 1–6.

Bullen, K.E. and Burke-Gaffney, T.N. (1958). Diffracted seismic waves near the PKP caustic. *Geophys. J., R. Astr. Soc.*, 1, 9–17.

Bullen, K.E. and Haddon, R.A. (1967). Derivation of an Earth model from free Earth oscillation data. *Proc. Nat. Acad. Sci., Wash.*, 58, 846–852.

Byerly, P. (1926). The Montana earthquake of June 28, 1925. *Bull. Seismol. Soc. Amer.*, **16**, 209–265.

Clark, S.P., Jr., and Ringwood, A.E. (1967). Density, strength and constitution of the mantle. In *The Earth's Mantle* (ed. T.F. Gaskell), Academic Press, London, pp. 111–124.

Gilvarry, J.J. (1957). Temperature in the Earth's interior. *J. Atm. Terr. Phys.*, **10**, 84–95.

Gutenberg, B. (1914). Über Erdbebenwellen. VII A. Beobachtungen an Registrierungen von Fernbeben in Göttingen und Folgerungen über die Konstitution des Erdkörpers. *Nachr. Ges. Wiss. Göttingen, Math.-Phys. Klasse*, pp. 1–52 and 125–176.

Gutenberg, B. (1951). $PKKP$, $P'P'$, and the Earth's core. *Trans. Amer. Geophys. Un.*, **32**, 373–390.

Gutenberg, B. (1959). *Physics of the Earth's Interior*. Academic Press, New York.

Gutenberg, B. and Richter, C.F. (1938). P' and the Earth's core. *Mon. Not. R. Astr. Soc., Geophys. Suppl.*, **4**, 363–372.

Haddon, R.A. and Bullen, K.E. (1969). An Earth model incorporating free Earth oscillation data. *Phys. Earth Planet. Interiors*, **2**, 35–49.

Halley, E. (1688). An account of the cause of the change of the variation of the magnetic needle, with an hypothesis of the structure of the internal parts of the Earth. *Royal Society papers*, 43–59.

Higgins, G. and Kennedy, G.C. (1971). The adiabatic gradient and the melting point gradient in the core of the Earth. *J. Geophys. Res.*, **76**, 1870–1878.

Jacobs, J.A. (1956). The Earth's interior. *Encyclopedia of Physics*, **47**, 364–406. Springer-Verlag, Berlin.

Jeffreys, H. (1926). The rigidity of the Earth's central core. *Mon. Not. R. Astr. Soc., Geophys. Suppl.*, **1**, 371–383.

Jeffreys, H. (1939a). The times of the core waves. *Mon. Not. R. Astr. Soc., Geophys. Suppl.*, **4**, 548–561.

Jeffreys, H. (1939b). The times of P, S and SKS and the velocities of P and S. *Mon. Not. R. Astr. Soc., Geophys. Suppl.*, **4**, 498–533.

Jeffreys, H. (1942). The deep earthquake of 1934 June 29. *Mon. Not. R. Astr. Soc., Geophys. Suppl.*, **5**, 33–36.

Jeffreys, Sir H. (1954). The times of P in Japanese and European earthquakes. *Mon. Not. R. Astr. Soc., Geophys. Suppl.*, **6**, 557–565.

Jeffreys, Sir H. (1958). The times of P up to $30°$ (second paper). *Geophys. J., R. Astr. Soc.*, **1**, 154–161.

Jeffreys, Sir H. (1970). *The Earth*, 5th ed. Cambridge University Press.

Jeffreys, H. and Bullen, K.E. (1933). Corrections to the times of the P wave in earthquakes. *Nature, Lond.*, **131**, 97.

Jeffreys, H. and Bullen, K.E. (1935). Times of transmission of earthquake waves. *Bur. Centr. Séism. Internat.* A, Fasc. 11, 202 pp.

REFERENCES

Jeffreys, H. and Bullen, K.E. (1940, 1967). *Seismological tables.* Brit. Ass., Gray-Milne Trust, 50 pp.

Lehmann, Inge (1934). Transmission times for seismic waves for epicentral distances around 20°. *Geodaet. Inst. Skr.*, **5**, 44 pp.

Lehmann, Inge (1936). P'. *Bur. Centr. Séism. Internat.* A, **14**, 3–31.

Lehmann, Inge (1959). Velocities of longitudinal waves in the upper part of the Earth's mantle. *Ann. Géophys.*, **15**, 93–118.

Lubimova, H.A. (1958). Thermal history of the Earth and the variable thermal conductivity of its mantle. *Geophys. J., R. Astr. Soc.*, **1**, 115–134.

Lubimova, H.A. (1967). Theory of the thermal state of the Earth's mantle. In *The Earth's Mantle* (ed. T.F. Gaskell), Academic Press, London, pp. 232–323.

Lubimova, H.A. (1969). Thermal history of the Earth. In *The Earth's Crust and Upper Mantle* (ed. P.J. Hart), American Geophys. Union, Washington, pp. 63–77.

Mohorovičić, A. (1909). Das Beben vom 8. x. 1909. *Jb. met. Obs. Zagreb*, **9**, 1–63.

Molodenski, M.S. (1953). Density and elasticity within the Earth. *Geophys. Inst. Trudy*, **19**, (146), 3.

Press, F. (1969). Earth models consistent with geophysical data. *Phys. Earth Planet. Interiors*, **3**, 3–22.

Takeuchi, H. (1950). On the Earth tide in the compressible Earth of varying density and elasticity. *Trans. Amer. Geophys. Un.*, **31**, 651–689.

Uffen, R.J. (1953). A method of estimating the melting-point gradient in the Earth's mantle. *Trans. Amer. Geophys. Un.*, **33**, 893–896.

Verhoogen, J. (1956). Temperatures within the Earth. *Phys. and Chem. of Earth*, **1**, 17–43.

Worzel, L.J. and Shurbet, G.L. (1955). Gravity interpretations from standard oceanic and continental sections. *Geol. Soc. Amer. Spec. Paper*, **62**, 87–100.

Zöppritz, K. (1907). Über Erdbebenwellen. II. Laufzeitkurven. *Nachr. Ges. Wiss. Göttingen, Math.-Phys. Klasse*, pp. 529–549.

CHAPTER 11

Evidence on compressibility in the Earth

The intimate connection between the incompressibility k and the density ρ, as expressed in the relation (7.15), shows that evidence involving k is very pertinent to the determination of the Earth's density variation. An illustration has already been provided in the use of information on ϕ, i.e. k/ρ, derived from seismic data in applying (10.7) to estimate density gradients in some regions of the Earth. In the present chapter, a quantity of additional evidence, theoretical and experimental, involving k is brought to bear.

Part of the most important evidence emerged in the first instance from a study of the variation of k with the pressure p as derived for the A-type Earth models, leading to an empirically based trial hypothesis on the variation of compressibility in the Earth over a wide range of pressure. Subsequent experimental and theoretical work strengthened essential features of the hypothesis and made it possible to assess numerically the reliabilities of particular aspects, resulting in sharpened determinations of k, ρ, and the rigidity μ in various parts of the Earth.

One of several far-reaching conclusions in which the hypothesis played a key role is (§11.7) that the Earth's inner core is solid (in the sense defined in §7.7).

The detailed evidence on k also enables the discussion in §10.8.4 on degrees of inhomogeneity in particular regions of the Earth to be carried considerably further. The results not only provide additional information on the suitability or otherwise of (10.7) in determining density gradients $d\rho/dz$, but also supply direct estimates of $d\rho/dz$ in some markedly inhomogeneous regions.

Along with the k-p hypothesis, evidence on k from a variety of further sources, including Birch's applications (1939, 1952) of finite-strain theory and laboratory evidence, are examined in the present

[Refs. on p. 223]

chapter and brought to bear where warranted. A review is included of B-type Earth models, which were substantially based on the k-p hypothesis.

The chapter is mainly concerned with regions sufficiently far below the Earth's surface to make the assumption of hydrostatic conditions adequate. Thus the relation (10.4) will be assumed to hold.

In contexts where it is necessary to discriminate between adiabatic and isothermal incompressibilities, k will denote the adiabatic and k' the isothermal incompressibility, as in §7.9.5.

11.1 Compression in the Earth

An idea of the magnitude of the compression f (as defined in §7.8.2) is needed intermittently, and it is convenient to give some results on f before proceeding to the main arguments on k.

By (7.28), taking logarithmic differentials, we have

$$\rho^{-1}\,d\rho = 3(1 + 2f)^{-1}\,df. \tag{11.1}$$

For a homogeneous material as defined in §10.1.4, Birch's adaptation (§7.9.7) of Murnaghan's finite-strain theory yields the relation (7.49) between p and f, and hence

$$p^{-1}\,dp = \{f^{-1} + 5(1 + 2f)^{-1}\}\,df. \tag{11.2}$$

Amendments which may be required to (7.49) are unlikely to be important for present purposes where only a rough first approximation is needed. It is also sufficiently accurate in the present section to treat the changes in p, ρ, and f as taking place adiabatically. We may then also write (see §7.5.1)

$$d\rho/dp = \rho/k. \tag{11.3}$$

[Strictly, Birch's formula (7.49) relates to isothermal conditions. Detail to be given in §§11.2.4 and 11.2.5 shows how departures from adiabatic conditions could be allowed for if required in using (11.2) and (11.3).] From (11.1), (11.2), and (11.3), we obtain (see Bullen, 1968a)

$$f = p(3k - 7p)^{-1}. \tag{11.4}$$

The relation (11.4), although derived above for a homogeneous material, is nevertheless also applicable at points P inside regions of

[Refs. on p. 223]

the Earth where the chemical composition is continuously changing. This is because dp, $d\rho$, and df in (11.1), (11.2), and (11.3) may be taken to relate, not necessarily to variations with respect to the depth z below the Earth's surface, but also to variations with respect to p such that the chemical composition of the material at P is stipulated to be unvaried.

The question of possible phase changes arising from changes in p can, however, be a source of complication in applying (11.4). Imagine a sample of material at zero pressure with the same composition as the material at P, and suppose the pressure is raised (adiabatically) to p. Then (11.4) gives that part of the total compression which does not arise from phase changes. If a significant part of the total density change brought about by the rise of the pressure from zero is due to phase changes, then (11.4) significantly underestimates the total compression. Thus values of f derived through (11.4) for points P of the Earth's interior must be regarded as minimum values. Inside the Earth's mantle, the possible occurrences of phase changes are expected to require only slight increases to the values of f given by (11.4). If the mantle-core transition should be predominantly associated with a phase transition (see §17.2.3), however, the values of f derived for the core could be appreciably too low.

Through (11.4), values of f were formally derived (Bullen, 1968a) for six different Earth models, including two models of type A. The results indicated that the values of f derived for A-type models are likely to be reliable (as minima) within 0·003 throughout the mantle and 0·02 inside the core. The writer's preferred estimates of the minimum f in the Earth are shown in Table 11.1.

Birch (1952) had earlier derived values of f for the mantle alone, but by a more complicated method which involves some approximations now known to be unnecessary and which is inapplicable to the core. His values also are minimum values which ignore phase changes. They agree closely in the mantle with the writer's results for the original Earth Model A, and in this respect supply a useful check on the reliability of (11.4).

The implications of phase changes incidentally need stressing in some closely related contexts. Using the concept of so-called 'uncompressed densities', many writers have presented unsound arguments which do not make sufficient (if any) provision for unknown phase changes during the process of decompression. In

[Refs. on p. 223]

particular, spurious support has been given in this way to some geochemical theories on the internal structure not only of the Earth but of other planets.

TABLE 11.1
Preferred estimates of the compression f in the Earth
(neglecting contributions from possible phase changes)

Depth (km)	f	Depth (km)	f
200	0·018	2200	0·103
400	0·029	2600	0·120
600	0·034	2900	0·132
800	0·041	3200	0·16
1000	0·050	3600	0·17
1400	0·070	4400	0·19
1800	0·087	6371	0·19

11.2 Variation of k in homogeneous regions of the Earth

The density ρ and the incompressibility k at depth z in the Earth depend in general on the variables p, τ, n_i, and q_i of §10.1.1. In the following sections, the notation $d\rho/dp$ is to be interpreted (differently from an interpretation in the third paragraph of §11.1) as $(d\rho/dz)/(dp/dz)$; similarly $dk/dp \equiv (dk/dz)/(dp/dz)$. Other symbols are as defined in subsections of §10.1.

The subsections §§11.2.1–11.2.7 are concerned principally with the development of expressions for dk/dp. Throughout these subsections except where stated, homogeneity as defined in §10.1.4 will be assumed; i.e. the n_i and q_i are treated as constant inside the region concerned.

11.2.1 *Formula for case where departures from adiabatic temperature gradients are neglected*

Taking $\phi = \alpha^2 - 4\beta^2/3$ (§10.1.2), where α and β are the P and S seismic velocities, we have

$$\phi = k/\rho, \tag{11.5}$$

$$\frac{dk}{dp} = \phi \frac{d\rho}{dp} + \rho \frac{d\phi}{dp}. \tag{11.6}$$

When the excess ϑ of $d\tau/dz$ over the adiabatic temperature gradient (§10.1.3) is neglected, (11.3) and (11.5) give

$$\frac{d\rho}{dp} = \phi^{-1}. \tag{11.7}$$

Then (11.6) and (10.4) give (Bullen, 1949)

$$\frac{dk}{dp} = 1 + g^{-1}\frac{d\phi}{dz}. \tag{11.8}$$

Considerable use has been made of the formula (11.8). [But see also §11.5.2 for a powerful generalization of (11.8).] The formula (11.8) enables dk/dp to be assessed in homogeneous regions of the Earth when values of $d\phi/dz$ and g are available, and can provide a test of homogeneity when independent information on dk/dp is also available. Illustrations are given in §§11.2.2 and 11.2.3.

11.2.2 Application to the region D

As seen in §§10.3.4 and 10.8.4, the region D' is probably fairly homogeneous. At the bottom of D', $d\phi/dz \approx 20$ m/s² and $g \approx 10$ m/s², so that (11.8) here gives $dk/dp \approx 3\cdot0$ (Bullen, 1946, 1949), in agreement with results in §10.7.2 for an A-type model.

The formula (11.8) was also prominent in the argument (Bullen, 1949) that the region D'' (2700 $<z<$ 2900 km) needs (see §10.3.4) to be treated distinctly from D'. On both the Jeffreys and Gutenberg velocity distributions (§9.6.2), $d\alpha/dz$ and $d\beta/dz$ fall rapidly and continuously to the order of zero inside D''. With $d\phi/dz \approx 0$, formal application of (11.8) would then give $dk/dp \approx 1$, whereas there is strong evidence (§§11.2.6, 11.4.4) that $dk/dp \approx 3$ in this part of the Earth. Hence the Jeffreys and Gutenberg velocities would entail considerable departures from (11.8), and therefore inhomogeneity, inside D''. By (11.6) and (10.4), the assumptions $d\phi/dz = 0$ and $dk/dp = 3$ formally give $d\rho/dz = 3g\rho/\phi$, i.e. a density gradient three times that for the case (10.7) of pure compression; the density increase inside D'' would as a consequence be increased from 0·1 to 0·3 g/cm³ (Bullen, 1949).

This estimate of density changes inside D'' could be affected by revisions of the values taken for $d\alpha/dz$ and $d\beta/dz$. It is fairly well established that $d\alpha/dz$ and $d\beta/dz$ both fall rapidly from D' to D''. But should one or both of $d\alpha/dz$ and $d\beta/dz$ differ significantly from zero inside D'', then $d\phi/dz$, i.e. $d(\alpha^2 - 4\beta^2/3)/dz$, would not be necessarily zero. In particular, should the fall in $d\beta/dz$ exceed that in $d\alpha/dz$ sufficiently, $d\phi/dz$ could remain, inside D'', equal to the value at the bottom of D'. In that event, (11.8) would be satisfied inside D'' with $dk/dp \approx 3$, and D'' would be indicated as homogeneous; instead of an abnormal density increase, there would then be a sharp fall in rigidity inside D''. Cleary (1969) suggested (see §12.1.4) that this may be the case. Most recent investigations favour the presence of inhomogeneity, but there may also be some fall in rigidity inside D''.

Throughout D, the seismic velocities α and β and their gradients appear to vary (effectively) continuously with z, and the same is likely to apply to chemical and physical properties inside D'' in particular In consequence, the thickness of D'' is not sharply determined and the currently assigned depth of 2700 km is only conventional.

The discussion on conditions inside D'' illustrates the sensitivity of (11.8) to questions of the normalcy of $d\phi/dz$, homogeneity, and values of dk/dp in a region (see also §11.5.2). Tests for homogeneity can be made more precise, however, and will be examined in further detail (§11.5), using the generalization of (11.8).

11.2.3 *Application to the region E*

Inside the region E, μ/k can be neglected, so that $d\phi/dz \approx d\alpha^2/dz$. To a first approximation, $d\alpha/dz$ is normal inside E, so that E may be treated as approximately homogeneous and (11.8) applied. With the Jeffreys and Gutenberg values of α, (11.8) gives dk/dp as ranging from about 2·8 at the top to 3·5 at the bottom of E. (A second approximation — see §12.2 — allows for the possibility that E may be mildly inhomogeneous.)

The small difference of 0·2 between the values of dk/dp indicated at the bottom of D' and top of E is less than the uncertainties in $g^{-1} d\phi/dz$, and so is not significantly different from zero.

11.2.4 *Allowance for temperature gradient*

In the notation of §§10.1.1—10.1.3, we have by (10.1, 4, 6, 11) for a homogeneous region: $d\rho/dz = P + S$; $dp/dz = g\rho$; $P = g\rho/\phi$; $S = -\gamma\phi\vartheta$. These relations yield

$$\frac{d\rho}{dp} = \phi^{-1}\left(1 - \frac{\gamma\phi\vartheta}{g}\right), \qquad (11.9)$$

[Refs. on p. 223]

and thence, by (11.6) and (10.13),

$$\frac{dk}{dp} = 1 + g^{-1}\frac{d\phi}{dz} - \omega, \qquad (11.10)$$

where the temperature correction term $\omega = \gamma\phi\vartheta/g = \lambda c\vartheta/g = -S/P$, and λ is Grüneisen's ratio (§10.1.3). In practice, numerical estimates of ω are brought to bear in determining dk/dp only in the lower mantle and core, where the main applications of (11.8) and (11.10) are made. Estimates of ω are immediately derivable from data in §10.8.2.

Near the bottom of D', $g^{-1}d\phi/dz \approx 2$ and $\omega \approx 0.04$. On these results, the temperature correction to the estimate of dk/dp obtained using (11.8) is less than 2 per cent. Higher up in D' corrections may reach 3 per cent. These corrections are quite minor, being no larger than the uncertainties in $g^{-1}d\phi/dz$.

Inside D'', (11.8) and (11.10) are not usually applied because of the likelihood of significant inhomogeneity. The value of ω in D'' is expected to be about the same as at the bottom of D'.

For the core, the detail in §10.8.2 makes it reasonable to neglect the temperature correction to dk/dp for the present.

11.2.5 *Isothermal and adiabatic incompressibility gradients*

In applying incompressibility theory to the Earth's interior, it is necessary for some purposes to assess differences between incompressibility-pressure gradients under isothermal and adiabatic changes.

By (7.42) and (10.12)

$$k - k' = \tau k'^2 \gamma^2/\rho c$$

$$\approx \tau\gamma\lambda k'. \qquad (11.11)$$

By (10.4), (10.8), and (10.10),

$$\left(\frac{\partial \tau}{\partial \rho}\right)_p \left(\frac{\partial \rho}{\partial s}\right)_p \frac{ds}{dp} = \frac{\vartheta}{g\rho}. \qquad (11.12)$$

Hence

$$\frac{dk}{dp} = \left(\frac{\partial k}{\partial p}\right)_s + \left(\frac{\partial k}{\partial s}\right)_p \frac{ds}{dp} \qquad (11.13)$$

$$= \left(\frac{\partial k}{\partial p}\right)_s + \frac{\vartheta}{g\rho}\left(\frac{\partial k}{\partial \tau}\right)_p. \qquad (11.13)$$

Using (11.5), the relation (11.10) then becomes

$$1 + g^{-1}\frac{d\phi}{dz} = \left(\frac{\partial k}{\partial p}\right)_s + \frac{\gamma\phi\vartheta}{g}\left\{1 + (\gamma k)^{-1}\left(\frac{\partial k}{\partial \tau}\right)_p\right\}. \quad (11.14)$$

Using (11.11) and (11.14), Birch (1952, p. 237) arrived at

$$1 + g^{-1}d\phi/dz = (\partial k'/\partial p)_\tau + \tau\gamma\lambda A + (\tau\gamma\lambda)^2 B + \gamma\phi\vartheta g^{-1} C, \quad (11.15)$$

where A, B, and C are certain dimensionless functions of k', p, τ, γ, c, and λ. From experimental evidence, he estimated that $A \approx -5$ and $C \approx -2$. The term containing B was neglected, $\tau\gamma\lambda$ being small (of order 0·05 — see below) and B being estimated to be of the same order as A and C. Thus (11.15) became

$$(\partial k'/\partial p)_\tau \approx 1 + g^{-1}d\phi/dz + 5\tau\gamma\lambda + 2\gamma\phi\vartheta g^{-1}. \quad (11.16)$$

Writing

$$\delta = dk/dp - (\partial k'/\partial p)_\tau, \quad (11.17)$$

we have from (11.10) and (11.16)

$$\delta \approx -5\tau\gamma\lambda - 3\gamma\phi\vartheta g^{-1}. \quad (11.18)$$

Data in §10.8.2 along with data on ϕ derived from Table 9.1 (§9.6.2) yield $\gamma\phi g^{-1} \approx (0\cdot14, 0\cdot09)$ km/deg and $\gamma\lambda \approx (2\cdot2, 0\cdot7) \times 10^{-5}$ deg^{-1} near the top and base, respectively, of the region D. Taking (for the purpose of a specimen calculation) $\tau = (2500°, 4000°)$ at the two levels and taking $\vartheta = 0\cdot5$ deg/km inside D then gives $\delta \approx (-0\cdot5, -0\cdot3)$, respectively. The largest uncertainties are in τ and ϑ, but the difference $|\delta|$ probably does not noticeably exceed 0·3 near the base of the mantle; higher up in the mantle; $|\delta|$ could reach about 0·7. Inside the core, similar calculations make it likely that $|\delta|$ is less than 0·3 [less than 0·1 if the results of Higgins and Kennedy (§10.8.2) are assumed, for the terms on the right side of (11.8) would then be of opposite signs].

In so far as the above examination uses the equations (11.7), (11.8), and (11.10), the results apply strictly only to homogeneous regions of the Earth. But since k varies only fairly slowly with chemical composition inside the Earth (see later), the differences δ are not likely to need any serious corrections for inhomogeneity.

11.2.6 *Approach through finite-strain theory*

Using his adaptation of Murnaghan's finite-strain theory, Birch sought to estimate dk/dp in homogeneous regions of the Earth by a process in which the formula (§7.9.7)

$$(\partial k'/\partial p)_\tau = (12 + 49f)/(3 + 21f) \qquad (11.19)$$

play a prominent part.

The formula (11.19), which relates to isothermal conditions, gives $(\partial k'/\partial p)_\tau$ equal to 4 at zero compression and decreasing monotonely as f increases. Inside each of the mantle and core, where f steadily increases as z increases (§11.1), (11.19) gives $(\partial k'/\partial p)_\tau$ as steadily decreasing. The values of f in Table 11.1 and (11.19) would give $(\partial k'/\partial p)_\tau$ as decreasing from about 3·6 to 3·2 inside the region D, and from about 3·2 to 3·0 units inside the core. [Birch himself applied (11.19) only inside the mantle.] These values have to be reduced a little to give the adiabatic gradient dk/dp; on the data in §11.2.4, Birch's theory would give $dk/dp \approx 3\cdot 0 - 2\cdot 8$ inside most of the Earth below 1000 km depth. Corrections to f for possible phase changes (see §11.1) would, if significant, reduce these estimates of dk/dp slightly.

The relation (11.19), along with (11.8) and thermodynamical extensions such as (11.15), was much used by Birch in investigating questions of homogeneity and other physical and chemical properties in particular regions of the Earth, account being taken of laboratory measurements on rocks.

Throughout the regions D' and E, the estimates of dk/dp derived through (11.19) agreed moderately well with the earlier estimates derived for A-type Earth models (see §10.7.2), and hence also with the estimates (§§11.2.2, and 11.2.3) derived by applying (11.8) to data on $d\phi/dz$. Thus the finite-strain theory supplied a tentative check that the assumption of homogeneity in these regions is at least a satisfactory first approximation.

The agreement on dk/dp is particularly close near the mantle-core boundary, where Birch's procedures yield $dk/dp \approx 2\cdot 9$. The writer's procedures had yielded $dk/dp \approx 3\cdot 0$ and $2\cdot 8$ at depth 2700 km and at the top of the core, and the k-p hypothesis (§11.4) indicates that dk/dp will not differ much from these values inside D''. The confirmation that $dk/dp \approx 3$ inside D'' is important in several applications of compressibility theory to the Earth; one example has already been given in §11.2.2.

Above the region D, (11.19) is applicable but rather more uncertainly (see §15.1.1); in particular, it gives some confirmation of

[Refs. on p. 223]

the now well-established conclusion that the region C is significantly inhomogeneous.

Inside the core, Birch did not apply (11.19) in any detail, preferring to use (11.8). There is some evidence (see §§11.4.5 and 15.1.1) that (11.19) may need sizeable corrections in the lower core.

For the purposes of the present chapter, Birch's finite-strain theory has value in the degree of confirmation it provides of broad conclusions otherwise derived on the variation of k and dk/dp inside the Earth, especially near the mantle-core boundary. It will be seen, moreover, in §§11.5.1 and 11.5.2 that, even when only rough estimates of dk/dp are available for a region, there is a powerful independent way of testing for departures from homogeneity.

11.2.7 *Models with constant* dk/dp

The Earth Model A results (§10.7.2) showed that dk/dp varies only slowly with p throughout most of the Earth (see also §§11.2.6 and 11.4). It is therefore useful for some purposes to consider simplified models (for applications inside limited or extended regions of the Earth and other planets) in which dk/dp is postulated to be constant.

The value of dk/dp in the Legendre-Laplace model (§6.4.2) is constant and equal to 2. Results for the Earth model A showed that this value is too small to fit the Earth.

For an Earth model with $dk/dp = 3$, the Emden equation (6.19) becomes

$$\frac{d}{dr}\left(r^2 \rho \frac{d\rho}{dr}\right) = -Ar^2\rho, \qquad (11.20)$$

which is solvable with the help of tables constructed by Emden, or nowadays an electronic computer. [The assumption $dk/dp = 3$ incidentally gives $k \propto \rho^3$ (see §6.4.2) and hence $a \propto \rho$ in a fluid region. Birch suggested that these simple proportionalities might prove useful in estimating ρ inside the core. Such a procedure does in fact yield a useful first approximation to the variations of ρ and k in the outer core, but the application of (10.7), even when uncorrected, gives better results.]

11.3 Some further implications of finite-strain theory

In order to facilitate comparisions to be made in §11.4, some further

[*Refs. on p. 223*]

formal results (Bullen, 1968b) entailed by Birch's finite-strain theory will now be derived (§ § 11.3.1–11.3.5). The results are of course subject to such errors as there may be in Birch's theory.

Differences between adiabatic and isothermal incompressibilities will here be disregarded as having only second-order effects in the applications of interest. This permits Birch's relations (7.49) and (7.52) to be written as

$$p = 3Kf(1 + 2f)^{\frac{5}{2}}, \qquad (11.21)$$

$$k = K(1 + 2f)^{\frac{5}{2}}(1 + 7f), \qquad (11.22)$$

where K denotes the incompressibility at zero strain, i.e. at $f = p = 0$. As p increases, f and k of course steadily increase for a given material.

11.3.1 *Variation of incompressibility ratios for different materials*

Let the subscripts 1 and 2 refer to two materials of different chemical composition whose compressions are f_1 and f_2 when at a common pressure p. Then (11.21) and (11.22) give

$$K_1 f_1 (1 + 2f_1)^{\frac{5}{2}} = K_2 f_2 (1 + 2f_2)^{\frac{5}{2}}, \qquad (11.23)$$

$$k_2/k_1 = (7 + f_2^{-1})/(7 + f_1^{-1}). \qquad (11.24)$$

It is convenient to introduce x, where

$$x = \frac{K_2/K_1 - k_2/k_1}{K_2/K_1 - 1}. \qquad (11.25)$$

Table 11.2 (Bullen, 1968b) gives values of f_2, k_2/k_1, x, and $(k_2 - k_1)/p$, computed using (11.21)–(11.25), for various assigned values of f_1 and K_2/K_1.

The values 0.14 and 0.20 for f_1 in Table 11.2 correspond approximately (see Table 11.1) to conditions at the mantle-core boundary and the Earth's centre, respectively. Thus, unless the contributions to compression from phase changes are substantial, the values assigned to f_1 in Table 11.2 approximately cover the range of interest in the Earth's interior. Also, the range of values assigned to K_2/K_1 is likely to be more than adequate for applications to the Earth's interior. In discussing essential implications of the table values, small compositional changes such as may occur inside the region D'' will be disregarded.

[Refs. on p. 223]

IMPLICATIONS OF FINITE-STRAIN THEORY

TABLE 11.2

Values of various quantities derived on Birch's finite-strain theory corresponding to assigned values of K_2/K_1 and f_1

f_1	(K_2/K_1)	1·00	1·20	1·40	1·60	1·80	2·00
0·00	f_2	0·000	0·000	0·000	0·000	0·000	0·000
0·05		0·050	0·043	0·038	0·034	0·030	0·028
0·14		0·140	0·124	0·112	0·102	0·094	0·087
0·20		0·200	0·179	0·163	0·150	0·139	0·130
0·00	k_2/k_1	1·00	1·20	1·40	1·60	1·80	2·00
0·05		1·00	1·12	1·24	1·36	1·48	1·60
0·14		1·00	1·06	1·13	1·19	1·25	1·31
0·20		1·00	1·05	1·09	1·14	1·18	1·22
0·00	x	(0·00)	0·00	0·00	0·00	0·00	0·00
0·05		0·40	0·40	0·40	0·40	0·40	0·40
0·14		0·67	0·67	0·68	0·69	0·69	0·69
0·20		0·76	0·76	0·77	0·77	0·77	0·78
0·05	$(k_2-k_1)/p$	0·00	1·09	2·15	3·22	4·29	5·36
0·14		0·00	0·30	0·60	0·89	1·17	1·45
0·20		0·00	0·19	0·37	0·55	0·73	0·90

11.3.2 Simplification

It can be deduced (Bullen, 1968b, p. 298) from (11.21) and (11.24) that

$$1 - x \rightarrow (1 + 2f_1)(1 + 7f_1)^{-2} \quad (11.26)$$

as $K_2/K_1 \rightarrow 1$. A remarkable feature of Table 11.2 is that, for the whole range of values shown for f_1, the ratio x is nearly constant for each assigned f_1 as K_2/K_1 varies from 1·0 to 2·0. This feature suggests that, for values of K_2/K_1 considerably in excess of unity, the expressions on the two sides of the arrow in (11.26) are nearly equal, and hence by (11.25) that

$$k_2/k_1 \approx 1 + (K_2/K_1 - 1)(1 + 2f_1)(1 + 7f_1)^{-2}. \quad (11.27)$$

A more detailed analytical scrutiny (*loc. cit.*, p. 300) showed that (11.27) is in fact highly accurate for $0 \leqslant f_1 \leqslant 0.27$ and $1.0 \leqslant K_2/K_1 \leqslant 2.0$, and thus for all values of f_1, f_2, and K_2/K_1 of likely interest in the Earth context.

The ratio K_2/K_1 is an index of the compositional difference between the two materials. When K_2/K_1 is assigned, (11.27) gives the

[Refs. on p. 223]

variation of k_2/k_1 with f_1 simply and immediately, without recourse to tiresome algebra otherwise needed in treating (11.21)–(11.24). Through (11.21), the variation is then given in terms of p.

11.3.3 *Trend of incompressibility ratios towards unity*

A result of some interest to the sequel is that $|k_2/k_1 - 1|$ diminishes steadily and indefinitely as p increases. This is seen on differentiating (11.24) and using (11.21). [The result is obvious from (11.27) within the range of validity of (11.27).] Thus Birch's theory incidentally entails that the incompressibilities of different materials tend toward equality as the pressure increases. The point was first made by Birch (1952) in a single specimen calculation which corresponds approximately to the entry $k_2/k_1 = 1.25$ in Table 11.2 for $K_2/K_1 = 1.80$ and $f_1 = 0.14$.

Table 11.2, among other things, exhibits in some detail this trend towards unity. For example, when p is increased from zero to the pressure at the mantle-core boundary (where $f = 0.14$), the excess of k_2/k_1 over unity is reduced by more than two-thirds for all values of K_2/K_1 in the table.

11.3.4 *Formal application to the neighbourhood of the mantle-core boundary*

Let the subscripts 1 and 2 now relate, in particular, to the materials at the base of the mantle and top of the core, respectively. Let Δk denote the jump $(k_2 - k_1)$ in k from mantle to core.

For $K_2/K_1 = 1.00, 1.20, 1.40$, and $f_1 = 0.14$, Table 11.2 gives $\Delta k/k_1 = 0, +0.06, +0.13$, respectively. Suppose the material at the base of the mantle is similar in composition to the material of the region B. Then the data in Table 10.2 (§10.7.2) give $K_1 \approx 1.15 \times 10^{11}$ N/m² (later amendments to Table 10.2 do not greatly affect this estimate of K_1). Suppose also that the material at the top of the core has an iron composition with $K_2 \approx 1.6 \times 10^{11}$ N/m². Then $K_2/K_1 \approx 1.4$ and Table 11.2 gives $\Delta k/k_1 \approx +0.13$.

Compositional changes inside the region C could slightly increase the foregoing estimate of K_1, while evidence (Bullen, 1952; MacDonald and Knopoff, 1958) that the representative atomic number for the outer core is somewhat less than that for pure iron would slightly reduce the estimate of K_2. But on the assumption

[Refs. on p. 223]

that the lower mantle and outer core consist predominantly of ultrabasic rock (or material of equivalent overall chemical composition) and iron, respectively, it does not seem feasible to lower the estimate of K_2/K_1 below 1·2.

On that assumption, Table 11.2 shows that Birch's finite-strain theory requires $\Delta k/k_1$ to be at least +0·06 and possibly as high as +0·13. This consequence of Birch's theory is important in tests against other approaches (see §11.4.1).

11.3.5 *Formal application to interior of core*

Let the subscripts 1 and 2 next relate to the materials at the base of the outer core and top of the inner core, respectively. (Any transition layers between outer and inner core are here being disregarded.) On currently preferred theories (but see also §17.2.4), the changes in composition from outer to inner core are comparatively slight, for example, from an alloy consisting predominantly of iron with some less dense material to fairly pure iron-nickel. The implied excess of K_2/K_1 over unity would then be quite small, probably appreciably less than 0·2. Since $f_1 \geqslant 0·19$ inside the lower core, Table 11.2 would then give $(k_2 - k_1)/k_1$ as appreciably less than 0·05. The most extreme compositional change on any theory that is not wholly implausible would give $K_2/K_1 \approx 1·4$, corresponding to a change from a modified form of ultrabasic rock in the outer core to iron-nickel in the inner core; then Table 11.2 gives $(k_2 - k_1)/k_1 \leqslant 0·09$.

Thus Birch's finite-strain theory would set a fairly severe upper bound to sudden changes of k inside the core. Although, as will be seen, Birch's theory may not be wholly reliable, it remains unlikely that any needed amendments will affect this particular result noticeably. (See also Cook, 1972, p. 324.)

11.4 Compressibility-pressure hypothesis

The original Model *A* calculations brought to notice (Bullen, 1946) a remarkable feature in the behaviour of k in the vicinity of the mantle-core boundary N. Whereas large discontinuous changes were indicated at N in the density ρ (from 5–6 to 9–10 g/cm^3) and in the rigidity μ (from 3·0 to 0·0 × 10^{11} N/m^2), the indicated changes in both k and dk/dp were so small as to be within the uncertainties of the determinations. The feature is exhibited in the table of results for

Model A'' (§10.7.2).

At the base of the mantle and top of the core, the original Model A calculations gave $k = 6.5$ and 6.2×10^{11} N/m^2, whence (in the notation of §11.3.4) $\Delta k/k_1 = -0.05$. For Model A'', $\Delta k/k_1 = +0.02$. For all other models later constructed by the writer's procedures (see Chapters 12–14), $\Delta k/k_1$ lies between -0.05 and $+0.02$.

As seen in §11.2.3, the Model A results also gave dk/dp nearly continuous in the vicinity of N.

Furthermore, laboratory results (e.g. of Bridgman, 1931) available at the time at pressures up to 10^{10} N/m^2, pointed to a trend toward equality in k and dk/dp for a wide class of materials, as p increases.

The overall evidence led to the formulation (Bullen, 1946, 1949) of an incompressibility-pressure (k-p) hypothesis to the effect that, throughout the Earth's lower mantle (below 1000 km depth) and core, irrespective of such variations of composition as may occur inside this entire region, k varies continuously and smoothly with p. The k-p hypothesis thus applies inside the pressure range $0.4 \leqslant p \leqslant 4.0 \times 10^{11}$ N/m^2, approximately. [In its original (1946) form, the k-p hypothesis envisaged the possibility that k might be largely independent of chemical composition at pressures of the order of 10^{11} N/m^2. Work of Feynman, Metropolis, and Teller (1949) indicated, however, that the hypothesis in that form was too wide (see Bullen, 1956, p. 83) and the hypothesis was then restricted to the materials predominating in the Earth at the pressures in question.]

The k-p hypothesis, intended initially as a trial to be tested against further evidence, soon received (see below) considerable support (in its restricted form) and has assisted in deriving close approximations to conditions inside the Earth at pressures beyond 0.4×10^{11} N/m^2. Its great utility is that it supplies, in effect, additional equations of condition (see e.g. Bullen, 1964) on ρ, k, and μ inside the Earth in a way that valuably supplements the more direct evidence from seismology. Among other things, it has assisted in estimating the density variation in certain regions where (10.7) is seriously inadequate, and where, as in the inner core, direct seismic data on S velocity values have been lacking (at least till recently – see §12.6.2).

Evidence bearing on the degree of reliability of the hypothesis is examined below: k is considered in §§11.4.1–11.4.3, and dk/dp in §§11.4.4 and 11.4.5.

[Refs. on p. 223]

11.4. Evidence on k at the mantle-core boundary

Both the Model A procedures and Birch's finite-strain theory give fairly small proportionate (positive or negative) jumps $\Delta k/k_1$ in k at N (§§11.3.4, 11.4). The jump on Birch's theory is, however, significantly greater than on the Model A procedures. Since a number of far-reaching conclusions have been drawn from Birch's theory, it is desirable to scrutinize its bearing on the k-p hypothesis in some detail, along with other pertinent evidence.

An important test comes from efforts to resolve the discrepancy between the maximum value +0·02 for $\Delta k/k_1$ indicated on the writer's approach through A-type Earth models and the minimum value +0·06 (§11.3.4) indicated on Birch's theory (as usually interpreted). An analysis (Bullen, 1968c; Bullen and Haddon, 1969) of the effects of admissible amendments to A-type models took account of the revised estimate of the Earth's moment of inertia I, of uncertainties inside D'', and of the extents of uncertainty in the estimated values of α, β, and ρ (especially in the upper mantle and lower core), in the representative value taken for the crustal thickness, and in the estimated core radius. The analysis indicated that $\Delta k/k_1$ is unlikely to lie far outside the range $-0\cdot 01$ to $+0\cdot 02$. [The revision of I (§5.7) required the original Model A estimate ($-0\cdot 05$) of $\Delta k/k_1$ to be increased to $-0\cdot 01$; uncertainties in $\Delta k/k_1$ arising from the other sources mentioned have smaller effects.] It was possible to arrive at +0·06 but only by assuming an unlikely and fairly extreme combination of errors, all affecting the estimation of $\Delta k/k_1$ in the same direction. Thus Birch's theory probably overestimates $\Delta k/k_1$ and the evidence derived through A-type models seems preferable in assessing the k-p hypothesis.

An incidental result of the analysis was that the circumstances for $\Delta k/k_1$ to be as high as +0·06 would (among other things) require the average density gradient in D' to be about 20 per cent less than that yielded by (10.7), i.e., would require the index η (§11.5.2) to be about 0·8. This would require the average super-adiabatic temperature gradient in D' to be at least 1·3 deg/km — greater should D' contain any inhomogeneities. Such a temperature gradient lies rather outside the bounds generally favoured for D', but the possibility needs noting.

It also needs to be noted that, in applying Birch's theory as above, it has been assumed that the lower mantle consists predominantly of material equivalent in composition to ultrabasic rock, and the outer

[Refs. on p. 223]

core of an iron alloy. Should the outer core consist predominantly of a high-pressure modification of the lower mantle material (see §17.2.3), the estimate of $\Delta k/k_1$ on Birch's theory could be much reduced. Difficulties over $\Delta k/k_1$ might also be noticeably reduced on the Fe_2O theory (§17.2.4).

The likelihood that Birch's theory on k needs some modification is, however, further evidenced on attempting to apply his equation (11.22) to the material at the base M of the Earth's mantle. Taking $K = 1\cdot15 \times 10^{11} N/m^2$ and $f = 0\cdot138$ (Birch's effective value at M) would give $k = 4\cdot2 \times 10^{11} N/m^2$, which is only two-thirds of the actual value (see Table 10.2) at M. Conversely, taking $k = 6\cdot5 \times 10^{11} N/m^2$ and $f = 0\cdot138$ at M would give $K = 1\cdot8 \times 10^{11} N/m^2$, which is too high on any plausible theory. Again, taking $K = 1\cdot15$ and $k = 6\cdot5 \times 10^{11} N/m^2$ would give $f = 0\cdot20$ at M, a result which would require much larger phase changes than are expected inside the mantle. On this argument, (11.22) would appear to be irreconcilable with conditions at M.

Also of interest are certain results on the varation of k with Z, where Z denotes the representative atomic number of a material. [For chemical compounds, Z is a weighted mean of the values of the atomic numbers of the constituent elements (see Knopoff and Uffen, 1954). For magnesium orthosilicate, iron, and nickel, $Z \approx 10$, 26, and 28, respectively.] Elsasser (1951) sought to provide evidence on the variation of ρ and p/k with p and Z at pressures of order $10^{11} N/m^2$ by interpolating between laboratory results at 10^{10} and theoretical results of Feynman (1949) beyond $10^{12} N/m^2$. The interpolated results proved to be seriously discordant with well-established evidence innate in the Model A calculations, and were substantially revised (Bullen, 1952) to achieve reconciliation with this evidence. (Although Elsasser accepted this revision straight away, a stream of other writers has continued to use his original interpolation and draw quite false conclusions.) The uncertainties of the interpolation were reduced by this revision but, principally because adequate evidence on phase changes for $10^{11} < p < 10^{13} N/m^2$ is still lacking, the uncertainties are still appreciable, especially for $10^{11} < p < 10^{12} N/m^2$. Assuming $10 \leqslant Z \leqslant 28$, the revised interpolation gave $\Delta k/k_1 \approx 0\cdot010 \Delta Z$, where ΔZ denotes the jump in Z at N. If Z is assumed equal to 12 in the region D, and 23 in the outer core (see Bullen, 1952; Knopoff and MacDonald, 1960), the procedure formally yields a value of $\Delta k/k_1$ near that on Birch's theory.

[*Refs. on p. 223*]

But the uncertainties forbid attaching any more weight to this result than to Birch's.

The more important results emerging from the above discussion are the following. The analysis of Earth models indicates that $\Delta k/k_1$ probably does not much exceed +0·02. The analysis does not closely determine a lower bound to $\Delta k/k_1$, but Birch's theory and the theory involving Z, in spite of their limitations, make it improbable that $\Delta k/k_1$ is significantly negative. Thus the overall evidence at present favours continuity or near-continuity of k at N.

11.4.2 *Implications on variation of k inside the core*

Any sudden changes in k inside the core are likely to be restricted to a small range of depth between the outer and inner core proper. The calculations in §§ 11.3.3 and 11.3.5 show that, because of the greater pressure and compression, Birch's theory implies smaller jumps in k in this part of the Earth than at N: the estimated proportionate jumps range from zero to an extreme upper limit of 0·09, depending on the extent of compositional change assumed from outer to inner core. The theory involving Z (see Bullen, 1952) gives similar results: for jumps in Z of 5 and 10 units, the proportionate jumps in k are indicated as 0·04 and 0·08, respectively.

It is likely that, as at N, both these theories overestimate jumps in k inside the core. But in any case it is important to note that the indicated upper bounds to the jumps are fairly small.

A further point is that any jump in ρ between the outer and inner core (see Chapter 12) is indicated as being appreciably smaller than the jump at N.

In view of the likely near-continuity of k throughout the whole lower mantle and at N, it would therefore be surprising if k should deviate far from continuity anywhere inside the core.

11.4.3 *Further evidence on degree of dependence of k on composition*

The degree of dependence of k on composition has been examined by other investigators, including Takeuchi and Kanamori (1966), Boschi and Caputo (1969), and Cook (1972). Boschi and Caputo showed values of k determined by Al'tschuler and others (1958) and McQueen and Marsh (1960) for some twenty metals, using shock-wave experiments. Significant variations of k with Z were indicated for these metals at pressures up to 4×10^{11} N/m^2, confirming the necessity of restricting the k-p hypothesis to materials of the type predominating in the Earth's deep interior. At the same time, the determinations of k by the different investigators show substantial discrepancies (up to 20–30 per cent) for the same material at the same pressure. Thus the experiments cannot be regarded as closely determining the variations of k with Z at core pressures.

Cook (1972) assembled evidence giving nearly the same values of k (all within 3 per cent of 6.5×10^{11} N/m^2) at the pressure at N for materials of the type occurring in the Earth such as iron oxide (FeO), stishovite (SiO$_2$), aluminium oxide (Al$_2$O$_3$), and magnesium oxide (MgO). But he suggested that the values of k may diverge when $p > 1.4 \times 10^{11}$ N/m^2. This suggested divergence rested on extrapolations in which several assumptions were implicitly made. Details in §11.3.3 on the trend of incompressibility ratios towards unity and in §11.4.1 on the dependence of k on Z make it rather difficult to sustain these suggested extrapolations throughout the range of pressures in the Earth's core. At the same time, Cook's evidence does tend to confirm the need for some caution in applying Birch's theory inside the core. Even if the least favourable interpretation of Cook's results were taken, however, the conclusion of §11.4.2 that any jump in k at the inner core boundary is slight would remain unaffected. [Incidentally, some of Cook's statements need amending. For example, he states that the writer's k-p hypothesis is directly applicable to the Moon and possibly Mars, but not other planets. This overlooks the fact that the k-p hypothesis has always been restricted to pressures greater than 0.4×10^{11} N/m^2 and so, as it stands, is not applicable to the Moon and Mars. Furthermore, the hypothesis is, with due caution, applicable to Venus (see Bullen, 1972). Equations of state for planets are discussed in Chapter 17.]

11.4.4 *Evidence on* dk/dp *at the mantle-core boundary*

At the bottom of the mantle, taking $f = 0.14$, Birch's finite-strain theory gives dk/d$p \approx 2.9$ (§11.2.6).

Consider now the greatest likely change in dk/dp at N when Birch's theory is assumed. Let $\Delta f/f_1$ be the proportionate jump in f at N. Then (11.4) yields

$$\Delta f/f_1 \approx -3\Delta k/(3k_1 - 7p_1), \qquad (11.28)$$

where p_1 denotes the pressure at N. Substituting $k_1 \approx 6.5 \times 10^{11}$ and $p_1 = 1.36 \times 10^{11}$ N/m² (see Table 10.2) into (11.28) gives $\Delta f/f_1 \approx -2.0 \Delta k/k_1$ and hence $|\Delta f| \leq 0.26 f_1$ on taking $|\Delta k|/k_1 \leq 0.13$ (see §§11.3.4, 11.4.1). Then, assuming (11.19) and writing X for $(\partial k'/\partial p)_\tau$, we obtain

$$\frac{3|\Delta X|}{X_1} = \frac{35|\Delta f|}{(12 + 49 f_1)(1 + 7 f_1)} \leq \frac{9 \cdot 1 f_1}{(12 + 49 f_1)(1 + 7 f_1)}.$$

The last expression has a maximum of about 0·035 (at $f_1 \approx 0.19$), so that $|\Delta X|/X_1 < 0.012$. Thus on Birch's theory $(\partial k'/\partial p)_\tau$ would change from mantle to core by at most the order of 1 per cent. Detail in §11.2.5 on δ indicates that the proportionate change in dk/dp would be of the same order. Thus the theory gives dk/dp as nearly continuous at N.

Although Birch's theory probably needs some amendment in respect of Δk (§11.4.1), any amendment to $\Delta(dk/dp)$ is likely to be proportionately much smaller since dk/dp varies much more slowly than k with p.

As already stated, evidence from A-type models independently indicates that $\Delta(dk/dp)$ is small, possibly zero, at N. Since also the estimates of dk/dp at N on different theories agree fairly well, the case for near-continuity is quite strong.

11.4.5 *Evidence on dk/dp inside the core*

Using the values of f in Table 11.1, and applying the correction δ of §11.2.5, Birch's formula (11.19) would give dk/dp as declining very slowly from its value 2·9 at the top of the core to about 2·8 at the centre, and so is compatible with close continuity of dk/dp throughout the core. Should the values of f in Table 11.1 need to be increased because of phase changes, the effect on dk/dp would be only slight: with $f = 0.3$ (a fairly extreme value), (11.17)–(11.19) would give $dk/dp \approx 2.7$.

At the top of the core, the estimate 2·8 of dk/dp derived through (11.8) in the first approximation (§11.2.3) agrees closely with the estimate 2·9 derived through (11.19). The estimate 2·8 might possibly need increasing by an amount not exceeding 0·3 should a suggestion of mild inhomogeneity in E (see §12.2) be substantiated; but the agreement would still be fairly good.

Below the top of the core, as z increases, the estimates of dk/dp

[Refs. on p. 223]

derived using (11.8) [or better, using the generalized relation (11.36)] become increasingly larger than those derived through Birch's theory. At $z = 3600$–4500 km, (11.8) gives $dk/dp \approx 3\cdot5$ when the Jeffreys or Gutenberg distribution of α is assumed, as against $2\cdot9$ on Birch's theory.

This discrepancy is possibly reducible by modifications to the assumed α (see §12.2). In theory, it could also be reduced by assuming (ad hoc) larger values of g in the outer core. This is most readily seen from the equation (11.36) (see §11.5.2) which shows that, for given $d\phi/dz$ and η, an inferred value of dk/dp can formally be reduced by raising the assumed g. Let m and σ be the mass and mean density within distance r of the Earth's centre; then $g = Gm/r^2 = 4\pi G r \sigma/3$. Thus raising g involves raising σ and hence raising the assumed densities in the lower core. Complete removal of the discrepancy in this way would, however, demand an appreciably higher central density ρ_0 than the overall evidence (§12.5), including Birch's own estimate of ρ_0 from shock-wave data, appears to permit. It is to be noted that the discrepancy would be increased, not reduced, by any inhomogeneity between 3600 and 4500 km depth since increasing η in (11.36) increases the inferred value of dk/dp.

Hence, inside the outer core, Birch's theory appears to give dk/dp too low by up to about 25 per cent. Independent estimates of Davies and Anderson (1971) and Cook (1972), so far as they go (numerous assumptions are made), appear to support this assessment. (See also §§12.2, 15.1.1).

Inside the lower core, it is difficult to assess dk/dp closely except through the k-p hypothesis itself. Errors in values given by Birch's theory may well exceed 25 per cent. The uncertainties attaching to the use of (11.8) also become fairly appreciable here because of uncertainties in both the distributions of α and the possible extent of inhomogeneity.

Using independent data from shock-wave experiments, (see §12.5.3), Birch (1961, 1963) presented curves which imply values of $\partial k/\partial p$ of order 6 for iron at $p \approx 3 \times 10^{11}$ N/m^2; $\partial k/\partial p$ here relates to Hugoniot thermodynamical conditions (see §12.5.3). Assuming that the data have been correctly interpreted (the point remains controversial), the result could be applied to estimating dk/dp inside the Earth's lower core, were a reliable means available for inferring the adiabatic k-p gradient from the Hugoniot. Investigations of Ahrens, Anderson, and Ringwood (1969) and Davies and Anderson (1971)

[Refs. on p. 223]

would suggest a value near 4. This would agree with a simple downward extrapolation from the outer core, using the k-p hypothesis. See also Steward (1973).

Some further discussion of Birch's theory is included in §15.1. See also Bullen (1970a).

11.4.6 *Further testing of the k-p hypothesis*

As will be seen, the k-p hypothesis leads to a number of far-reaching formal consequences, the most important of which is that the Earth's inner core is solid (§11.7). By examining independent evidence on these consequences, additional tests can be supplied on the degree of reliability of the hypothesis.

Runcorn (1974) has suggested a basis for the k-p hypothesis in terms of fairly fundamental ideas concerning the close packing of ions in the deep mantle and core and the lack of strong dependence on the chemistry of the repulsive forces between the ions.

11.5 Theory for inhomogeneous regions

Much of the formal theory in the preceding sections relates primarily to homogeneous regions of the Earth. But some regions cannot now be reasonably treated as homogeneous even to a first approximation. Even where the assumption of homogeneity does give a useful first approximation, the stage has now been reached where a more general theory is necessary. Some steps in that direction will now be indicated.

The simplifying assumption will be made that the representative atomic number Z is (for our purpose) a sufficient index of chemical composition, and Z will be taken to vary continuously with the depth z inside any one region. Except where mentioned, terms involving ϑ will be ignored as small compared with the main effects of inhomogeneity to be here examined. Thus two independent variables, p and Z, will now be involved. Both these variables are functions of z in applications to the Earth's interior.

Conventions on symbolism for differentials will correspond to the procedure in §11.2. Thus if χ is any function of z, $d\chi/dp$ will denote $(d\chi/dz)/(dp/dz)$. The notation $\partial\chi/\partial p$ and $\partial\chi/\partial Z$ will relate to changes at constant composition and pressure, respectively; all partial derivatives with respect to other variables, e.g. $\partial p/\partial k$, will relate to

[*Refs. on p. 223*]

changes at constant pressure.

The relations (11.5) and (11.6) of §11.2.1 continue to hold, but the left side of (11.7) has now to be replaced by $\partial \rho/\partial p$. Thus we now have

$$\frac{\partial \rho}{\partial p} = \phi^{-1} = \frac{\rho}{k}, \quad (11.29)$$

$$dk = \phi d\rho + \rho d\phi. \quad (11.30)$$

By (11.29) and (11.30),

$$\partial k/\partial p = 1 + \rho \partial \phi/\partial p. \quad (11.31)$$

We continue to use the hydrostatic relation

$$dp = g\rho dz. \quad (11.32)$$

11.5.1 Amended relation for density gradient

By (11.29), (11.30), and (11.32), we have

$$\frac{dk}{dp} = \frac{\phi}{g\rho}\frac{d\rho}{dz} + g^{-1}\frac{d\phi}{dz}. \quad (11.33)$$

Hence (Bullen, 1963)

$$\frac{d\rho}{dz} = \left(\frac{dk}{dp} - g^{-1}\frac{d\phi}{dz}\right)\frac{g\rho}{\phi}. \quad (11.34)$$

By (11.8), the equation (11.34) reduces to the Williamson-Adams equation (10.7) in a homogeneous region. When only meagre information was available on dk/dp, it was appropriate to assume homogeneity, at least for trial purposes, in estimating $d\rho/dz$ inside any region of the Earth, and to make adjustments as contradictions arose, as in the case of the region C (§10.5). As has been seen, the assumption of homogeneity continues to yield useful approximations to $d\rho/dz$ in some regions. The growth of information on dk/dp has now, however, reached the point where the Williamson-Adams equation must be regarded as superseded by (11.34) and related theory. Even where (10.7) gives a useful first approximation result, it is important to have a means of finding how far the coefficient of $g\rho/\phi$ in (11.34) differs from unity. Sometimes (see e.g. §14.7) the coefficient can be estimated independently of knowledge of dk/dp. A number of aspects of (11.34) will be examined in the following subsections.

[Refs. on p. 223]

11.5.2 *The index* η

It is convenient to write (11.34) in the form

$$d\rho/dz = \eta g\rho/\phi, \qquad (11.35)$$

where

$$\eta = \frac{dk}{dp} - g^{-1}\frac{d\phi}{dz}. \qquad (11.36)$$

It is readily deduced that

$$\eta = (k/\rho)d\rho/dp = (d\rho/dp)/(\partial\rho/\partial p), \qquad (11.37)$$

$$\eta - 1 = \frac{\partial\rho/\partial Z}{\partial\rho/\partial p}\frac{dZ}{dp}. \qquad (11.38)$$

For the particular case of a homogeneous region, $\eta = 1$. Then (11.35) reduces to the Williamson-Adams equation

$$d\rho/dz = g\rho/\phi, \qquad (11.39)$$

(11.36) reduces to (11.8), and (11.37) gives $k = \rho dp/d\rho$.

In general, the contribution to $d\rho/dz$ due to changes of chemical composition, i.e. $(\partial\rho/\partial Z)dZ/dz$, is positive. Hence, by (11.38), $\eta \geqslant 1$ in general. (In regions where ϑ is significant, η can however, be less than unity and even negative — see later.)

A rough specimen claculation shows that η is very sensitive to changes of composition (see also §11.2.2). Imagine a hypothetical continuous change of composition, inside a region R of thickness 100 km, which would contribute an increase of 1 g/cm^3 in ρ, independently of the increase due to pressure; this would give $(\partial\rho/\partial Z)dZ/dz \approx 10^{-7} \text{ g/cm}^4$. Taking R to be inside the region D, Table 10.2 (§10.7.2) would give $(\partial\rho/\partial p)dp/dz \approx 5 \times 10^{-9} \text{ g/cm}^4$. By (11.38), η would then be of order 20 inside R. This sensitivity makes η an important index of departures from chemical homogeneity inside the Earth.

The index η is particularly useful in practice because, through (11.36), it depends only on dk/dp, g, and $d\phi/dz$ — all of which can be numerically estimated within usable limits inside much of the Earth. Numerical evidence on dk/dp has been discussed in §§11.2 and 11.4 (see also §11.5.4). Estimates of g are available from Earth models such as those discussed in §10.7.2; for present purposes,

[*Refs. on p. 223*]

these estimates are more than sufficiently accurate, as found on examining g in a wide variety of Earth models. Estimates of $d\phi/dz$ are obtained through (10.3) from seismic data on α and β. It is principally because of the evidence now available on η that (10.7) must be regarded as superseded.

Since the Emden equation (6.19) incorporates the Williamson-Adams equation, it also is superseded. Taking (11.35) and (11.37) in place of (6.14,15), we obtain

$$\frac{d}{dr}\left\{r^2 k\rho^{-2}\eta^{-1}\frac{d\rho}{dr}\right\} = -4\pi G r^2 \rho \tag{11.40}$$

as the generalization of (6.17). Assuming $dk/dp = n$ where n is constant (as in §6.4.2), and using (11.37), gives $dk/k = n'd\rho/\rho$, where $n' = n/\eta$. In most regions of the Earth (the principal exception is the region C), η can be treated as constant to a useful first approximation. We then have $k = C\rho^{n'}$, and hence

$$\frac{d}{dr}\left\{r^2 \rho^{n'-2}\frac{d\rho}{dr}\right\} = -A'^2 r^2 \rho, \tag{11.41}$$

where $A' = \sqrt{(4\pi G\eta/C)}$, as a generalization of (6.19). In (11.41), we still have an Emden equation, so that techniques based on Emden equation theory can still, with suitable caution, be used in estimating density-depth relations even where departures from the Williamson-Adams equation are appreciable (Bullen, 1970b). But the replacement of n and A in (6.19) by n' and A' can have substantial numerical effects. It is to be noted that (11.41), like (6.19), assumes a linear dependence of k on p: neither of these Emden equations would be reliable where marked departures from this linearity occurred.

11.5.3 *Temperature allowance*

The discussion in §§11.5–11.5.2 ignores deviations from adiabatic temperature gradients. In the present subsection, ρ and k will be treated as functions of ϑ as well as of p and Z. On repeating the derivations of the above equations with allowance made for dependence on ϑ, it is found that all of (11.29)–(11.37) and (11.40) remain valid, $\partial/\partial p$ being interpreted as relating to changes with respect to p at constant composition under adiabatic conditions. In particular, the important formula (11.36) holds unmodified.

But η now depends on ϑ through dk/dp. The dependence can be made explicit as follows. From theory in §§10.1.1 and 10.1.3, we now have (Bullen, 1967)

$$\frac{d\rho}{dz} = \frac{g\rho}{\phi} + \frac{\partial \rho}{\partial Z}\frac{dZ}{dz} - \gamma\rho\vartheta, \qquad (11.42)$$

and hence, by (11.35), (11.29), and (11.32),

$$\eta - 1 = \frac{\phi}{g\rho}\frac{\partial \rho}{\partial Z}\frac{dZ}{dz} - \frac{\gamma\phi\vartheta}{g} \qquad (11.43)$$

$$= \frac{\partial \rho/\partial Z}{\partial \rho/\partial p}\frac{dZ}{dp} - \frac{\gamma\phi\vartheta}{g}. \qquad (11.44)$$

For some purposes, it is convenient to have available the corresponding relation for dk/dp, namely,

$$\frac{dk}{dp} = 1 + g^{-1}\frac{d\phi}{dz} + \phi\frac{\partial \rho}{\partial Z}\frac{dZ}{dp} - \frac{\gamma\phi\vartheta}{g}, \qquad (11.45)$$

which is yielded by (11.36) and (11.43). The relation (11.38) is now inadequate and has to be replaced by (11.44).

It has been pointed out (§10.84) that, in general, the presence of a positive super-adiabatic gradient and chemical inhomogeneity affect $d\rho/dz$ in opposite directions. The relation (11.43) shows the effect in terms of the index η which now becomes an index of the combined effects on $d\rho/dz$ of inhomogeneity and super-adiabatic temperature gradient. For η to be equal to unity, i.e. for the Williamson-Adams equation (11.39) to be relevant, either there must be no deviation from uniform chemical composition and phase and no deviation from an adiabatic temperature gradient, or there must be an accidental combination of all three effects causing them to cancel one another out.

For ease of exposition, the above discussion has been carried out taking Z as the sole chemical variable. The essential results are unaffected when the theory is further generalized by taking, as in §10.1, a set of variables in place of Z. Likewise the theory can be simply adapted to allow for phase changes as well. All that is formally necessary is to add, in each equation in which Z appears explicitly, terms analogous to those in Z. Explicit allowance for phase is needed only in contexts where an attempt is made to discriminate between contributions from changes in chemical composition and phase.

[Refs. on p. 223]

11.5.4 *Allowance for variation of k with composition*

When (11.36) is applied in practice, the values substituted for dk/dp are frequently not the values of dk/dp, but of $\partial k/\partial p$. In that event [e.g. when Birch's relation (11.19) is used and the correction δ applied], dependence of k on composition is being ignored. At the pressures prevailing in the Earth's lower mantle and core, the discussions in §11.3 and 11.4 indicate that the variation of k with Z at constant pressure is in fact fairly slight for the range of materials involved. It is nevertheless desirable, so far as can be done, to assess the possible effect of this dependence in calculations involving η. Dependence on ϑ will be ignored in §§11.5.4 and 11.5.5.

Let the function j be defined by

$$j = \{(k/\rho)(\partial \rho/\partial k) - 1\}^{-1}. \tag{11.46}$$

In general, ρ increases with Z in the context of interest, so that j can be a useful measure of the degree of dependence of k on Z. If k were entirely independent of Z, then $\partial k/\partial \rho$ and j would be zero.

Let η_0 be the value yielded for η through (11.36) when dependence of k on composition is ignored. Thus

$$\eta_0 = \frac{\partial k}{\partial p} - \frac{1}{g}\frac{d\phi}{dz} \tag{11.47}$$

and

$$\eta - \eta_0 = \frac{dk}{dp} - \frac{\partial k}{\partial p} = \frac{\partial k}{\partial Z}\frac{dZ}{dp}. \tag{11.48}$$

Then, by (11.29) and (11.38),

$$\frac{\eta - 1}{\eta - \eta_0} = \frac{k}{\rho}\frac{\partial \rho}{\partial k} \tag{11.49}$$

and, by (11.46),

$$\eta - \eta_0 = j(\eta_0 - 1) \tag{11.50}$$

(see Bullen, 1965a). If numerical evidence on j can be provided, (11.50) can be used to assess the correction $(\eta - \eta_0)$.

Precise theory for estimating $\partial \rho/\partial k$ and j is not available, so that only a general idea of the correction $(\eta - \eta_0)$ can be indicated. For $p = 3 \times 10^{11}$ N/m², the details of the interpolation referred to in

§11.4.1 formally yielded (see Bullen, 1952) $\partial\rho/\partial Z \approx 0\cdot 3$ and $\partial k/\partial Z \approx 8 \times 10^9$, where ρ and k are in g/cm³ and 10^{11} N/m². Taking values of ρ and k at the bottom of the region E from Table 10.2 would then give $j \approx 0\cdot 3$. The interpolation involves a rather gross simplification of the relation between the dependences of ρ and k on composition at pressures not greater than the order of 10^{11} N/m². But it is likely that, where j differs markedly from 0·3 inside the Earth's lower mantle and core, this will usually be due to $\partial\rho/\partial k$ being markedly greater than the interpolation value, in which case j will be smaller than 0·3. For example, at the mantle-core boundary artificially treating (for present purposes only) the changes in ρ as rapid but continuous, j would be indefinitely small.

When j is assumed equal to 0·3, (11.50) formally gives $\eta - \eta_0 = 0, 0\cdot 3, 0\cdot 6, 1\cdot 5, 3, 6, 9$ when $\eta_0 = 1, 2, 3, 6, 11, 21, 31$; for larger η_0, $\eta \to 1\cdot 3\eta_0$. In regions which deviate only slightly from homogeneity, it is of course to be expected that the corrections $(\eta - \eta_0)$ will be small. But in regions where the deviations are sizeable, the evidence on j, to the extent that it can be relied on, indicates that allowance for variation of k with composition in applications of (11.36) may require corrections in η up to 30 per cent. The uncertainty in $(\eta - \eta_0)$ is, however, less than the general uncertainty of estimates of $d\rho/dz$ in markedly inhomogeneous regions and therefore does not affect the application of (11.36) very seriously.

11.5.5 *Homogeneity and normal seismic velocity variation*

For a purpose connected with seismic ray theory, a rough classification was made (§9.3) into normal and abnormal variation of the seismic velocities α and β with the depth Z. It is to be expected that regions of normal variation will (apart from effects of super-adiabatic gradients, here ignored) be near-homogeneous. The links provided between η and $d\phi/dz$ in (11.36) and between η_0 and $d\phi/dz$ in (11.47) enable some further light to be thrown on this question. It is to be noted that the variation of ϕ, i.e. of $\alpha^2 - 4\beta^2/3$, is here involved, not $d\alpha/dz$ and $d\beta/dz$ separately as in §9.3.

Consideration will be confined to the lower mantle and core, where there is good evidence (§§11.4.4, 11.4.5) that $\partial k/\partial p$ lies approximately between 3 and 4. Substituting (say) $\partial k/\partial p = 3\cdot 4 \pm 0\cdot 6$ and $\eta_0 = 1$ into (11.47) gives

$$d\phi/dz = 2\cdot 4g \pm 25\%. \qquad (11.51)$$

[Refs. on p. 223]

A useful guide as to whether a region is close to homogeneity is given by testing the observed seismic velocity variations against (11.51). The relation (11.51) can indeed be regarded as giving a serviceable definition of normal variation of ϕ. Such a definition is not identical with that in §9.3, but since both definitions are intended only for broad purposes, the formal differences between them do not matter very much.

It needs to be noted that, while observed variations of ϕ which conform to (11.51) are compatible with near-homogeneity, they establish near-homogeneity only if $(\eta - \eta_0)$ is small when $\eta_0 \approx 1$. Detail in §11.5.4 shows that this is strongly expected to be the case but rests on the assumption that the coefficient j is not large. Should j be abnormally large, i.e. should $(\rho/k)\partial k/\partial p$ be near unity, inside some particular range of depth, there could be significant inhomogeneity in spite of normal variation of ϕ. Such an occurrence would require an unusual combination of circumstances and would probably be confined to very limited ranges of depth in the Earth, but the possibility cannot be quite ruled out. Thus the presence of normal seismic velocity variation in a region is strong, but not quite conclusive, evidence of near-homogeneity.

11.6 Degrees of inhomogeneity in particular regions of the Earth

The values of η in the individual regions $A-G$ (§10.3) will now be examined, along with related discussions on chemical homogeneity or inhomogeneity, phase, and temperature gradient.

11.6.1 *Region A*

The region A is too complex for theory on η to add usefully to the information already set down in §10.3.1.

11.6.2 *Region B*

The apparent normalcy of the Jeffreys (1939) values of $d\alpha/dz$ and $d\beta/dz$ in the region B (§10.3.2) was originally taken (Bullen, 1936, 1940) as evidence of near-homogeneity and applied in constructing the A-type models. Birch (1952, 1961) brought two arguments to bear giving general support to the near-homogeneity of the region B: the first was based on (11.8), along with evidence from finite-strain theory and laboratory evidence on the order of magnitude of the temperature correction terms which appear in (11.16); the second was based on the empirical relation (10.14).

Departures from normalcy of the velocity gradients in B have, however, since been indicated (§10.3.2 – see also §12.1.1). Birch's evidence on near-homogeneity in B makes it likely that these departures are predominantly associated with super-adiabatic temperature gradients, at least in the outermost 200 km or so. By (11.43), this would indicate that η is significantly less than unity inside much of B, a conclusion independently supported by evidence to be given in §14.7. So far, it has not been possible to assess η closely inside B through use of the theory in preceding sections. The principal reasons are that assessments of $d\alpha/dz$, $d\beta/dz$, and ϑ are still fairly uncertain at most depths and that complexities due to lateral variations are most acute in this part of the Earth. The best assessments so far of η in B have been yielded with the help of free Earth oscillation data (Chapter 14).

11.6.3 *Region C*

The procedure used in §10.7.1 to determine $d\rho/dz$ inside the region C is independent of the use of (11.35) and (11.39) and involves no prior estimate of η. Subsequent application of (11.35) yields values of η somewhat greater than 2 near the top of C and steadily diminishing with z to about unity at the bottom. The early evidence (Bullen, 1936) that C is significantly inhomogeneous has been supported in all subsequent work, including work of Birch using the approaches mentioned in §11.6.2. It remains the case that (11.39) is seriously inaccurate inside C.

Following the writer's work, Jeffreys (1936) suggested that the 20° discontinuity might be associated with a transformation of ordinary olivine to a high-pressure modification, and Bernal (1936) quoted evidence on a transformation of magnesium germanate in support of this view. Ringwood (1958) demonstrated the transformation of fayalite (iron orthosilicate) to a spinel structure, with a 12 per cent increase in density at a pressure of 0.4×10^{11} N/m² and and temperature of 600°C, and constructed geochemical models compatible with continuous phase transformation inside C. See also Akimoto and Ida (1966). Birch (1961) questioned whether the binary system ($Mg_2SiO_4 - Fe_2SiO_4$) is sufficient to account for the indicated

[Refs. on p. 223]

density changes inside C, and suggested that at least a ternary system is required. Attempts to work out the fine structure of the region C (and other parts of the Earth's interior) through arguments based on geochemistry and results from laboratory experiments are currently fashionable, but a tendency for workers in these fields to be black and white rather than probability minded in their arguments makes it difficult to attach appropriate degrees of precision to their detailed findings. Nevertheless, their work justifies a general statement that the inhomogeneity of C probably involves some small but significant changes of composition as well of phase.

Reference has been made in §§ 10.3.2 and 10.3.3 (see also § 12.1.2) to evidence requiring some revision of the delineation of the regions B and C, including subdivisions. This revision affects the distributions of η slightly.

11.6.4 *Region D*

The treatment of the region D as near-homogeneous in the original Model A calculations also received general support in Birch's 1952 studies. Following the subdivision into D' and D'' (§ 11.2.2 – Bullen 1946, 1949), D' became the one part of the mantle in which the velocity variations could be treated as approximately normal and η taken nearly equal to unity. Some investigations (see e.g. Evernden and Clark, 1970) have indicated significant departures from normality inside a few narrow ranges of depth in D', but confirmation of this must await finer determinations of $d\alpha/dz$ and $d\beta/dz$. A radically different suggestion that $\eta \approx 0$ inside much of D' arose in the course of work of Landisman, Satô, and Nafe (1965) (see § 14.6.2) on free Earth oscillations. It remains just possible that the average value of η inside D' is as small as 0·8 (see § 11.4.1), but a value less than 0·8 is now very unlikely. The value 0·8 would not preclude (11.39) from providing a rough first approximation to $d\rho/dz$.

The values of α and β as functions of p inside D' do not fit a simple extrapolation from data on any silicate rock occurring at the Earth's surface, but would be compatible with the presence of a high-pressure modification of olivine or an equivalent composition in the form of closely packed oxides of magnesium, silicon, and iron. Birch has suggested that there might also be a slightly increasing proportion of iron with increase of depth, but involving an overall change in the mean atomic weight of only about 0·5.

[*Refs. on p.* 223]

As pointed out in §11.2.2, abnormalities in $d\alpha/dz$ and $d\beta/dz$ indicate that the region D'' is either significantly inhomogeneous or a region of rather rapidly declining rigidity. But the effect on density would be only slight, involving a value of η not greater than 3 and hence an addition of at most 0·2 g/cm³ to the assessed density at the bottom of D''.

Birch suggested some accumulation of free iron inside D'' to account for the possibly increased value of η. Others, for example Ringwood (1958), suggested that some core material may have diffused upward across the mantle-core boundary. (For some further detail, see B, §13.9.2.)

11.6.5 *Region E*

As mentioned in §11.2.3, the region E, the outer core proper, is probably fairly homogeneous, with the possibility of some mild inhomogeneity, especially in the outermost part. (See also §12.2.1.) The currently preferred estimates of η inside E range between 1 and 1·4 (Bullen, 1969a).

Using (ρ, p) tables for the A-type and related models, Bullen (1952), Knopoff and MacDonald (1960), and others investigated the mean representative atomic number for E. The estimates for E are all less than the values of Z for iron (26) and nickel (28) by at least a few units. Independently of theory on Z, Birch (1961) inferred from his formula (7.51) and shock-wave data that the mean density of E is about 15 per cent less than that for iron. Thus it became fairly well established that the outer core contains a significant quantity of material less dense than iron. Most investigators have assumed that E is composed of an alloy of iron with elements such as silicon (MacDonald and Knopoff, 1958; Ringwood, 1959), carbon, sulphur, and possibly even some hydrogen (see Birch, 1952). Birch (1972) found that a silicon-iron alloy, with about 15 per cent of silicon by weight, would fit his preferred evidence. Another view is that E consists principally of a phase-modification of the lower mantle material (see §17.2.3). More recently it has been suggested that E is composed of the iron oxide Fe_2O, which is unstable at pressures less than core pressures (see §17.2.4). Endeavours to estimate the composition of E depend basically on geophysically derived estimates of ρ, and have not as yet fed back useful evidence on the Earth's density distribution.

[*Refs. on p. 223*]

11.6.6 *The lower core*

The early P velocity distributions obtained by Jeffreys and Gutenberg (§10.3.6) between the outer and inner core proper were later so substantially revised that it is now inappropriate to apply them to close determinations of density and incompressibility in the lower core. There is indeed now a considerable likelihood that there is no transition layer F between the outer and inner core proper. The revisions are discussed in §§12.3, 12.7, and 15.6. It is nevertheless instructive, as well as of some historical interest, to set down certain formal consequences of applying theory on η to the early velocity distributions.

Inside the region F, the conventional velocity distribution chosen by Jeffreys (§10.3.6) gave $d\alpha^2/dz \approx -0.15$ km/s². Taking values of g and $\partial k/\partial p$ within the permissible limits, treating F as fluid, and using (11.36) would then give $\eta \approx 30$ (possibly several units greater on allowing for the correction $\eta - \eta_0$). Thus the Jeffreys velocity distribution would imply very substantial inhomogeneity inside F and a density gradient some 30 times that given by the Williamson-Adams equation (11.39). Let $\Delta\rho$ and ΔZ be the increases in ρ and Z between the top and bottom of F. A value of order 30 for η would give $\Delta\rho \approx 3$ g/cm³ and $\Delta Z \approx 12$. (The estimate of ΔZ is very rough because of the uncertainties in data on $\partial\rho/\partial Z$.)

Inside the region G, the inner core proper (§10.3.7), $d\alpha^2/dz$ as given by Jeffreys (as well as by most later investigators) is smaller than normal. If it is assumed that $d\beta^2/dz$ is correspondingly small, application of (11.36) then gives $\eta \approx 4$, which would indicate moderate inhomogeneity, with the density increasing by 2–3 g/cm³ through G. (This result is independent of whether the inner core is assumed solid or fluid.) The inhomogeneity if real could be partly due to a changing iron-nickel ratio.

The foregoing results, which depend very much on values assumed for $d\phi/dz$, were prominent in an early estimate (Bullen, 1950) of order 18 g/cm³ — more than 5 g/cm³ above the minimum possible (§10.6) — for the Earth's central density ρ_0. As will be seen (§12.5), corrections to $d\alpha/dz$ inside F and other evidence require this estimate of ρ_0 to be substantially reduced.

Gutenberg gave $d\alpha^2/dz$ as large and positive over a range of depth approximating to the region F (see §10.3.6). On the theory of §11.5.2, this would require η and $d\rho/dz$ to be negative unless the function j [equation (11.50)] were to behave here very surprisingly (Bullen, 1963). Thus Gutenberg's velocity distribution in F is *a priori* very improbable. Inside G, Gutenberg's results agreed with those of Jeffreys in giving $d\alpha^2/dz$ smaller than normal.

The above details, apart from their historic interest, illustrate further both the sensitivity of η to departures from homogeneity and the very great extent to which (11.39) can be in error when the departures are significant. The results also highlight the degree of dependence of estimates of density variation on the observational values of $d\phi/dz$ in particular regions of the Earth. (See also §12.7.2).

11.7 Solidity of the inner core

The outstanding observed feature characterizing the change from the outer to the inner core proper is a sizeable net rise in the P seismic velocity α. Otherwise, it is not possible to interpret the main branches in the travel-time curves for core phases that have been found by all investigators since Lehmann's initial work in 1936. The net rise is of the order of 10 per cent, and its main part is either discontinuous or spread over a fairly narrow range of depth. For broad purposes, it may be assumed that α^2 jumps by about 20 per cent inside a small range of depth just above the inner core.

By (8.6),

$$\alpha^2 \rho = k + 4\mu/3, \qquad (11.52)$$

where μ denotes the rigidity. Thus a sudden rise in α^2 must be associated with a sudden corresponding rise in k and/or μ, since ρ cannot (significantly) decrease with depth. On the k-p hypothesis, a sizeable jump in k is forbidden. On the seismic evidence that α^2 jumps by about 20 per cent, the hypothesis therefore entails (Bullen, 1946, 1950) that μ is significantly greater in the inner core than in the outer, and thus that the inner core is solid (in the sense defined in §7.7).

The evidence in §11.4.2, while allowing for a limited departure from continuous variation of k with p at the inner core boundary, L say, gives fairly strong support to this conclusion. Thus even if Z were to jump by 5 units at L, the jump in k would not be expected to exceed one-fifth of 20 per cent. Even in the hypothetical extreme case of a change from a modified form of lower mantle material to iron-nickel at L, the jump in k would not be expected to exceed 9 per cent, and could be much less.

Other evidence comes from thermodynamical considerations. Birch's work in 1940 led him to mention the possibility of an inner core composed of crystalline iron. Simon (1953), assuming a model core consisting wholly of iron, inferred that a transition from liquid to solid iron would take place at the inner core boundary if the temperature were about 3600°C, a temperature well compatible with other evidence. On a similar assumption, Jacobs (1954) considered curves representing the adiabatic temperature and the melting-point in terms of depth inside the Earth, and inferred that a crossing of the curves at the inner-core boundary is consistent with all the evidence on temperature. He contemplated an earlier molten Earth which started solidifying from the centre and from the mantle-core boundary, upward in both cases, leaving a molten outer core trapped between. Calculations carried out by Lubimova (1956) also gave the inner core a temperature below melting point. Kennedy and Kraut (see Slichter, 1966) arrived empirically at a linear relation between the compression and melting-point temperature for various materials. This led them to a tentative estimate of 3700°C for the temperature at the inner core boundary, a result which is very close to Simon's estimate and obtained by a quite different method. (See also Higgins and Kennedy, 1971; Birch, 1972.)

Al'tschuler and Kormer (1961) inferred from shock-wave studies at pressures of order $3 \times 10^{11} \text{N/m}^2$ that the inner core is solid.

In a study of the modes of convection currents needed in the outer core on the dynamo theory of the Earth's main magnetic field, Chandrasekhar (1952) found necessity for a boundary of the type that a solid inner core would provide. Evidence from the Earth's magnetic field also led Knopoff and MacDonald (1958) to the view that the inner core is unlikely to be liquid.

A fairly direct test of the question would be provided if a phase such as *PKJKP* (see §9.4.2) involving S waves in the inner core, could be detected. However, there is a problem of exciting S waves, sufficiently large to be observable, from incident P waves in the outer core. Calculations (Bullen, 1951) showed that, with the largest earthquakes, the phase *PKJKP* could reach the border of observability over a limited range of epicentral distance on assuming the most favourable case of a single sharp boundary between outer and inner core. The phase has not so far been reliably detected at ordinary seismological observatories, but the increased resolving power of large seismic array stations has raised the probability of detection (see §12.6.2). Another direct seismic test is possible through examining amplitudes of the phase *PKiKP* (corresponding to waves reflected upward at the inner core boundary — see §9.4.2). Caloi (1961) detected such waves and inferred strong support for the solidity of the inner core. (See also Melik-Gajkazan, 1955; Bolt and O'Neill, 1965.)

Additional evidence on the solidity of the inner core is discussed in §§12.6 and 14.9.

11.8 Earth models of type *B*

The k-p hypothesis was made a central feature in constructing a second type of Earth model called the *B* type: k and dk/dp were formally taken to vary smoothly and continuously with p below 1000 km depth. In the original Model *B* (Bullen, 1950), the older value 0·3335 was used for the Earth's moment of inertia coefficient y, and the 1939 Jeffreys values of α and β were used below 1000 km depth. Other *B*-type models incorporating substantially revised observational data were later constructed (Bullen, 1965b; Bullen and Haddon, 1967, 1968).

11.8.1 *Construction of Model B*

The equation (11.52) provides one equation of condition on the values of ρ, k, and μ inside the Earth, the reliability depending only on the reliability of the observational values used for the *P* velocity α. A second equation of condition is provided for the mantle by using $\mu/\rho = \beta^2$ and the observational values of β, and for the outer core *E* by taking $\mu = 0$ (§§10.2, 10.35). These conditions were applied in constructing Model *B*, as well as Model *A*.

Further procedures for Model *B* involved principally the k-p hypothesis and the equivalent of using equations (11.35) and (11.36). [The original Model *B* was constructed before the forms (11.35) and (11.36) had been explicitly set down]. Application of (11.35) and (11.36) also involves knowledge of α and β.

For the regions D' and E, (11.35) was used with η taken equal to unity.

Inside D'', $d\phi/dz$ was taken as zero, so that the k-p hypothesis gave $\eta \approx 3$, as in §11.2.2, suggesting a moderate accumulation of denser matter near the bottom of the mantle (see also §11.6.4).

[*Refs. on p. 223*]

With k taken continuous at the mantle-core boundary N, knowledge of α and β on the two sides of N formally determined the ratio of the jump in ρ at N.

The k-p hypothesis required the inner core of Model B to be solid, in accordance with the argument in §11.7.

Assessments of η in the lower core (§11.6.6) were supplied through (11.36) and the assumed smooth variation of dk/dp. (Needed values of g were obtained by forming successive approximations to the density distribution in the lower core: any assigned distribution of ρ determines a unique distribution of g.) For the inner core, no direct observational values of β are available and it was originally assumed that $d\phi/dz$, like $d\alpha^2/dz$, is here small, whence $\eta \approx dk/dp$. The relation (11.35) was applied in F and G, using the values of η as thus derived.

The above procedures do not formally fix the values of any density jumps at the $E-F$ and $F-G$ boundaries, and nothing has been said so far about details in the regions A, B, and C. But it transpired that, along with the assumed values of M (the Earth's mass) and y, the procedures virtually determined values of ρ to close precision throughout nearly the whole Earth. Little flexibility was left in respect of possible density jumps inside the core or changes inside the regions B and C from the value of $d\rho/dz$ derived at the top of D'.

As a consequence of this limited flexibility, it was necessary to extrapolate the density distribution in D' upward nearly to the top of the region B with little change in the density gradient, in contrast to the comparatively high rate of diminution of ρ with decreasing z that occurs inside the region C of Model A. Thus the mean density of the regions B and C in the original Model B had to be taken significantly greater than in Model A. Within the small range of flexibility permitted, Model B was selected as the parametrically simplest model meeting the above requirements: ρ was taken continuous throughout the core; the smooth upward extrapolation of ρ from D' was taken to 80 km depth; and the Model A distribution of ρ was adopted above 80 km, resulting in a discontinuous jump from 3·36 to 3·87 g/cm^3 at this depth. Other density results for the original Model B are: 4·41 g/cm^3 at 1000 km depth; a jump from 5·57 to 9·74 g/cm^3 at the mantle-core boundary; nearly 18 g/cm^3 at the centre.

The values of g and p were derived from the values of ρ in the usual way, using (10.4,5). Values of k and μ in the mantle and of k in the outer core were deduced from the values of ρ, assuming the

Jeffreys values of α and β. Values of k and μ below the outer core were derived through the $k\text{-}p$ hypothesis.

11.8.2 *Critique of Model B*

Model B was set up primarily to test the $k\text{-}p$ hypothesis. Since Model B in a sense grew out of Model A, the differences between the two models are of course fairly small at most levels. But some of the differences are significant.

The first main difference is that the inner core of Model B is solid. In this respect, because of the extent of support from further evidence, Model B was an improvement on the original A models, the latter having been constructed before the $k\text{-}p$ hypothesis had been formulated. The applications of the hypothesis inside the inner core have, however, only small repercussions outside, so that the A models can be readily adjusted to incorporate the indicated solidity of the inner core.

A second difference is the excess of the density gradient in the region D'' of Model B over that in Model A. Because D'' occupies only a comparatively small range of depth, the effect of the difference is fairly slight outside D'' and adjustments to the A models can again be readily made to allow for it. The possible alternative (§ 11.2.2) that the rigidity diminishes inside D'' has to be kept in mind.

An apparently more serious difference was the excess of the original Model B over the Model A densities inside the upper mantle. This excess, which is about 0.5 g/cm³ at 80 km and remains positive down to 600 km depth, was a fairly direct consequence of reducing the change $|\Delta k|$ in k at N to zero in going from Model A to Model B. Geochemical studies did not support the high upper mantle densities of the original Model B, and several investigators took this as crucial evidence against the $k\text{-}p$ hypothesis. However, the 1963 revision of y from 0.3335 to 0.3308 (§ 5.7) entirely altered the situation. For Earth models constructed before 1963, including the original Model B, the revision of y required some transfer of mass from the mantle to the core, and an effect (see also § 12.7.2) was to increase the flexibilities mentioned in § 11.8.1 sufficiently to allow a B-type model to have the same density distribution in the upper mantle as an A-type model. Thus evidence on densities in the upper mantle no longer discriminates against the B-type models or the $k\text{-}p$ hypothesis.

Another feature of Model B is that it provided the first

[*Refs. on p. 223*]

observationally based estimate of the central density ρ_0. (The model A calculations had provided only a minimum estimate — §10.6). Although the original Model B estimate of about 18 g/cm³ for ρ_0 is now superseded as a consequence of revised and new observational evidence, the underlying theory is unaffected (see §12.5).

Prior to 1963, the differences between Models A and B had served to indicate broadly the uncertainty in the determination of the Earth's density distribution down to around 5000 km depth. The subsequent revisions, however, brought the two types of model into quite close agreement, the A-type models now having Δk insignificantly different from zero at N (§11.4.1). Unless future evidence should unexpectedly disturb the present findings on Δk, the need to discriminate between the A and B types has therefore largely lapsed.

The general principles underlying the construction of Model B have thus been fairly well substantiated, although the numerical detail has had to be revised to meet new observational evidence. The original Model B was superseded by a set of B-type models (Bullen and Haddon, 1967, 1968) which incorporated the revised y, etc. These models are described in §12.7.2.

11.8.3 *Empirical relations between k and p*

For the whole of the Earth's interior below 1000 km depth, a quadratic formula for k in terms of p represents the Model B results within 2 per cent accuracy (Bullen, 1950). The formula, adjusted (Bullen, 1968d) to allow for later revisions of the observational data, is

$$k = 2\cdot34 + 3\cdot00p + 0\cdot10p^2, \tag{11.53}$$

where the units are 10^{11} N/m².

In view of misapplications that have been made by several investigators, it needs to be stressed that (11.53) is relevant only to the pressure range $0\cdot4 < p < 4\cdot0 \times 10^{11}$ N/m² (approx.). Above 1000 km depth, the variation of k with p cannot be assumed to have settled down to moderately smooth behaviour: if applied to the top of the region B, (11.53) would give about twice the actual value of k.

Some investigators have preferred to represent the dependence of k on p in terms of separate linear relations for the lower mantle and core. The relations

$$k = 2\cdot29 + 3\cdot16p, \quad k = 1\cdot84 + 3\cdot44p \tag{11.54}$$

(Bullen, 1968) give the closest linear fits for the lower mantle and

[Refs. on p. 223]

core of a B-type model. For some other examples of linear (k,p) relations, see Cook (1972) and §15.1.2.

The relations (11.53) and (11.54) provide convenient summaries which can be used for trial purposes in certain outside contexts, e.g. those parts of the interiors of terrestrial planets where the pressures lie inside the stipulated range.

REFERENCES

Ahrens, T.J., Anderson, D.L., and Ringwood, A.E. (1969). Equations of state and crystal structures of high-pressure phases of shocked silicates and oxides. *Rev. Geophys.*, **7**, 667–707.

Akimoto, S., and Ida, Y. (1966). High-pressure synthesis of Mg_2SiO_4 spinel. *Earth Planet. Sci. Letters*, **1**, 358.

Al'tschuler, L.V., and others (1958). *Zurn. Eksp. Teor. Fiz.*, **36**, 606.

Al'tschuler, L.V., and Kormer, S.B. (1961). On the internal constitution of the Earth. *Bull. Acad. Sci. U.S.S.R., Geophys. Ser.* (English trans.), pp. 18–21.

Bernal, J.D. (1936). Hypothesis on 20° discontinuity. *Observatory*, **59**, 268.

Birch, F. (1939). The variation of seismic velocities within a simplified Earth model, in accordance with the theory of finite strain. *Bull. Seismol. Soc. Amer.*, **29**, 463–479.

Birch, F. (1952). Elasticity and constitution of the Earth's interior. *J. Geophys. Res.*, **57**, 227–286.

Birch, F. (1961). Composition of the Earth's mantle. *Geophys. J., R. Astr. Soc.*, **4**, 295–311.

Birch, F. (1963). Some geophysical applications of high-pressure research. In *Solids under Pressure*, McGraw-Hill, New York. pp. 137–162.

Birch, F. (1972). The melting relations of iron, and temperatures in the Earth's core. *Geophys. J. R. Astr. Soc.*, **29**, 373–387.

Bolt, B.A. and O'Neill, M.E. (1965). Times and amplitudes of the phases PKiKP and PKIIKP. *Geophys. J. Roy. Astr. Soc.*, **9**, 223–231.

Boschi, E. and Caputo, M. (1969). Equations of state at high pressure and the Earth's interior. *Riv. Nuovo Cim.*, **1**, 441–513.

Bridgman, P.W. (1931). *The Physics of High Pressure*. Bell, London.

Bullen, K.E. (1936). The variation of density and the ellipticities of strata of equal density within the Earth. *Mon. Not. R. Astr. Soc., Geophys. Suppl.*, **3**, 395–401.

Bullen, K.E. (1940). The problem of the Earth's density variation. *Bull. Seismol. Soc. Amer.*, **30**, 235–250.

Bullen, K.E. (1942). The density variation of the Earth's central core. *Bull. Seismol. Soc. Amer.*, **32**, 19–29.

Bullen, K.E. (1946). A hypothesis on compressibility at pressures of the order of a million atmospheres. *Nature, Lond.*, **157**, 405.

Bullen, K.E. (1949). Compressibility-pressure hypothesis and the Earth's interior. *Mon. Not. R. Astr. Soc., Geophys. Suppl.*, **5**, 355–368.

Bullen, K.E. (1950). An Earth model based on a compressibility-pressure hypothesis. *Mon. Not. R. Astr. Soc., Geophys. Suppl.*, **6**, 50–59.

Bullen, K.E. (1951). Theoretical amplitudes of the seismic phase PKJKP. *Mon. Not. R. Astr. Soc., Geophys. Suppl.*, **6**, 163–167.

Bullen, K.E. (1952). On density and compressibility at pressures up to thirty million atmospheres. *Mon. Not. R. Astr. Soc., Geophys. Suppl.*, **6**, 383–401.

Bullen, K.E. (1956). Seismology and the broad structure of the Earth's interior. *Phys. & Chem. of the Earth*, **1**, 68–93.

Bullen, K.E. (1963). An index of degree of chemical inhomogeneity in the Earth. *Geophys. J., R. Astr. Soc.*, **7**, 584–592.

Bullen, K.E. (1964). New evidence on rigidity in the Earth's core. *Proc. Nat. Acad. Sci., Wash.*, **52**, 38–42.

Bullen, K.E. (1965a). On compressibility and chemical inhomogeneity in the Earth's core. *Geophys. J., R. Astr. Soc.*, **9**, 195–202.

Bullen, K.E. (1965b). Models for the density and elasticity of the Earth's core. *Geophys. J., R. Astr. Soc.*, **9**, 233–252.

Bullen, K.E. (1967). Note on the coefficient η. *Geophys. J., R. Astr. Soc.*, **13**, 459.

Bullen, K.E. (1968a). Compression in the Earth. *Geophys. J., R. Astr. Soc.*, **16**, 31–36.

Bullen, K.E. (1968b). Dependence of compressibility and compression on chemical composition in finite-strain theory. *Phys. Earth Planet. Interiors*, **1**, 297–301.

Bullen, K.E. (1968c). Incompressibility at the Earth's mantle-core boundary. *Proc. Nat. Acad. Sci., Wash.*, **60**, 752–757.

Bullen, K.E. (1968d). Empirical equations of state for the Earth's lower mantle and core. *Geophys. J., R. Astr. Soc.*, **16**, 235–238.

Bullen, K.E. (1969a). Compressibility-pressure gradient and the constitution of the Earth's outer core. *Geophys. J., R. Astr. Soc.*, **18**, 73–79.

Bullen, K.E. (1969b). Seismic and related evidence on compressibility in the Earth. In *The Application of Modern Physics to the Earth and Planetary Interiors* (ed. S.K. Runcorn), Wiley-Interscience, London, pp. 287–297.

Bullen, K.E. (1969c). The interiors of the planets. *Ann. Rev. Astron. Astrophys.*, **7**, 177–200

Bullen, K.E. (1970a). Comparison of sources of evidence on the variation of incompressibility in the Earth's deeper interior. *Phys. Earth Planet. Interiors*, **3**, 36–40.

Bullen, K.E. (1970b). Note on application of Emden's equation to planetary interiors. *Mon. Not. R. Astr. Soc.*, **149**, 51–52.

Bullen, K.E. (1972). Compressibility and planetary interiors. *Phys. Earth Planet. Interiors*, **6**, 131–135.

REFERENCES

Bullen, K.E. and Haddon, R.A. (1967). Earth models based on compressibility theory. *Phys. Earth Planet. Interiors*, **1**, 1–13.

Bullen, K.E. and Haddon, R.A. (1968). Corrections to three Earth models. *Phys. Earth Planet. Interiors*, **1**, 401–402.

Bullen, K.E. and Haddon, R.A. (1969). Upper bound to change in incompressibility at the Earth's inner core boundary. *Geophys. J., R. Astr. Soc.*, **17**, 179–183.

Caloi, P. (1961). Seismic waves from the outer and inner core. *Geophys. J., R. Astr. Soc.*, **4**, 139–150.

Chandrasekhar, S. (1952). The thermal instability of a fluid sphere heated within. *Phil. Mag.*, **7**, 1317–1329.

Cleary, J. (1969). The S velocity at the core-mantle boundary, from observations of diffracted S. *Bull. Seismol. Soc. Amer.*, **59**, 1399–1405.

Cook, A.H. (1972). The dynamical properties and internal structures of the Earth, the Moon and the planets. *Proc. R. Soc. Lond.* A, **328**, 301–336.

Davies, G.F. and Anderson, D.L. (1971). Revised shock-wave equations of state for high-pressure phases of rocks and minerals. *J. Geophys. Res.*, **76**, 2617–2627.

Elsasser, W.M. (1951). Quantum-theoretical densities of solids at extreme compression. *Science*, **113**, 105–107.

Evernden, J.F. and Clark, D.M. (1970). Study of teleseismic P. *Phys. Earth Planet. Interiors*, **4**, 1–31.

Feynman, R.P., Metropolis, N. and Teller, E. (1949). Equations of state of elements based on the generalized Fermi-Dirac theory. *Phys. Rev.*, **75**, 1561–1572.

Higgins, G. and Kennedy, G.C. (1971). The adiabatic gradient and the melting point gradient in the core of the Earth. *J. Geophys. Res.*, **76**, 1870–1878.

Jacobs, J.A. (1954). Temperature distribution within the Earth's core. *Nature, Lond.*, **173**, 258.

Jeffreys, H. (1936). Hypothesis on 20° discontinuity. *Observatory*, **59**, 268.

Jeffreys, H. (1939a). The times of P, S and SKS and the velocities of P and S. *Mon. Not. R. Astr. Soc., Geophys. Suppl.*, **4**, 498–533.

Jeffreys, H. (1939b). The times of the core waves. *Mon. Not. R. Astr. Soc., Geophys. Suppl.*, **4**, 548–561, 594–615.

Knopoff, L. and MacDonald, G.J.F. (1960). An equation of state for the Earth. *Geophys. J., R. Astr. Soc.*, **3**, 68–77.

Knopoff, L. and Uffen, R.J. (1954). The density of compounds at high pressures and the state of the Earth's interior. *J. Geophys. Res.*, **59**, 471–484.

Landisman, M., Satô, Y. and Nafe, J. (1965). Free vibrations of the Earth and the properties of its deep interior regions. Part 1: Density. *Geophys. J., R. Astr. Soc.*, **9**, 439–502.

Lubimova, H.A. (1956). Thermal history of the Earth and its geophysical effect. *Dokl. Akad. Nauk, U.S.S.R.*, **107**, (11), 55–58.

MacDonald, G.J.F. and Knopoff, L. (1958). The chemical composition of the outer core. *Geophys. J., R. Astr. Soc.*, **1**, 284–297.

McQueen, R.G. and Marsh, S.P. (1960). Equation of state for nineteen metallic elements from shock-wave experiments to two megabars. *J. App. Phys.*, **31**, 1253–1269.

Melik-Gajkazan, I.A. (1955). *Akad. Nauk SSSR, Geophys. Inst., Trudy*, **26**, (153), 117.

Ringwood, A.E. (1958). The constitution of the mantle – I, II and III. *Geochim. et Cosmochim. Acta,* **13**, 303–321; **15**, 18–29, 195–212.

Ringwood, A.E. (1959). On the chemical evolution and densities of the planets. *Geochim. et Cosmochim. Acta,* **15**, 267–283.

Runcorn, S.K. (1974). A physical interpretation of Bullen's compressibility-pressure hypothesis. Proceedings of NATO Conference, Newcastle-upon-Tyne, April 1974. In course of publication.

Simon, Lord (F.E.) (1953). The melting point of iron at high pressures. *Nature, Lond.,* **172**, 746.

Slichter, L.B. (1966). A glimpse at the geophysical scene. *Trans. Amer. Geophys. Un.,* **47**, 346–354.

Stewart, R.M. (1973). Composition and temperature of the outer core. *J. Geophys. Res.,* **78**, 2586–2597.

Takeuchi, H. and Kanamori, H. (1966). Equations of state of matter from shock wave experiments. *J. Geophys. Res.,* **71**, 3985–3994.

CHAPTER 12

Some second approximations

Any of the original A-type and B Earth models provides a first approximation to the distributions of the main internal physical properties of the Earth, namely, the density ρ, pressure p, gravitational intensity g, incompressibility k, and rigidity μ. Various amendments, mostly of the nature of small corrections to the original models, will now be considered, leading to second (not final) approximations.

In the present chapter, most of the amendments to be considered arise from certain revisions of seismic bodily-wave travel-time (T, Δ) data and related effects on the P and S velocities α and β. Travel-time data continue to be substantially the most reliable primary source of evidence on α throughout most of the Earth, and a highly important source on β inside the mantle. The estimation of β, however, now leans rather heavily on other data as well, including data from observations of surface seismic waves and free Earth oscillations. Evidence from these sources will be mentioned, but considered in more detail in later chapters.

The present chapter also takes account of results in Chapter 11 on incompressibility and the coefficient η, and of results from shock-wave experiments in which transitory pressures exceeding a million atmospheres are reached. These results supply, among other things, new evidence on the Earth's central density ρ_0 and on rigidity in the lower core.

Details are given of an Earth model B_2 constructed according to the procedures of the original Model B except that certain revised observational data, including the revised estimate 0·3308 of the Earth's moment of inertia coefficient, have been incorporated.

12.1 P and S velocities in the mantle

After 1940, a vast accumulation of new data on seismic travel times

[Refs. on p. 255]

and velocity distributions led to numerous attempts to improve the J.B. tables and the 1939 Jeffreys velocities. The attempts fall into two classes: those concerned with the most suitable representations to take for spherically symmetrical Earth models; and those concerned with lateral deviations (§10.3.2) — i.e. with representations for particular geographic regions below which the values of α and β for given depths z deviate detectably from the average.

The new evidence includes results from studies of seismic waves through the Earth from chemical and nuclear explosions (§8.1.1), where there is an advantage over natural earthquake studies in that source locations and origin times can be available to high precision. Other important evidence is provided by array stations — networks of stations all with similar recording equipment spread systematically over large areas (525 stations spread over 10,000 square miles in the case of the Montana array). Among other things, array stations supply more direct information on $dT/d\Delta$ (see e.g. Toksöz and Wiggins, 1969) than that derived through the procedures of §9.5.

The following subsections are concerned with the estimation of α and β in the mantle.

12.1.1 *The outermost 200 km*

The 1939 Jeffreys velocity model (§9.6.2) has P_n and S_n velocities (the velocities at the top of the region B) equal to 7·75 and 4·35 km/s. These values fit well the data for some geographical areas, e.g. Japan, but for many other areas the velocities are now known to be distinctly greater. For Europe, for example, Jeffreys (1947) inferred a P_n velocity equal to $8·09 \pm 0·08$ km/s with $dT/d\Delta = 13·7$ s/deg at the top of the region B, in contrast to 14·3 s/deg in the 1939 model.

A general idea of the global spread of values of P_n and S_n velocities is given in Table 12.1, due to Brune (1969). (See also Ritsema, 1969; Huestis, Molnar, and Oliver, 1973.) Using evidence from seismic surface as well as bodily waves, Brune classified P_n and S_n velocities according to the type of overlying crust. (The type referred to as 'basin-range' is characterized by recent breaking of the crust into basins and ranges, and is the seat of numerous volcanic extrusions and earthquake foci.) The table also shows Brune's preferred estimates of the corresponding crustal thicknesses.

Since deep-ocean basins and continents cover most of the Earth's surface, the table would favour, for the Earth as a whole, models in

Table 12.1
Brune's summary of P_n and S_n velocities

Crustal type	Crustal thickness (km)	P_n velocity (km/s)	S_n velocity (km/s)
Continental shield	35	8·3	4·7
Mid-continent	38	8·2	4·6
Alpine	55	8·0	
Basin-range	30	7·8	4·4
Great island arc	30	7·4—7·8	4·3
Deep-ocean basin	11	8·1—8·2	4·5
Mid-ocean ridge	10	7·4—7·6	4·3

which the P_n and S_n velocities are about 3-4 per cent greater than in the 1939 model. Raising the P_n and S_n velocities requires compensating reductions in the average P and S velocity gradients in the outermost 200 km (at least) of the mantle, in order to meet the requirements of the overall travel-time data. The likely reductions of gradient are sufficiently large to preclude the velocity variations from being treated as normal in this part of the Earth, though the extent of abnormality has not yet been finely resolved.

As far back as 1926, Gutenberg, arguing from evidence on bodily-wave amplitudes, had suggested that there is a 'low velocity layer' inside the outermost 100—200 km of the mantle, the velocity gradients actually becoming negative for a range of depth. The presence of such a layer, especially should the detail involve violation of the relation (9.17), would make it difficult to delineate the velocity distributions at all accurately, using only bodily-wave arrival times.

Later work substantiated Gutenberg's suggestion in certain particulars, though not completely. For example, taking a conventional crust ($0 < z < 35$ km, $\beta = 3.55$ km/s), Lehmann (1961) arrived at a model for Europe and north-eastern America in which β increases from 4·60 to 4·65 km/s inside the region B down to $z = 120$ km, is equal to 4·3 km/s inside a low S velocity zone ($120 < z < 220$ km), and jumps to 4·7 km/s at $z = 220$ km. For the European sub-crust, she gave (1959) α as constant, however, equal to 8·1 km/s, down to $z = 220$ km, there jumping to 8·4 km/s. These models are by no means uniquely determined. [See Bullen (1961) for some of the cautions needed in drawing inferences from bodily-wave data in this

[Refs. on p. 255]

part of the Earth.] But they are in line with most recent bodily-wave evidence in indicating that, under most geographical areas, $d\beta/dz$, but not $d\alpha/dz$, becomes negative over a range of depth inside the uppermost 200 km of the mantle. (There is evidence that $d\alpha/dz$ is also negative under some limited geographical areas.) Summaries of recent evidence on low velocity layers have been supplied, among others, by Lehmann (1967, 1970) and D.L. Anderson (1967). The main seismological evidence appears to be sufficiently met by assuming that $d\mu/dz < 0$ inside part of the outermost 200 km, a condition which is feasible in the light of evidence on temperature gradients (§ 11.6.2).

Lateral variations of α and β below the crust have long been indicated in seismic surface wave studies (see Chapter 13), and Jeffreys later (1952) showed that lateral variations are statistically significant in bodily-wave observations. A quantity of detail has since been added, especially through analyses of nuclear explosion records, including notably the Bikini explosion of 1946 (Gutenberg and Richter, 1946; Bullen, 1948), the Gnome explosion of 1961 (Carder et al., 1962), and several later explosions. Table 12.1 gives an indication of lateral variations just below the crust. The variations on the whole diminish with increasing depth but are still detectable down to several hundred kilometres depth; Molnar and Oliver (1969) have summarized important sections of the evidence. Detailed information on the variations is, however, so far available only in a few limited geographical areas. Thus lateral variations contribute a large part of the as yet unresolved uncertainties in travel times for the outer mantle.

12.1.2 *Remainder of the mantle*

The unresolved complications inside the outermost 200 km of the mantle accentuate the difficulties of improving estimates of the velocity distributions at greater depths, especially inside the next several hundred kilometres. This is principally because, for a ray which does not penetrate very far below the 200 km level, the portions which lie above this level are mostly substantial.

For the whole outermost 900 km of the Earth, the P and S velocity distributions now appear to be somewhat more complicated than the 1939 distributions, though it remains the case that the region C is characterized by velocity gradients greater than normal.

Carder (1965) inferred from analyses of nuclear explosion records that rapid changes in $d\alpha/dz$ occur at 370–400 and 630–650 km depth. Niazi and Anderson (1965), using array-station data, gave 320 and 640 km. Many subsequent studies confirmed the existence of these two rapid changes and are consistent with taking the higher one near $z = 400$ km, as in § 10.3.2. The presence of the second rapid change requires C to be subdivided into C' and C'', the separating boundary being generally taken near $z = 650$ km.

The effect of the revisions inside the regions B and C is fairly small on α and β, being less at most depths than the amendments already indicated for P_n and S_n. The average velocity gradients over (say) 400 km ranges of depth likewise need only fairly small amendments. On the other hand, amendments to $d\alpha/dz$ and $d\beta/dz$ for some smaller ranges of depth are quite large, with the consequence that it is now unsafe to assume that α and β vary normally anywhere inside the outermost 900 km; it is likely that several types of abnormality occur.

Between 900 and 2700 km depth (the region D'), suggested amendments to α, β, $d\alpha/dz$, and $d\beta/dz$ have been mostly fairly slight, and the evidence continues to favour variations of α and β with z that are close to normal. Considerable uncertainties remain in the region D'' (the lowest 200 km of the mantle); these will be discussed in §§ 12.1.4, 12.4 and later.

12.1.3 *Recent estimates of P velocities*

Among innumerable recent attempts to produce amended P travel times and velocities for the mantle, the most noted was that of a committee of twelve chaired by E. Herrin (1968), making extensive use of modern computing facilities. The results may be taken as a fairly representative example of much recent work. (See also Hales, Cleary, and Roberts, 1968.) Although subsequent testing has indicated that the Herrin model falls short of original aspirations on a number of points, the model is important as the fruit of a recent major effort on the part of a large group of investigators.

Table 12.2 shows the Herrin velocities at particular depths, along with their excesses over the 1939 Jeffreys velocities (§ 9.6.2). The crustal velocities are conventional and are included in the table solely because of their involvement with the computation of subcrustal velocities. The detail shown below the crust is broadly compatible

[*Refs. on p. 255*]

with the foregoing discussion on the extent of likely amendments needed to the 1939 mantle P velocities and provides a general indication of the order of the uncertainties still remaining. It is to be noted, however, that near the mantle-core boundary N the uncertainty in estimates of α is greater than indicated in the table. For example, Taggart and Engdahl (1968) using data on travel times of the phase PcP gave $\alpha = 13 \cdot 40$ km/s at N. A slight negative P velocity gradient inside part at least of the region D'' was also suggested in work of Sacks (1967) and others on diffracted P waves.

It is interesting to note that a comprehensive study by Gilbert, Dziewonski, and Brune (1973) indicated that, in certain important respects, the J.B. P travel-time tables for the mantle are more compatible with free Earth oscillation data (Chapter 14) than are various more modern tables.

12.1.4 *Recent estimates of S velocities*

Many studies of S travel times have been principally directed at deriving improved distributions of β in the upper mantle. Among the more important are those of Nishimura, Kishimoto, and Kamitsuki (1958), Lehmann (1961), Arnold, Jeffreys, and Shimshoni (1963), Jeffreys (1966), Ibrahim and Nuttli (1967), Nuttli (1969), and Kovach and Robinson (1969). Doyle and Hales (1967) gave revised S travel-time data for $28° < \Delta < 82°$, enabling the revision of β to be carried into the lower mantle. Cleary, Porra, and Read (1967) and Cleary (1969), using observations of diffracted S waves for $\Delta \geqslant 100°$ and other travel-time data, inferred that $d\beta/dz$ is markedly negative inside D''. Cleary tentatively gave $\beta'' = 6\cdot 8$ km/s, where β'' is the S velocity at the bottom of the mantle. Some (not all) investigators supported Cleary's general findings: Bolt, Niazi, and Somerville (1969) gave $\beta'' = 7 \cdot 0 \pm 0 \cdot 1$ km/s; Randall (1971) gave $6 \cdot 8$ km/s.

Hales and Roberts (1970a) in a study of the β distribution for the whole mantle, presented two models resulting from different smoothing treatments of their data. The mean of the two distributions (labelled β_{HR}) and a distribution of Randall (1971) (labelled β_R) are shown in Table 12.3, along with differences from the Jeffreys 1939 distribution. The data of Hales and Roberts came from records at North American stations of selected events. Randall's data have some special interest in that they comprise all S phases reported in the International Seismology Summary and Centre bulletins for the set

Table 12.2

Comparison of the Herrin (α_H) and Jeffreys 1939 (α_J) P velocities

Depth (km)	α_H (km/s)	$\alpha_H - \alpha_J$
0	6·00	+0·43
15	6·00	+0·43
15	6·75	+0·25
40	6·75	
40	8·05	+0·28
100	8·12	+0·17
200	8·32	+0·06
300	8·68	+0·10
400	9·13	+0·20
600	10·20	−0·05
800	11·08	−0·02
1000	11·44	+0·02
1200	11·72	+0·01
1400	12·05	+0·06
1600	12·32	+0·06
1800	12·55	+0·02
2000	12·78	−0·01
2200	13·03	0·00
2400	13·25	−0·02
2600	13·49	−0·01
2700	13·60	−0·04
2894	13·64	0·00

of earthquakes used by Herrin and colleagues (§ 12.1.3). Randall adhered closely to the statistical principles formulated by Jeffreys (1948) — see also Arnold (1968) — in treating the data. The values in Table 12.3 for the crust are only conventional.

The two sets of results have been included in the table since Hales and Roberts, but not Randall, included a low S velocity layer in the upper mantle — while Randall, but not Hales and Roberts, included a drop in β near the bottom of the mantle. The values in brackets were derived by interpolation from a table in Randall's paper. (Randall foreshadowed some possible modification below 2000 km depth because of certain complexities connected with the phase SKS.) The results in Table 12.3 indicate broadly the extent of revision of the original Jeffreys β distribution that may be required on evidence as

[*Refs. on p. 255*]

Table 12.3

Comparison of the Hales-Roberts (β_{HR}), Randall (β_R), and Jeffreys 1939 (β_J) S velocities

Depth (km)	β_{HR}	$\beta_{HR} - \beta_J$ (km/s)	β_R	$\beta_R - \beta_J$
0	3;50	+0·14	3·40	+0·04
15	3·50	+0·14	3·40	+0·04
15	3·50	−0·24	3·80	+0·06
40	4·60	+0·23	3·80	
40	4·60	+0·23	4·35	−0·02
100	4·60	+0·15	4·36	−0·09
180	4·15	−0·42	4·45	−0·12
200	4·60	0·00	4·47	−0·13
300	4·70	−0·06	4·67	−0·09
400	5·13	+0·19	4·95	+0·01
600	5·46	−0·20	5·63	−0·03
800	6·22	+0·09	6·22	+0·09
1000	6·34	−0·02	6·38	+0·02
1200	6·50	0·00	6·49	−0·01
1400	6·62	0·00	6·63	+0·01
1600	6·73	0·00	6·73	0·00
1800	6·83	0·00	6·82	−0·01
2000	6·93	0·00	(6·93)	(0·0)
2200	7·01	−0·01	(7·02)	(0·0)
2400	7·09	−0·03	7·12	0·00
2660	7·19	−0·05	7·24	0·00
2894	7·30	0·00	6·78	−0·52

yet available. Quite possibly, amendments larger than the differences in the table may be required at some depths, there being a variety of problems in correctly identifying the times of onset of S phases on seismograms. (See **B**, §10.6.) Comparisons with β distributions derived for the mantle by some other authors are exhibited in the paper of Hales and Roberts.

12.1.5 *Supplementary data from amplitude observations*

Auxiliary information on bodily-wave travel times and velocity (v,r) distributions in the Earth comes from studies of recorded wave amplitudes. Formulae connecting amplitudes with (T,Δ) and (v,r) relations are given in **B**, Chapter 8. The formulae show that large and

[Refs. on p. 255]

small amplitudes are generally associated with large and small values of $d^2T/d\Delta^2$, respectively, i.e. with large and small curvatures of the (T, Δ) curve. A decrease in dv/dz over a range of values of z contributes to a decrease in the (T, Δ) curvature over a range of values of Δ (see §9.3.4). Thus, as one example of the use of amplitude data, if contributions to the amplitudes from other effects could be neglected, a decrease in dv/dz could be revealed in a study of the variation of amplitudes with Δ. Other examples are the occurrence of large amplitudes at certain types of cusp in (T, Δ) curves (§§9.3.2, 9.3.3).

In practice, it has in the past proved difficult to draw from amplitude data inferences that are not much more reliably determined by data on phase onset times. This is because the recorded amplitudes depend sensitively on a variety of additional factors, including uncertainties in geological and crustal structures for some distance below recording stations. Quite subtle changes, vertical and lateral, in assumed velocity distributions can also affect the calculated amplitudes noticeably. The role of amplitude studies has therefore usually been subordinate to travel-time studies — principally to suggest where intensive travel-time studies might be made, rather than to add significant numerical detail to the final (v, r) results. (Cf. Gutenberg's low velocity layer, §12.1.1.)

Latterly, however, amplitude studies have been rising in importance, largely as a consequence of the application of modern communication theory to seismology, coupled with the advent of array stations. (See e.g. Robinson, 1967; Båth, 1974.) The processes of combining observations taken at array stations have resulted in sharpened accuracy in measuring amplitudes of arriving waves for selected seismic phases, and various devices, theoretical and practical, assist in filtering out extraneous contributions to the amplitudes. The stage has now been reached where bodily-wave amplitude studies sometimes enable important discriminations to be made between otherwise equally plausible velocity distributions. An important illustration was provided in work of Buchbinder and Sacks (see §12.3.2) on the core P velocity distribution.

Attempts to estimate the sizes of sudden density changes in the Earth using seismic data are discussed in §15.5.

12.2 Structure of the outer core

12.2.1 *Suggestion of mild inhomogeneity in outer core*

The original Model A and B procedures treated the outer core E as a

[Refs. on p. 255]

single homogeneous layer, used the Jeffreys 1939 distribution of α, and took η equal to unity throughout E. Later work (Bullen, 1969) indicated that the Jeffreys and Gutenberg distributions imply some mild inhomogeneity inside the outermost 700 km of E. The notation E_1 will be used for this outermost subregion of E, and E_2 for the subregion below.

The subdivision into E_1 and E_2 arose in the course of an attempt to reconcile Birch's (1952) estimates of dk/dp with seismological evidence. Birch himself had inferred reasonable agreement inside the region E between his relation (11.19) and a selection of the seismological evidence. This apparent agreement had been achieved by substituting into (11.8) an average value for $d\alpha^2/dz$ for E derived from Gutenberg's (1951) P velocity distribution. Later evidence indicated that undue weight had been given to Gutenberg's results in the lower part of E_2, and a re-examination (Bullen, 1969, 1970) became necessary. The more general formula (11.36) was used in place of (11.8).

It emerged that for $2900 < z < 4500$ km, Gutenberg's values of α accord reasonably well with those of Jeffreys. (For $4500 < z < 5000$ km, Birch's estimates of dk/dp are significantly discordant with the Jeffreys values of $d\alpha^2/dz$.) Both sets of velocity data give $g^{-1}d\alpha^2/dz$, and therefore $(dk/dp - \eta)$ [by (11.36), β being taken zero inside E], as increasing by 0·8-0·9 between the top and the bottom of E_1. These data therefore actually imply that (i) there are significant departures from Birch's formula (11.19) at the pressures and compressions of the outer core; or (ii) E_1 is significantly inhomogeneous; or both. For E_2, the data give $(dk/dp - \eta)$ as nearly constant.

From experiments on a wide range of compressed rocks and metals, O.L. Anderson (1968) inferred that $(\partial^2 k'/\partial p^2)_\tau$ is undetectably different from zero at pressures equal to those in the Earth's core; the behaviour of d^2k/dp^2 would not be expected to be greatly different (see §11.2.5) for homogeneous materials. On Anderson's evidence, the stated seismic data are thus compatible with uniform composition and phase throughout E_2, but not E_1. The seismic data also entail a value for dk/dp in E_2 appreciably larger than that given by (11.19). Hence, on the data used, it would appear that both (i) and (ii) apply.

The Jeffreys data on $d\alpha^2/dz$ give $(dk/dp - \eta)$ as varying from about 1·8 at the top of E_1 to 2·6 at the bottom. (The Gutenberg data give values about 0·2 units less throughout most of E_1.) When E_1 is

[Refs. on p. 255]

no longer assumed homogeneous, there is no strong reason (apart from Birch's theory) to expect dk/dp to be closely constant throughout E_1. It then becomes a problem to determine how the observationally-based variation of $g^{-1}d\alpha^2/dz$ should, through (11.36), be distributed between dk/dp and η. On the Jeffreys data, the procedure which gives the least deviation from Birch's formula happens also to give dk/dp practically continuous at N, with dk/dp varying from about 3·1 to 3·6 and η from 1·4 to 1·0 between the top and bottom of E_1; throughout E_2, $dk/dp \approx 3\cdot 6$ and $\eta \approx 1$. (On the Gutenberg data, the same procedure would give $dk/dp \approx 3\cdot 4$ inside E_2, and a similar distribution of η in E_1 and E_2 to that on the Jeffreys data.)

The degree of inhomogeneity that would be entailed by $\eta = 1;4$ is fairly mild because η is so sensitive to variations of composition. (Cf. numerical detail on η in §§11.6.3, 11.6.4, 11.6.6.) The inhomogeneity could be due to a slight continuous change of composition or of phase inside E_1. The presence of phase changes would seem slightly the more probable (Bullen, 1969).

12.2.2 *Assessment in the light of recent evidence*

The suggestion that E_1 may be slightly inhomogeneous depends predominantly on the reliability of the data on $d\alpha^2/dz$. Since α drops sharply (from the order of 13·6 to the order of 8 km/s) across the mantle-core boundary N, there is a sizeable range of depth immediately below N, including all the region E_1, in which rays which are of P type above N cannot have their lowest points. The estimation of α inside this range therefore cannot be made directly from observations of the phases P and PKP. Since the value, α' say, of α at the top of the core exceeds that of β at the bottom of the mantle, the distribution of α in the region E_1 can, however, be derived by analysing the onset times of the phases S, SKS, $SKKS$, etc. Other methods involve subtracting travel times of the phase PcP from PKP at equal values of the ray parameter $p = dT/d\Delta$ (§§9.1.2, 9.1.4), or subtracting ScS from SKS. Data on all the phases mentioned were taken into account in deriving the 1939 values of α immediately below the mantle, but the precision with which some of the onset times involved can be measured is rather below that for the main P and PKP phases. It was therefore suggested in the 1969 paper that additional data on $d\alpha^2/dz$ inside E_1 should be examined with a view to testing further the suggestion that E_1 is slightly

[Refs. on p. 255]

inhomogeneous.

A revision by Randall (1970) yielded values of α exceeding those of Jeffreys inside E_1. The excess is 0·16 for α' and diminishes with increasing depth to 0·03 km/s at $z = 3600$ km. (Randall's α', 8·26 km/s, exceeded Gutenberg's by about 0·26 km/s.) Application of (11.36) to Randall's distribution gives $(dk/dp - \eta)$ increasing fairly steadily from 1·6 to 2·3 from 2900 to 3600 km depth and so supports the presence of mild homogeneity in E_1.

Below 3600 km, Randall's distribution gave $(dk/dp - \eta)$ as continuing to rise until a maximum of about 3·1 is reached at $z = 4300$ km, suggesting a possible continuation of mild inhomogeneity to this depth. Inside the range $4300 < z < 4800$ km, the distribution had $d\alpha^2/dz$ rapidly diminishing, the implied value of $(dk/dp - \eta)$ falling to about $-1·5$ at 4800 km; thus greater departures from homogeneity would be implied inside this last range of depth, with η possibly reaching the order of 4 to 5.

Unpublished work of Haddon (briefly outlined in Bullen, 1972) showed that if Randall's value of α' is assumed and if an appropriately adjusted simple variation of α is taken in place of Randall's more complex variation over the whole range $2900 < z < 4800$ km, then the discrepancy between the inferred distribution of $(dk/dp - \eta)$ and Birch's estimate of the distribution of dk/dp could be much reduced. However, if α' is taken to be 8·10 km/s or less, it is difficult to remove the discrepancy.

More recently still, Dziewonski and Gilbert (1972), taking account of free Earth oscillation evidence, arrived at two Earth models in which $\alpha' = 8·24$ and 8·27 km/s, in close agreement with Randall's estimate. (Later — see §16.4 — they gave 8·12 km/s.) A model of Jordan and Anderson (1974) gave $\alpha' = 8·02$ km/s; these authors also considered the region E_1 to be inhomogeneous.

Mention may be made of a different type of investigation by Buchbinder (1968). Using observational data from eight nuclear explosions and three natural earthquakes, Buchbinder inferred that the direction of initial motion of recorded *PcP* pulses is reversed near $\Delta = 32°$ and that the amplitudes of the phase *PcP* pass through a minimum near this Δ. On the usual theory of energy partitioning of seismic waves at a simple boundary (**B, Chapter 6**), this set of observations, if taken at face value, would entail that α' is appreciably less than 8·0 km/s and that the density is nearly continuous across the mantle-core boundary N. Buchbinder fitted his interpretation

[*Refs. on p. 255*]

in a series of models in which α' lies between 7·2 and 7·5 km/s and the ratio of the density change across N lies between 1·00 and 1·05. On the Model A procedures, these circumstances could actually be accommodated by including at the top of the outer core a thin layer inside which the main physical properties change very abnormally rapidly. But the entailed values of k and dk/dp inside such a layer deviate much too far from both the k-p hypothesis and Birch's finite-strain theory to be at all plausible. It seems probable that some other complexity in the vicinity of N, involving less variation in k and dk/dp will ultimately account for Buchbinder's observations (see §15.5). The fact that the observations have not yet been satisfactorily interpreted has led some commentators to attach some weight to the implication on α', at least to the extent of considering it rather unlikely that α' exceeds 8·1 km/s.

A summary of the present position is that the overall evidence favours the presence of mild inhomogeneity inside the region E_1 at least, and that Randall's investigation makes it necessary to keep open the possibility that the inhomogeneity may extend more than 700 km into the core. But it remains desirable to sharpen the precision with which α' and $d\alpha/dz$ are determined in the outermost 1000 km of the core.

12.3 Structure of the lower core

The largest revisions to the Jeffreys 1939 velocity distribution have been in the vicinity of the region F (§10.3.6). This region was originally delineated as lying between 4980 and 5120 km depth and was characterized by a strong negative velocity gradient which, by (11.36), would entail an abnormally large value of η, of order 30 units (§11.6.6). The 1939 distribution was not intended to be other than provisional but, apart from a small modification by Jeffreys himself (1942) in which the estimated inner-core radius, R_i say, was somewhat reduced, no significant revision was attempted till the 1960s.

Meanwhile a quantity of onsets on seismograms, arriving in advance of the times expected according to the J.B. tables for the phase *PKIKP*, had been detected. These onsets, which are referred to as *precursors* to *PKIKP* and are not allowed for in the 1939 velocity distribution, are described in §12.3.1. Various efforts to interpret the precursors are discussed in §12.3.2.

[*Refs. on p. 255*]

12.3.1 *Precursors to the phase PKIKP*

Corresponding to the J.B. tables and the Jeffreys velocity model, the earliest onsets on seismograms over the range of epicentral distances $100° < \Delta < 142°$ are associated with the phase *PKIKP*.

Precursors to these onsets for $118° < \Delta < 140°$ had been noticed by Gutenberg (1957) during scrutinies of numerous records of natural earthquakes. These precursors showed a wide observational scatter and had small amplitudes.

An analysis of seismic records of certain nuclear explosions (Bullen and Burke-Gaffney, 1958) gave unmistakable evidence of early arrivals of a *P* phase at epicentral distances Δ between $137°$ and $142°$. At Pretoria ($\Delta = 137°.4$), Kimberley ($139°.6$), and Tamanrasset ($140°.6$), the arrivals were 11, 9, and 6 seconds, respectively, earlier than *PKIKP*.

Since 1958, additional *PKIKP* precursor readings have been gathered by various investigators.

12.3.2 *Interpretations of the PKIKP recursors*

Gutenberg (1957) attributed the precursors to selective wave dispersion near the inner-core boundary, but this suggestion did not find general favour. Most investigators at that time preferred to interpret the precursors in terms of diffraction — diffraction near a caustic for precursors near $142°$, and an unspecified form of diffraction for those at shorter distances.

Soon afterwords, a series of attempts was made to obtain improved velocity models for the lower core in which the *PKIKP* precursors were principally interpreted in terms of ray theory, rather than diffraction.

Using travel-time data for some deep-focus earthquakes in the vicinity of Fiji, New Hebrides, and the Celebes Island, Nguyen Hai (1961, 1963) arrived at a core velocity distribution which deviated markedly from the 1939 distribution in an intermediate zone extending from nearly 4400 km depth to the boundary of the inner core, R_i being taken as 1236 km. The velocity α was taken continuous throughout the outer core and intermediate zone. Between 4380 and 4480 km depth, α fell from 9·91 to 9·86 km/s and then rose to 9·98 km/s. The velocity gradient was fairly steady from there to the inner core boundary except that $d\alpha/dz$ turned mildly negative over a short range of distance just above the inner core. At the inner

[*Refs. on p. 255*]

core boundary, α jumped from 10·27 to 10·80 km/s and then followed a Gutenberg-type inner-core distribution (§10.3.5), with $\alpha = 11\cdot24$ km/s at the centre.

Bolt (1962, 1964) interpreted the *PKIKP* precursors wholly in accordance with ray theory and introduced a new branch into the core travel-time curves which went roughly through the middle of the scattered Gutenberg readings (§ 12.3.1). He postulated a model four-layer core with separating boundaries at $z \approx 4560$, 4710, and 5160 km, and gave $R_i \approx 1210-1220$ km. It is convenient to refer temporarily (for calculations based on Bolt's velocity model) to these four layers as E', E'', F, and G (see Bullen, 1965). Bolt adhered to the 1939 velocities throughout E'; the gradient $d\alpha/dz$ was taken zero throughout each of E'', F, and G; α was taken continuous at the $E'-E''$ boundary, jumping from 10·03 to 10·31 km/s at the $E''-F$ boundary, and to 11·23 km/s at the $F-G$ boundary.

Adams and Randall (1963, 1964), like Nguyen Hai, terminated the outer core proper near 4400 km depth. Like Bolt, they gave two transition layers but with α discontinuous at each of the internal boundaries (at $z = 4390$, 4890, and 5115 km). The jumps in α at the three boundaries were from 9·87 to 10·00, 9·98 to 10·41, and 10·15 to 11·16 km/s. They gave $d\alpha/dz$ mildly negative both in the lower half of their upper transition layer and throughout their lower transition layer. Outside the transition layers they adhered to the 1939 distribution.

The distributions of Bolt, Nguyen Hai, and Adams and Randall between 4000 and 5600 km depth are exhibited in Fig. 12.1, along with the 1939 and the currently preferred distributions (see §16.4). The really important change from the 1939 model was the removal of the strong negative velocity gradient in the region F. The absence of strong negative velocity gradients anywhere in the core has been sustained in most later investigations. As will be seen, this change from 1939 had a marked effect on the estimation of densities below the outer core proper. For a useful summary on core velocity models around this time, see Engdahl (1968). Over the period 1964—70, Bolt's velocity model was, because of its comparative parametric simplicity, frequently assumed in estimating densities in the lower core (see later).

Haddon (1970) could not reconcile Bolt's velocities with certain free Earth oscillation periods (see later) and so sought an improved model, still adhering to ray theory with the precursors. He managed

[Refs. on p. 255]

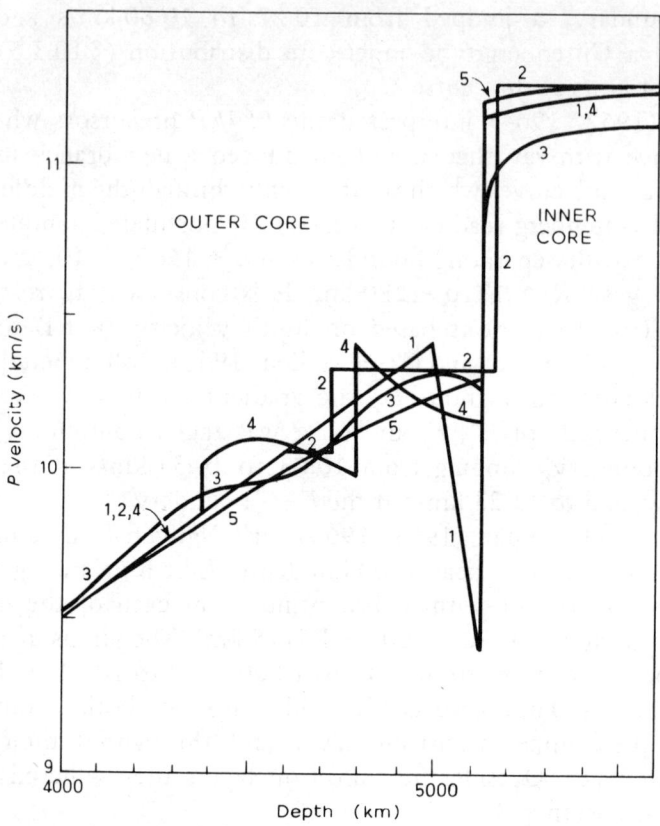

Fig. 12.1. Proposed P velocity distributions in the vicinity of the inner-core boundary: (1) Jeffreys (1939); (2) Bolt (1962); (3) Nguyen Hai (1963); (4) Adams and Randall (1964); (5) currently preferred distribution

to fit all the available bodily-wave travel-time and oscillation data by constructing a model with a five-layer core (see Bullen, 1972) in which a prominent feature was a low P velocity layer 200 km thick just outside the inner core. In this model, the outer core proper was terminated at 4300 km depth and R_i taken as 1216 km. To a slight extent, Haddon's model reverted towards the 1939 model but his transition zone was more extended and more complicated. Yanovskaya (1972) independently gave a preferred series of core α distributions remarkably similar to Haddon's and including a low-velocity layer like Haddon's just outside the inner core (she also gave a second series in which the main low-velocity layer occurred some 400 km higher than in Haddon's model). Yanovskaya's work showed clearly that it was not possible to infer, from the free Earth oscillation and bodily-wave travel-time data available and as interpreted at the time, a narrowly determined α distribution for $4300 < z < 5150$ km.

Work of Buchbinder (1971) and Sacks and Saa (1971) indicated that, in spite of success in other respects, models of the Haddon and Yanovskaya types could not be fully reconciled with amplitude data on bodily-wave core phases. Using amplitude and travel-time data, these investigators independently arrived at much simpler velocity models. Each of their preferred models had $R_i = 1216$ km (with uncertainty stated as not less than 10 km), and had α varying smoothly from the inner-core boundary to the mantle-core boundary, except for very slight discontinuities. The model of Sacks and Saa showed a single such discontinuity at $z \approx 4770$ km, with α jumping by about 0·05 km/s from the value 10·1 km/s on the upper side; in the 60 km just outside the core, $d\alpha/dz$ was abnormally small; at the inner-core boundary, α jumped from 10·4 to 11·2 km/s. Buchbinder's model showed two discontinuities above the inner core at $z = 4550$ and 4850 km, the velocities here being 9·99 and 10·15 and the velocity jumps 0·01 and 0·02 km/s, respectively; the jump at the inner-core boundary was from 10·26 to 10·84 km/s, and the central velocity was 11·28 km/s.

Haddon tested the models of Buchbinder, Sacks and Saa, and could not reconcile them with the overall data, including oscillation data. He was unable to secure a fully satisfactory core velocity distribution so long as simple ray theory was used in interpreting the precursor observations [see Bullen and Haddon (1973) for a summary of the background leading to this development]. Among other problems was how to account for an observed curvature in the trend of the (T, Δ) precursor curves. By apparent exhaustion of the possibilities, Haddon (1972) was led to infer that the precursors are likely to be associated, not with the presence of transition layers of the Bolt type, nor with any discontinuities inside the outer or inner core, but with scattering near the mantle-core boundary. Some details of Haddon's scattering theory will be presented in § 15.6. Haddon's theory is compatible with a simple two-layer core of the type originally envisaged by Lehmann. Independently of Haddon's theory, a number of other recent Earth models have only a single

[Refs. on p. 255]

discontinuity inside the core. On the other hand, some investigators still prefer models of the Buchbinder-Sacks type. For example, Bertrand and Clowes (1974) gave a preferred model in which α jumps discontinuously by 0·06, 0·09, 0·77 km/s at z = 4490, 4790, 5130 km, respectively.

A number of attempts have been made to estimate the sharpness of the inner-core boundary. For example, Engdahl, Flinn, and Romney (1970) used array-station data on the phase *PKiKP* (see §11.7) to infer that the boundary is fairly sharp. Their findings were compatible with $R_i \approx 1216$ km. On the other hand, Sacks used a different approach (see §12.6.2) to infer that the boundary is probably not sharp.

More recently, Gilbert, Dziewonski, and Brune (1973) inferred from free Earth oscillation and other data that R_i probably lies between about 1230 and 1250 km. (See also §§14.8, 16.4.) Engdahl, Flinn, and Massé (1974) subsequently inferred a value 1220−1230 km from *PcP* and *PKiKP* travel-time data.

12.4 Radius of the Earth's core

Gutenberg's 1914 estimate of the depth z_c of the mantle−core boundary N was 2900 km, the uncertainty being 50−100 km. Jeffreys (1939) gave z_c = 2898 ± 3 km, or R_c = 3473 ± 3 km, where R_c is the radius of the core. On various subsequent occasions, Gutenberg assumed values of z_c up to 2920 km.

Later investigations pointed to values of R_c somewhat greater than 3473 km. Kogan (1960), Carder (1964), and Buchbinder (1965) found values greater by 13, 30, and 10 km, respectively, to be compatible with their (independent) studies of *PcP* waves generated by nuclear explosions. Sacks (1966, 1967) made a spectral analysis of *P* wave onsets for $70° < \Delta < 165°$ and inferred that a rapid decrease in the amplitudes of short-period *P* waves which sets in somewhere between $\Delta = 93°$ and $96°$ is consistent with simple diffraction round N. In the absence of other complications, this would give R_c exceeding the 1939 value by about 60 km. In the light of other evidence this increase is too great, so that other complications must occur near N. Dorman, Ewing, and Alsop (1965), taking account of free Earth oscillation data available at the time, produced a model in which R_c = 3483 km; but this model has α and β falling sharply inside the lowest 30 km of the mantle to 10·3 and 5·4 km/s at N, and

[*Refs. on p. 255*]

so has the defect that it implies an unsatisfactory value ($< 4 \times 10^{11}$ N/m^2) of k at N.

Bullen and Haddon (1967a, 1967b), using a more comprehensive set of free Earth oscillation and bodily-wave data, inferred that $R_c = 3490 \pm 5$ km and constructed models which incorporated this result and avoided all obviously unsatisfactory features (see § 14.6.2). The value of R_c could be reduced by about $2x$ km if η were reduced by $0.1x$ from the value unity that had been taken for η inside the region D'; reducing η by 0.1 would, on the data in § 10.8.2, require a super-adiabatic temperature gradient of nearly 1 deg/km in the lower mantle. The value of R_c was also subject to some small additional uncertainty connected with uncertainties in the distributions of α, β, and ρ in the region D''. Press (1968) also applied free Earth oscillation data to infer that R_c is appreciably greater than 3473 km (see § 15.3.2).

Intensive efforts followed to resolve the discrepancy between the increased estimate of R_c and the earlier estimate, 3473 ± 3 km. In a reassessment of his original bodily-wave evidence, Jeffreys (see Haddon and Bullen, 1969) inferred an upper bound of 3480 km for R_c, and for a time the results of a variety of investigators seemed to indicate that the value could not be much greater. For example, Taggart and Engdahl (1968), using recent bodily-wave data, including nuclear explosion data on *PcP*, gave 3477 ± 2 km. Using array-station measurements, Johnson (1969) gave 3481 km. Using S and *ScS* data, Hales and Roberts (1970b), however, gave 3490 ± 5 and 3486 ± 5 km for two models. Engdahl and Johnson (1972) inferred from nuclear explosion data that the estimate 3477 km of Taggart and Engdahl needs to be increased by 5–15 km.

Jeffreys (1967) suggested that allowances for effects of damping on free Earth oscillation periods might resolve the discrepancy. Until these allowances have been reliably calculated (the problem is difficult), special caution is indeed needed in drawing fine inferences from the oscillation periods. At the same time, a tentative calculation led Haddon (see Haddon and Bullen, 1969) to infer that damping has less effect on the estimated R_c than other sources of uncertainty.

At one stage, Muirhead and Cleary (1969) appeared to have resolved the discrepancy by adopting Cleary's suggestion (§ 12.1.4) that $d\beta/dz$ becomes markedly negative inside the region D''. They showed that the Bullen-Haddon procedures would then permit R_c to be as small as 3478 km. Evidence of small lateral variations in α

[Refs. on p. 255]

(of order 1 per cent) in the lowermost few hundred kilometres of the mantle (see e.g. Julian and Sengupta, 1973) have, however, cast some doubt on Cleary's interpretations.

More recently, several investigations using extended free Earth oscillation data as well as bodily-wave data (see §14.8) have favoured values of R_c near 3486 km. A value of this order would now appear to be fairly widely favoured were it not for the remaining uncertainty over damping.

12.5 The Earth's central density

12.5.1 *Lower bound*

The estimate (§10.6) of a lower bound a little greater than 12 g/cm^3 for the central density ρ_0 still stands. This estimate ignores possible effects of changes in chemical composition or phase inside the core; such changes would increase the estimated ρ_0.

It is of some interest that a rough estimate of the minimum ρ_0 may be arrived at using

$$\rho = \sigma(1 + 2f)^{3/2} \qquad (12.1)$$

(§7.3.2), where σ here denotes density at zero pressure. Substituting $\sigma = 7\cdot88$ g/cm^3 (the value for iron at zero pressure) and $f = 0\cdot19$ (the minimum compression at the Earth's centre — §11.1) into (12.1) formally yields $\rho_0 = 12\cdot8$ g/cm^3. This gives an estimate of the minimum ρ_0 if the material at the centre of the Earth consists predominantly of iron or denser alloy. The presence of some nickel would slightly increase the estimate. Any phase changes in the material would increase the value required for f, and hence also ρ_0.

12.5.2 *Preliminary estimates*

The value of ρ_0 in the original Model A (1942) is 17·3 g/cm^3 (§10.7.1), but this value was quite arbitrarily taken. The first calculated value of ρ_0, nearly 18 g/cm^3 (the original Model B value, §§11.6.6 and 11.8.1), happened to be fairly close to the Model A value. [It was for this reason that Model A had been used till around 1960, in preference to Model A' (§10.7.1) which had the minimum ρ_0.] As mentioned in §§11.8 and 11.8.1, the calculation giving $\rho_0 \approx 18$ g/cm^3 (i) assumed the 1939 core distribution of α, and

(ii) assumed that $d\phi/dz \approx d\alpha^2/dz$ inside all regions of the core, where $\phi = \alpha^2 - 4\beta^2/3$. [The assumption (ii) permits the presence of solidity and therefore non-zero values of β inside regions of the core, but treats variation of β inside these regions as negligible.]

In a series of subsequent calculations (Bullen, 1962–7), the assumption (i) was first modified to take account of the removal (§ 12.3.2) of the strong negative P velocity gradient in the region F: $d\alpha^2/dz$ was now taken zero inside F instead of the original Jeffreys value of -0.15 km/s^2 (§ 11.6.6). By (11.36), this change reduced the estimate of η inside F from the order of 30 units to the order of dk/dp, i.e. to 3 or 4 units. By (11.35), this in turn reduced $d\rho/dz$ inside F by a factor of about 10 and hence permitted a reduction in the estimated ρ_0 to about 15 g/cm^3 (Bullen, 1962, 1964a). A second step was a revision of the assumption (ii) (see §§ 12.6, 12.7), enabling ρ_0 to be reduced to 13 g/cm^3 or less.

12.5.3 *Evidence from shock-wave data*

Independently of seismology, Birch (1961, 1963) sought to estimate ρ_0 from data on shock-wave experiments at pressures reaching the order of $(3-4) \times 10^6$ atmospheres. In ordinary seismic wave transmission at points not close to the source, the speed u of particles of the medium is small compared with the speed v of advance of the waves. With shock waves, however, u is comparable with v and, to a good approximation (see Birch, 1966),

$$p = \sigma u v, \qquad (12.2)$$

$$\rho = \sigma v/(v - u). \qquad (12.3)$$

Substituting into (12.2) and (12.3) values of u and v observed in shock-wave experiments on a material yields the *Hugoniot* or shock-wave equation of state, a relation between p and ρ for the material. In the Hugoniot conditions, the temperature rise is greater than in adiabatic conditions, and the problem of inferring the corresponding adiabatic or isothermal relation from the Hugoniot is difficult (see references in § 11.4.5).

Birch assembled a quantity of shock-wave data, including data of Al'tschuler and others (see Al'tschuler and Kormer, 1961) and McQueen and Marsh (1960), constructed curves giving the 'sound velocity' $\sqrt{(\partial p/\partial \rho)}$ in terms of ρ for a number of metals, and compared the results with curves showing the observed variation of α

in the core. He assumed the difference between the Hugoniot and adiabatic compression curves to be insignificant for the purposes of this comparison. The comparison, along with other evidence on the composition of the inner core, led Birch to infer that $\rho_0 \leqslant 13$ g/cm^3. Later questionings of the degree of reliability of geophysical inferences from the shock-wave observations (see §15.1.1) resulted in confirmation that ρ_0 probably does not exceed 13 by more than 0·5 g/cm^3, but a somewhat greater excess is not precluded.

12.5.4 *Other evidence*

Engdahl, Flinn, and Romney (1970) pointed out that their observations (§12.3.2) of the phase *PKiKP* set new constraints on ρ, k, and μ in the lower core. Bolt and Qamar (1970) used the data of Engdahl, Flinn, and Romney to infer that $\rho_0 \leqslant 14\cdot25$ g/cm^3, though the data seem rather meagre for that purpose. Thus, for the present, it is appropriate to rely principally on Birch's evidence. Bolt and Qamar also used amplitude data on the phases *PKiKP* and *PcP* to infer that $\rho_0 \geqslant 12\cdot7$ g/cm^3, but the limited data used seem again hardly sufficient to preclude values down to the previous reliably established minimum value of ρ_0 (§10.6). In any case, most recent work on Earth models independently yields values of about 12·5 g/cm^3 or more for ρ_0 (see later).

12.6 Further evidence bearing on rigidity in lower core

The theory on η (§§11.5–11.5.5) will now be applied to the pair of findings that dα/dz is nowhere markedly negative in the core (§12.3.2) and that ρ_0 does not much exceed 13 g/cm^3 (§12.5.3). The results substantially modify the original Model *A* and *B* distributions of ρ, k, p, etc. in the lower core and bear on the question of the inner core's rigidity (§11.7). Some additional evidence is given in §12.6.2.

12.6.1 *Connection between inner-core rigidity and central density*

As already mentioned, the assumptions (i) and (ii) of §12.5.2 permitted a value of about 15 g/cm^3 for ρ_0 when (i) was modified to remove the strong negative *P* velocity gradient that had previously been indicated in the region *F*. This was approximately the minimum permissible value of ρ_0 if the original Model B procedures were to be entirely followed. So it became desirable to see whether the procedures could be adapted to yield agreement with $\rho_0 \leqslant 13$ g/cm^3. It was found (Bullen, 1964a, 1964b) that this could be done by departing from the assumption (ii) that $d\phi/dz \approx d\alpha^2/dz$ in the lower core. In outline the argument is as follows.

Since $\rho_0 > 12$ g/cm^3, taking $\rho_0 \leqslant 13$ g/cm^3 would leave only a little room inside the lower core for any excess density gradient over that due to simple adiabatic compression, i.e. would not permit η to exceed unity by very much inside most of the lower core. Specific quantitative examination indicated that η would lie between 1 and 2 inside the inner core. The relation $\phi = \alpha^2 - 4\beta^2/3$ entails that $d\phi/dz = -4d\beta^2/3dz$ when $d\alpha/dz = 0$. This applies in, for example, Bolt's velocity model for the regions E'', F, and G; Bolt's model will be tentatively assumed in the present argument. Then (11.36) gives

$$d\beta^2/dz = \tfrac{3}{4}g(\eta - dk/dp). \qquad (12.4)$$

Substituting $\eta = O(1\text{-}2)$ and $dk/dp = O(3\text{-}4)$ in (12.4) gives $d\beta/dz < 0$. Thus the present assumptions on α and ρ_0 together entail that $d\beta/dz$ is significantly different from zero in the inner core, and so require a marked deviation from the assumption (ii).

When $d\beta/dz < 0$ throughout a range of depth, β itself must depart from zero inside that range; and the same applies to μ since $\mu = \rho\beta^2$. Thus if the results of Birch and Bolt could both be accepted, new evidence would be provided on the presence of solidity in the inner core. A substantial increase to Bolt's value of $d\alpha/dz$ in the inner core would, however, reduce the weight of this new evidence; such a possibility cannot be ruled out (but would not affect other evidence on the inner core solidity).

Again on the formal assumption $d\alpha/dz = 0$, an explicit expression for $d\mu/dz$ in the lower core may be derived from (12.4) using $\beta^2 = \mu/\rho$, $\phi = k/\rho$, and

$$d\rho/dz = \eta g\rho/\phi. \qquad (12.5)$$

The result is

$$\frac{d\mu}{dz} = \frac{\mu}{\rho}\frac{d\rho}{dz} + \frac{3}{4}g\rho\left(\eta - \frac{dk}{dp}\right) \qquad (12.6)$$

$$= g\rho\left\{\frac{\eta\mu}{k} + \frac{3}{4}\left(\eta - \frac{dk}{dp}\right)\right\}. \qquad (12.7)$$

[Refs. on p. 255]

A relation equivalent to (12.7) can alternatively be derived on writing

$$k + 4\mu/3 = \alpha^2 \rho, \qquad (12.8)$$

differentiating with respect to p, treating α as constant, and using (12.4). This gives

$$d\mu/dp = \eta\mu/k + \tfrac{3}{4}(\eta - dk/dp), \qquad (12.9)$$

where d/dp is interpreted as usual as $(dp/dz)^{-1}d/dz$.

The relations (12.7) and (12.9) show directly that the assumptions $d\alpha/dz = 0$, $\eta = O(1\text{-}2)$, and $dk/dp = O(3\text{-}4)$ entail the presence of rigidity in the inner core, for (12.7) and (12.9) are then incompatible with μ and $d\mu/dz$ both being zero. On the same assumptions, and assuming also that the rigidity does not have an improbably high value, it would follow further from (12.7) that $d\mu/dz$ is negative in the inner core. Thus the present assumptions would entail not only a change to solidity below the fluid outer core, but also a mild trend back towards fluidity with increasing depth inside the inner core. Some numerical detail is given in §12.7.

12.6.2 Additional evidence

Independent evidence bearing on the solidity of the inner core was supplied by Sacks (1971) in the course of a spectral analysis of data on P waves inside the core and their attenuation. He inferred that the value of the attenuation parameter Q (§7.4) is of order 200 in the outermost part of the inner core proper, rising rapidly to the order of 600 lower down. The contrast with the value (an order of magnitude higher) in the fluid outer core proper led him to suggest that the top of the inner core may be in a state of partial melting; this could be the prelude to the setting in of inner-core rigidity. On Sacks's findings, the inner-core boundary would probably not be sharply defined, in contrast to the conclusion (§12.3.2) of Engdahl and others. Sacks pointed out in this connection that, on his analysis, S waves in the inner core would probably be difficult to detect directly.

The prospect of such direct detection has been enhanced through the advent of modern seismic array-station and filtering techniques, and Julian, Davies, and Sheppard (1972) reported observations of the phase $PKJKP$ (see §11.7) through the use of these techniques. Their evidence was rather limited, and further checks are desirable (see later). Additional support for the solidity of the inner core came from recent analyses of free Earth oscillation data (see §14.9). These analyses implied an average value of about 3·5 km/s for β in the inner core as against 3·0 inferred by Julian, Davies, and Sheppard. It was suggested that the latter investigators might have observed *SKJKP* or *PKJKS*, not *PKJKP*. The foregoing results may be compared with the estimate (Bullen and Haddon, 1967c) that β ranges from about 3·9 to 2·9 km/s inside the inner core (see Table 12.4, §12.7.2).

12.6.3 *Note on scientific inference*

Most of the evidence in §§11.7, 12.6.1, and 12.6.2 on the solidity of the inner core is indirect. Questions have frequently been raised as to the (so-called) 'physical justification' of inferences drawn thus indirectly. Put that way, the questions tend to misrepresent the character of scientific inference. As pointed out in §6.1.1, all scientific inferences are inductive inferences drawn from observational evidence — in the present case, the seismic and other data mentioned. If some well-established set of physical results turns out to be relevant to a proposition, that is part of the evidence needing to be taken into account: such evidence may in fact sometimes markedly increase or decrease the probability (or degree of reliability) attachable to an inference, the probability being a function of the total evidence used (§6.1.1). But if no evidence of this type is available, that does not preclude an assessment of probability from being made, nor preclude the probability from being sometimes high. Notwithstanding the difficulties of providing a more direct physical 'justification', it seems appropriate to assess the probability that the inner core is solid as indeed quite high, on the total evidence now available.

12.7 Improved *B*-type models

12.7.1 *Some calculations relating to the lower core*

The seismic velocity revisions made in the lower core during 1962–4 (§12.3.2) were followed by recalculations of the distributions of ρ, k, and μ. Bolt's (1962) *P* velocity distribution is formally assumed for convenience in the following detail, which illustrates some of the

[*Refs.* on p. 255]

types of inference that could be drawn from lower core velocity distributions around that time. Throughout the whole core, dk/dp is assumed continuous, of order 3–4. Units are understood to be g/cm^3 for ρ and 10^{11} N/m^2 for k and μ. The subscript zero indicates values at the Earth's centre.

At the E'–E'' boundary (in the nomenclature of §12.3.2), Bolt's distribution has α continuous and $d\alpha/dz$ changing discontinuously from a normal value to zero. Since $\alpha^2 \rho = k + 4\mu/3$, continuity of α is strong (not quite conclusive) evidence of continuity of $\rho, k,$ and μ (see §10.1.4). It would then follow that the region E'' (4560–4710 km), like E', is probably fluid and, further, that $d\phi/dz = d\alpha^2/dz = 0$ in E''. Taking $dk/dp = 3$–4 would then, by (11.35, 36), give an increase of 0.3–0.4 in ρ inside E'', i.e. three to four times the contribution due to simple adiabatic compression. Thus E'' would be moderately inhomogeneous.

At the E''–F boundary, Bolt gave a small jump in α^2 (from 100·6 to 106·4 km^2/s^2). Literal application of the k-p hypothesis would then require the region F (4710–5160 km) to have some rigidity; but the jump in α^2 is here sufficiently small to be attributed, within permissible limits of deviation from the hypothesis, to a jump in k instead of μ, so that F could well be fluid. On the other hand, the size of the jump in α^2 at the F–G boundary (from 106·4 to 126·1 km^2/s^2) remains in keeping with the argument in §11.7 that the region G, the inner core proper, is significantly solid.

Further results of interest emerge from scrutiny of a series of lower-core models (Bullen, 1965) derived by applying (11.35, 36) and related theory on k and p to Bolt's distribution. [The numerical values appearing below have been adjusted to fit the revised value 0·3308 (§5.7) of the Earth's moment of inertia coefficient; the older value had been used in the 1965 paper.]

The smallest central density ρ_0 for the whole series of models was nearly 12·8. The excess over the absolute minimum was principally a consequence of the inferred inhomogeneity inside E''. The models with the smallest ρ_0 had a slight drop in k at the E''–F boundary. When k was made continuous, the smallest ρ_0 rose to 13·0 g/cm^3. The smallest k_0 was 14·6. For models with $\rho_0 \leqslant 13$ g/cm^3, the rigidity μ in the inner core ranged from at least 1·9 at the top to 1·1 at the bottom; higher values of μ were not precluded. For models with an assumed entirely fluid core, the minimum ρ_0 was 14·9; the model with this minimum had $k_0 = 18·8$, with k jumping discontinuously by 6 and 18 per cent at the E''–F and F–G boundaries, respectively. For a model in which k was assumed continuous and $d\beta/dz$ zero in the lower core, ρ_0 was a little greater than 15 g/cm³ and μ_0 of order 3.

An outstanding result was that Bolt's velocity distribution would demand a central density of nearly 15 g/cm³ if the core were wholly fluid. This shows forcefully that a wholly fluid core is seriously incompatible with Bolt's (and similar) α distributions and Birch's conclusion that $\rho_0 \leqslant 13$ g/cm³. Also, the rigidity in the inner core would, on the stated assumptions, be at least equal to the mean rigidity of the mantle, or half the rigidity near the base of the mantle.

12.7.2 The Model B_2

The results of §12.7.1 were extended (Bullen and Haddon, 1967c, 1968) to give a series of fourteen revised B-type models for the whole Earth. The original Model B procedures were followed, but the revised estimate of the moment of inertia coefficient y was incorporated. The seismic velocities used were (with certain inconsequential adjustments) the Jeffreys 1939 velocities as modified by Bolt in the lower core. Throughout the whole region consisting of the lower mantle and core, k and dk/dp were taken as continuous, with dk/dp of order 3–4 in the principal models. Inside the core, ρ was taken as continuous. (Any discontinuities in ρ in the core would raise the implied central densities ρ_0.)

The revision gave an important change in the upper mantle density distribution. Whereas in constructing the original model B it had been found necessary to extrapolate the density inside the region D' linearly upward to 80 km depth (§11.8.1), the reduced value of y permitted the new B-type models to have density distributions similar to those of A type in the upper mantle. Thus, as stated in §11.8.2, the revision of y removed an earlier principal difficulty with B-type models.

Two classes of revised B-type models were constructed, with ρ_0 around 15 and 13 g/cm³, respectively. So long as weight is given to Birch's inference that $\rho_0 \leqslant 13{\cdot}0$ g/cm³, the second class is to be preferred. The main characteristics of a representative model called B_2 are shown in Table 12.4. (The model here called B_2 differs slightly from the model B_2 of the 1967 paper in that the core radius is taken as 3485 km.) Its region B has a density distribution of

[Refs. on p. 255]

Table 12.4
Properties of Model B_2

Region	Depth (km)	ρ (g/cm^3)	p	k (10^{11} N/m^2)	μ	g (m/s^2)
---	0		0·000			9·822
A		(2·84)		(0·65)	(0·36)	
---	33	3·32	0·009	1·15	0·63	9·845
B						
---	245	3·51	0·080	1·45	0·77	9·911
C						
---	984	4·49	0·382	3·44	1·81	9·932
D'	2000	5·06	0·86	5·04	2·42	10·02
D''	2700	5·40	1·24	6·20	2·85	10·53
	2886	5·69	1·35	6·54	3·04	10·77
---	2886	9·95	1·35	6·54	0	10·77
E'	4000	11·39	2·48	10·34	0	7·95
E''	4560	11·87	2·94	11·98	0	6·33
---	4710	12·30	3·05	12·42	0	5·87
F	4710	12·30	3·05	12·42	0·53	5·87
	5160	12·74	3·34	13·60	0·00	4·35
---	5160	12·74	3·34	13·60	1·90	4·35
G	6371	13·03	3·68	15·04	1·11	0

Region	Depth (km)	α (km/s)	β	E (10^{11} N/m^2)	σ
---	0				
A		(6·30)	(3·55)	(0·91)	(0·267)
---	33	7·75	4·35	1·60	0·270
B					
---	245	8·40	4·67	1·96	0·277
C					
---	984	11·42	6·35	4·63	0·276
D'	2000	12·79	6·92	6·27	0·293
D''	2700	13·61	7·26	7·41	0·301
	2886	13·64	7·30	7·88	0·299
---	2886	8·12	0	0	0·5
E'	4000	9·53	0	0	0·5
E''	4560	10·05	0	0	0·5
---	4710	10·05	0	0	0·5
F	4710	10·33	2·07	1·54	0·48
	5160	10·33	0·00	0·00	0·50
---	5160	11·25	3·86	5·44	0·43
G	6371	11·25	2·91	3·25	0·46

[*Refs. on p. 255*]

Model A type with η taken equal to unity. No particular significance attaches to the value 245 km given for the depth of the B–C boundary: inspection of the fourteen revised models shows that this value is sensitive to quite minor changes in the assumptions made. Values of η in other regions are: C, order 1–2; D', 1; D'', 3; E', 1; E'', order 3–4; F, order 1–2; G, 1. The procedures formally determined values of the gradients $d\mu/dz$ and $d\beta/dz$ inside the regions F and G, but only minimum values (the values shown in the table) for μ and β themselves. The slight rigidity formally indicated inside F is probably devoid of physical significance (§12.7.1). Additional details on Model B_2, and details on other models showing the effects of various postulated deviations from the assumptions for B_2, are given in the paper cited.

The purpose of Table 12.4 is to exhibit a B-type model. (The presentation of this type of model was deferred from §11.8 to the present section because of the sizeable changes from the upper mantle and lower core of the original Model B.) The general principles followed in constructing B-type models continue to be strongly relevant to the problem of estimating the Earth's density distribution, but later observational data (Chapters 14–16) have demanded a number of changes from Model B_2. Furthermore, should, for example, Haddon's scattering theory of $PKIKP$ precursors (§15.6) be substantiated, the variations of properties required below 4500 km depth would be simpler than in Table 12.4.

On comparing Table 10.2 for Model A'' with Table 12.4 for B_2 it is seen that (apart from the lower core) the differences are small. To some extent this is because the two models have a number of assumptions in common, but it is pertinent to note that the differences are substantially less than those between the original Models A and B. As pointed out earlier, the improved agreement is principally due to the changes in the assumed value of y.

REFERENCES

Adams, R.D. and Randall, M.J. (1963). Observed triplication of PKP. *Nature* (Lond.), **200**, 744.

Adams, R.D. and Randall, M.J. (1964). The fine structure of the Earth's core. *Bull. Seismol. Soc. Amer.*, **54**, 1299–1313.

Al'tschuler, L.V. and Kormer, S.B. (1961). On the internal constitution of the Earth. *Bull. Acad. Sci. USSR, Geophys. Ser.* (English trans.), 18–21.

Anderson, D.L. (1967). Latest information from seismic observations. In *The Earth's Mantle*, 355–420, Academic Press, London.

Anderson, O.L. (1968). On the use of ultrasonic and shock-wave data to estimate compressions at extremely high pressures. *Phys. Earth Planet. Interiors*, **1**, 169–176.

Arnold, E.P. (1968). Smoothing travel-time tables. *Bull. Seismol. Soc. Amer.*, **58**, 1345–1351.

Arnold, E.P., Jeffreys, Sir H. and Shimshoni, M. (1963). S in three European earthquakes. *Geophys. J., R. Astr. Soc.*, **8**, 12–16.

Archimbeau, C.B., Flinn, E.A. and Lambert, D.G. (1967). Fine structure of the upper mantle. *J. Geophys. Res.*, **74**, 5825–5865.

Båth, M. (1974). *Spectral Analysis in Geophysics.* Elsevier, Amsterdam.

Bertrand, A.E.S. and Clowes, R.M. (1974). Seismic array evidence for a two-layer core transition zone. *Phys. Earth Planet. Interiors*, **8**, 251–268.

Birch, F. (1952). Elasticity and constitution of the Earth's interior. *J. Geophys. Res.*, **57**, 227–286.

Birch, F. (1961). Composition of the Earth's mantle. *Geophys. J., R. Astr. Soc.*, **4**, 295–311.

Birch, F. (1963). Some geophysical applications of high-pressure research. In *Solids under Pressure*, 137–162, McGraw-Hill, New York.

Birch, F. (1966). Compressibility; elastic constants. In *Handbook of Physical Constants* (ed. S.P. Clark, Jr.), 97–173, Geological Society of America, New York.

Bolt, B.A. (1962). Gutenberg's early *PKP* observations. *Nature* (Lond.), **196**, 121–124.

Bolt, B.A. (1964). The velocity of seismic waves near the Earth's center. *Bull. Seismol. Soc. Amer.*, **54**, 191–208.

Bolt, B.A., Niazi, M. and Somerville, M.R. (1970). Diffracted *ScS* and the shear velocity at the core boundary. *Geophys. J., R. Astr. Soc.*, **19**, 299–305.

Bolt, B.A. and Qamar, A. (1970). Upper bound to the density jump at the boundary of the Earth's inner core. *Nature* (Lond.), **228**, 148–150.

Brune, J.N. (1969). Seismic waves and crustal structure. In *The Earth's Crust and Upper Mantle*, 230–242, American Geophysical Union, Washington.

Buchbinder, G.G.R. (1965). *PcP* from the nuclear explosion Bilby, September 13, 1963. *Bull. Seismol. Soc. Amer.*, **55**, 441–461.

Buchbinder, G.G.R. (1968). Properties of the core-mantle boundary and observations of *PcP*. *J. Geophys. Res.*, **73**, 5901–5923.

Buchbinder, G.G.R. (1971). A velocity structure of the Earth's core. *Bull. Seismol. Soc. Amer.*, **61**, 429–456.

Bullen, K.E. (1948). The Bikini bomb and the seismology of the Pacific region. *Nature* (Lond.), **161**, 62.

Bullen, K.E. (1961). Seismic travel-times and velocity distributions. *Bur. Centr. Séism. Internat.*, A, **21**, 7–13.

Bullen, K.E. (1962). Earth's central density. *Nature* (Lond.), **196**, 973.

REFERENCES

Bullen, K.E. (1964a). Rigidity and density in the Earth's core. *Nature* (Lond.), **201**, 807.

Bullen, K.E. (1964b). New evidence on rigidity in the Earth's core. *Proc. Nat. Acad. Sci. USA*, **52**, 38–42.

Bullen, K.E. (1965). Models for the density and elasticity of the Earth's lower core. *Geophys. J., R. Astr. Soc.*, **9**, 233–252.

Bullen, K.E. (1969). Compressibility-pressure gradient and the constitution of the Earth's outer core. *Geophys. J., R. Astr. Soc.*, **18**, 73–79.

Bullen, K.E. (1970). Comparison of sources of evidence on the variation of incompressibility in the Earth's deeper interior. *Phys. Earth Planet. Interiors*, **3**, 36–40.

Bullen, K.E. (1972). Some problems connected with constructing Earth models. *Proc. Conference on Solid Earth Problems, Buenos Aires, 1970*, II, 63–77.

Bullen, K.E. and Burke-Gaffney, T.N. (1958). Diffracted seismic waves near the PKP caustic. *Geophys. J., R. Astr. Soc.*, **1**, 9–17.

Bullen, K.E. and Haddon, R.A. (1967a). Earth oscillations and the Earth's interior. *Nature* (Lond.), **213**, 574–576.

Bullen, K.E. and Haddon, R.A. (1967b). Derivation of an Earth model from free oscillation data. *Proc. Nat. Acad. Sci. USA*, **58**, 846–852.

Bullen, K.E. and Haddon, R.A. (1967c). Earth models based on compressibility theory. *Phys. Earth Planet. Interiors*, **1**, 1–13.

Bullen, K.E. and Haddon, R.A. (1968). Corrections to three Earth models. *Phys. Earth Planet. Interiors*, **1**, 401–402.

Bullen, K.E. and Haddon, R.A. (1973). Some recent work on Earth models, with special reference to core structure. *Geophys. J., R. Astr. Soc.*, **35**, 31–38.

Carder, D.S. and others (1962). The Gnome explosion. *Bull. Seismol. Soc. Amer.*, **52**, 977–1077.

Carder, D.S. (1964). Travel times from the central Pacific nuclear explosions and inferred mantle structure. *Bull. Seismol. Soc. Amer.*, **54**, 2271–2294.

Carder, D.S. (1965). Upper mantle from travel time studies. *U.S. Program for International Upper Mantle Project, Progress Report*, p. 9.

Cleary, J. (1969). The S velocity at the core-mantle boundary, from observations of diffracted S. *Bull. Seismol. Soc. Amer.*, **59**, 1399–1405.

Cleary, J., Porra, K. and Read, L. (1967). Diffracted S. *Nature* (Lond.), **216**, 905–906.

Dorman, J., Ewing, J. and Alsop, L.E. (1965). Oscillations of the Earth: new core-mantle boundary model based on low-order free vibrations. *Proc. Nat. Acad. Sci. USA*, **54**, 364–368.

Doyle, H.A. and Hales, A.L. (1967). An analysis of the travel times of S waves to North American stations, in the range $28°$ to $82°$. *Bull. Seismol. Soc. Amer.*, **57**, 761–771.

Dziewonski, A.M. and Gilbert, F. (1972). Observations of normal modes from 84 recordings of the Alaskan earthquake of 1964 March 28. *Geophys. J., R. Astr. Soc.,* **27**, 393–446.

Engdahl, E.R. (1968). Core phases and the Earth's core. Ph.D. thesis, Saint Louis University.

Engdahl, E.R., Flinn, E.A. and Romney, C. (1970). Seismic waves reflected from the Earth's inner core. *Nature* (Lond.), **228**, 852–853.

Engdahl, E.R. and Johnson, L.E. (1972). A new *PcP* data set from nuclear explosions on Amchitka Island. *Trans. Amer. Geophys. Un.,* **53**, 1045.

Gilbert, F., Dziewonski, A.M. and Brune, J. (1973). An informative solution to a seismological inverse problem. *Proc. Nat. Acad. Sci. USA,* **70**, 1410–1413.

Gutenberg, B. (1926). Untersuchungen zur Frage, bis zu welcher Tiefe die Erde kristallin ist. *Z. Geophys.,* **2**, 24–29.

Gutenberg, B. (1951). $PKKP$, $P'P'$, and the Earth's core. *Trans. Amer. Geophys. Un.,* **32**, 373–390.

Gutenberg, B. (1957). The 'boundary' of the Earth's inner core. *Trans. Amer. Geophys. Un.,* **38**, 750–753.

Gutenberg, B. and Richter, C.F. (1946). Seismic waves from atomic bomb tests. *Trans. Amer. Geophys. Un.,* **27**, 776.

Haddon, R.A. (1970). [See Bullen (1972).]

Haddon, R.A. (1972). Corrugations on the mantle-core boundary or transition layers between inner and outer cores? *Trans. Amer. Geophys. Un.,* **53**, 600.

Haddon, R.A. and Bullen, K.E. (1969). An Earth model incorporating free Earth oscillation data. *Phys. Earth Planet. Interiors,* **2**, 35–49.

Hales, A.L., Cleary, J. and Roberts, J.L. (1968). Velocity distributions in the lower mantle. *Bull. Seismol. Soc. Amer.,* **58**, 1975–1989.

Hales, A.L. and Roberts, J.L. (1970a). The travel times of S and SKS. *Bull. Seismol. Soc. Amer.,* **60**, 461–489.

Hales, A.L. and Roberts, J.L. (1970b). Shear velocities in the lower mantle and the radius of the core. *Bull. Seismol. Soc. Amer.,* **60**, 1427–1436.

Herrin, E. and 12 others (1968). Seismological tables for P phases. *Bull. Seismol. Soc. Amer.,* **58**, 1193–1241.

Huestis, S., Molnar, P. and Oliver, J. (1973). Regional Sn velocities and shear velocity in the upper mantle. *Bull. Seismol. Soc. Amer.,* **63**, 469–475.

Ibrahim, A.K. and Nuttli, O. (1967). Travel-time curves and upper mantle structure from long period S waves. *Bull. Seismol. Soc. Amer.,* **57**, 1063–1092.

Jeffreys, H. (1942). The deep earthquake of 1934 June 29. *Mon. Not. R. Astr. Soc., Geophys. Suppl.,* **5**, 33–36.

Jeffreys, H. (1947). Seismic waves in western and central Europe. *Mon. Not. R. Astr. Soc., Geophys. Suppl.,* **5**, 105–119.

Jeffreys, Sir H. (1948). *Theory of Probability* (2nd Ed.), Cambridge University Press.

Jeffreys, H. (1952). The times of P up to $30°$. *Mon. Not. R. Astr. Soc., Geophys. Suppl.,* **6**, 348–364.

REFERENCES

Jeffreys, Sir H. (1966). Revision of travel times. *Geophys. J., R. Astr. Soc.*, **11**, 5–12.

Jeffreys, Sir H. (1967). Radius of the Earth's core. *Nature* (Lond.), **215**, 1365–1366.

Johnson, L.R. (1969). Array measurements of P velocities in the lower mantle. *Bull. Seismol. Soc. Amer.*, **59**, 973–1008.

Jordan, T.H. and Anderson, D.L. (1974). Earth structure from free oscillations and travel times. *Geophys. J., R. Astr. Soc.*, **36**, 411–459.

Julian, B.R., Davies, D. and Sheppard, R.M. (1972). PKJKP. *Nature* (Lond.), **235**, 317–318.

Julian, B.R. and Sengupta, M.K. (1973). Seismic travel time evidence for lateral inhomogeneity in the deep mantle. *Nature (Lond.)*, **242**, 443–447.

Kogan, S.D. (1960). Travel times of longitudinal and transverse waves, calculated from data on nuclear explosions made in the Marshall Islands. *Bull. Acad. Sci. USSR, Geophys. Ser.* (English Transl.), No. 3, 10pp.

Kovach, R.L. and Robinson, R. (1969). Upper mantle structure in the basin and range province, western North America, from the apparent velocities of S waves. *Bull. Seismol. Soc. Amer.*, **59**, 1653–1665.

Lehmann, Inge (1959). Velocities of longitudinal waves in the upper part of the Earth's mantle. *Ann. Géophys.*, **15**, 93–118.

Lehmann, Inge (1961). S and the structure of the upper mantle. *Geophys. J., R. Astr. Soc.*, **4**, 124–138.

Lehmann, Inge (1967). Low-velocity layers. In *The Earth's Mantle*, 41–61, Academic Press, London.

Lehmann, Inge (1970). The 440-km discontinuity. *Geophys. J., R. Astr. Soc.*, **21**, 259–372.

McQueen, R.G. and Marsh, S.P. (1960). Equation of state for nineteen metallic elements from shock-wave measurements to two megabars. *J. App. Phys.*, **31**, 1253–1269.

Molnar, P. and Oliver, J. (1969). Lateral variations of attenuation in the upper mantle and discontinuities in the lithosphere. *J. Geophys. Res.*, **74**, 2648–2682.

Muirhead, K.J. and Cleary, J. (1969). Free oscillations of the Earth and the D'' layer. *Nature* (Lond.), **223**, 1146.

Nguyen Hai (1961). Propagation des ondes longitudinales dans le noyau terrestre d'après les séismes profunds des îles Fidji. *Ann. Géophys.*, **17**, 60–66.

Nguyen Hai (1963). Propagation des ondes longitudinales dans le noyau terrestre. *Ann. Géophys.*, **19**, 285–346.

Niazi, M. and Anderson, D.L. (1965). Upper mantle structure of western North America from apparent velocities of P waves. *J. Geophys. Res.*, **70**, 4633–4640.

Nishimura, E., Kishimoto, Y. and Kamitsuki, A. (1958). On the nature of the $20°$ discontinuity in the Earth's mantle. *Tellus*, **10**, 137–144.

Nuttli, O. (1969). Travel times and amplitudes of S waves from nuclear explosions in Nevada. *Bull. Seismol. Soc. Amer.*, **59**, 385–398.

Press, F. (1968). Earth models obtained by Monte Carlo inversion. *J. Geophys. Res.*, **73**, 5223–5234.

Randall, M.J. (1970). SKS and seismic velocities in the outer core. *Geophys. J., R. Astr. Soc.*, **21**, 441–445.

Randall, M.J. (1971). A revised travel time for S. *Geophys. J., R. Astr. Soc.*, **22**, 229–234.

Ritsema, A.R. (1969). Seismology and upper mantle investigations. In *The Earth's Crust and Upper Mantle* (ed. P.J. Hart), 110–115, American Geophysical Union, Washington.

Robinson, E.A. (1967). *Statistical Communication and Detection*, 362 pp., Griffin, London.

Sacks, I.S. (1966). Diffracted wave studies of the Earth's core: 1, amplitudes, core size, and rigidity. *J. Geophys. Res.*, **71**, 1173–1181.

Sacks, I.S. (1967). Diffracted P-wave studies of the Earth's core: 2, lower mantle velocity. *J. Geophys. Res.*, **72**, 2989–2994.

Sacks, I.S. (1971). Anelasticity of the inner core. *Carnegie Inst. of Washington Year Book, 1971*, 416–419.

Sacks, I.S. and Saa, G. (1971). The structure of the transition zone between the inner core and the outer core. *Carnegie Inst. of Washington Year Book, 1971*, 419–426.

Taggart, J.N. and Engdahl, E.R. (1968). Estimation of PcP travel times and the depth to the core. *Bull. Seismol. Soc. Amer.*, **58**, 1293–1303.

Toksöz, M.N. and Wiggins, R.A. (1969). Seismic arrays and the structure of the Earth's interior. *Nerem Record*, 170–171.

Yanovskaya, T.B. (1972). Determination of a set of velocity-depth curves for the transition zone in the Earth's core from travel times of successive arrivals of PKP waves. *Geophys. J., R. Astr. Soc.*, **29**, 227–235.

CHAPTER 13

Evidence from seismic surface waves

So far, the observational seismological evidence brought to bear in estimating the density ρ, incompressibility k, rigidity μ, etc. below the Earth's surface has been mainly confined to bodily waves. The present chapter is concerned with evidence from surface waves.

The basic theory of the transmission of seismic surface waves has been set down in §8.5 for the simple cases of Rayleigh and Love waves, including the derivation of the equations (8.34) and (8.42) for the wave or 'phase' velocity in the two cases, The theory shows that Love waves are dispersed (§§8.5.3–8.5.4), as are all observed seismic surface waves, whether of SH or P-SV type. Stoneley (1925) drew attention to the importance of distinguishing between the phase velocities v and the group velocities V in practical observations. For an approximately sinusoidal wave group, κ will denote the wave number, λ the wave length, and τ the period. Thus $\lambda = 2\pi/\kappa$ and

$$\tau = 2\pi/\kappa v; \tag{13.1}$$

v and V are functions of κ, and hence of τ and λ.

For surface-wave observations to be useful in investigating properties below the Earth's surface, the wave motion must not die off too rapidly down to the depths of interest. Scrutiny of the wave forms involved, e.g. the form $C \exp\{\iota\kappa(sx_3 + x_1 - vt)\}$ in (8.37), shows that the rate of diminution of the wave amplitudes with depth decreases as κ decreases. Hence the greater the wave length $2\pi/\kappa$ and period $2\pi/\kappa v$, the greater is the depth to where the properties of the medium significantly affect the surface wave velocities, i.e. the greater is the extent of the effective wave guide below the surface; hence also the greater is the depth to which useful information bearing on ρ, k, and μ can be supplied. A rough working rule is that observations of surface waves of length λ can supply useful information on the structure down to a depth of about $\frac{1}{3}\lambda$.

[Refs. on p. 283]

Until fairly recent times, the available instruments did not record seismic waves with periods much exceeding a minute. With v of the order of 4 km/s, this meant that the observations could give information only on the crust and its near vicinity. With the much longer periods that are now reliably recorded (see §8.5.5), information can now be provided on subsurface structure down to several hundred kilometres depth.

As will be seen, a principal section of the information is in the form of equations of condition which ρ and other unknowns have to satisfy. It is supplementary to other information and not sufficient in itself to determine the structure uniquely. It supplements evidence from bodily waves in two important respects: first, the equations of condition are sometimes determined to better precision than bodily-wave data; secondly, whereas the bodily-wave evidence is confined to yielding values of k/ρ and μ/ρ (see Chapters 9 and 10), the surface-wave evidence yields equations involving ρ, k, and μ in other combinations. In these respects, the surface-wave evidence contributes to the primary task of arriving at the most suitable spherically symmetrical representation of the Earth's internal density distribution.

Surface waves that are predominantly influenced by, and which throw light on, subcrustal structure have been called *mantle waves*. The mantle waves are of course also influenced to some degree by the crustal structure.

Because they are limited to the outer part of the Earth, observed seismic surface waves show marked dependence on geographical region. The crustal structures in different geographical regions are substantially different, and the differences are now known to extend well below the crust. Most earlier surface-wave studies were crustal and limited to what were assumed to be 'pure paths', i.e. paths for which a fairly uniform subsurface structure was assumed to be sustained throughout. Later studies, especially studies of mantle structure, have attempted to interpret observations over 'mixed paths' which are often longer and include two or more pure-path segments. To a first approximation, the paths may be classified into continental-shield type and ocean-basin type but, as will be seen, this does not meet all requirements.

In the overall density investigation, sorting out the geographical regional variations is a secondary task in which surface-wave evidence is particularly relevant. Efforts have been made to infer results for

[*Refs. on p. 283*]

pure paths from observations for mixed paths. Model structures are postulated for regions that appear to be not too mixed and are tested against the observations in the usual way with a view to narrowing the range of plausible structures. The regional results derived in this way contribute towards determining the full three-dimensional density distribution in the outer part of the Earth, but there is still a long way to go (see §16.1.2).

The present chapter gives an indication of the contributions so far made by studies of surface-wave observations to both the primary and secondary density tasks.

Except where otherwise stated, the model structures to be considered will contain plane horizontal boundaries and properties will vary only with the depth z (or x_3).

13.1 Underlying principles in applying surface-wave data

The present section indicates in a general way the type of procedure used in deriving information about the Earth from seismic surface-wave observations. The illustration to be given for the simple case of Love waves serves as a guide to more complicated procedures considered later.

13.1.1 *Construction of dispersion curves*

As a result of dispersion, a typical single seismogram taken at a sufficient distance from an earthquake source shows surface waves as a series of approximately sinoidal wave groups (see §8.5.4) with (in general) more or less gradually changing periods. For a particular group, which will have travelled with the group velocity $V(\kappa)$ from the source (see B, §5.3.2), the period τ can be read from the seismogram.

The origin time and location of the source being otherwise determined, the observations thus yield a numerical tabular relation between τ and V. Let this be represented by

$$\phi_1(\tau, V) = 0, \qquad (13.2)$$

where ϕ_1 is regarded as a known function.

The group-velocity equation (8.47), namely,

$$V = d(\kappa v)/d\kappa, \qquad (13.3)$$

and (13.1) then give
$$\phi_1(2\pi/\kappa v, d(\kappa v)/d\kappa) = 0, \tag{13.4}$$
and hence
$$\phi_2(\kappa, v, A) = 0, \tag{13.5}$$
where A is an integration constant and ϕ_2 may also be regarded as a known function.

Dispersion curves (§8.5.6) connect pairs of variables such as v, V, κ, and τ. The observational relation (13.2) gives a dispersion curve connecting V with τ or κv. If A is known, (13.5) gives a dispersion curve connecting v with κ, and (13.1), (13.2), and (13.5) then enable curves to be drawn connecting V and κ, or v and τ, etc. The curves derived in this way through (13.2) are observational dispersion curves.

As stated in §8.5.6, theoretical dispersion curves can be derived for specified model structures representing the outer part of the Earth. The process of derivation for a particular case is described in §13.1.2.

By comparing corresponding observational and model dispersion curves (or using equivalent procedures) the suitabilities of particular model representations can be tested, and in this way light thrown on a subsurface structure through which the observed waves have passed.

13.1.2 *Illustration for case of simple Love waves*

Essential features in applying dispersion curves to the problem of subsurface structure are well illustrated in the case of simple Love waves. For this case (§8.5.3), a model structure is postulated which consists of a uniform outer layer L' of constant thickness H' lying in welded contact on a uniform semi-infinite medium L, the boundaries of L and L' being plane. Consideration is restricted to the displacement component u_2 and corresponding SH waves. Unprimed symbols relate to properties of L — the density ρ, rigidity u, and S velocity β; primed symbols relate to L'.

The phase-velocity equation (8.42) has the form
$$\phi_3(\kappa H', v/\beta', \beta/\beta', \rho/\rho') = 0, \tag{13.6}$$
since $\mu = \rho\beta^2$. The form of the function ϕ_3 is definite since the character (in this case the Love character) of the structure has been specified. If numerical values of H', β, β', and ρ/ρ' are specified (as additionally assigned features of the model), (13.6) gives a theoretical dispersion curve connecting κ and v. Useful information about the

[*Refs. on p. 283*]

structure may be derived by comparing this curve directly with the observational curve obtained through (13.5).

13.1.2.1

The equations through which the model and the observational dispersion curves are derived can be brought to bear in a variety of ways.

Suppose now that not all of H', β, β', and ρ/ρ' are assigned. The variables κ and v can be eliminated from (13.1), (13.5), and (13.6), resulting in a relation of the form

$$\phi_4(\tau, H', \beta, \beta', \rho/\rho', A) = 0, \qquad (13.7)$$

where ϕ_4 is also a definite function. The observations embodied in (13.2) can thus be used to deliver a set of equations of condition (13.7) on H', β, β', ρ/ρ', and A, one equation for each observed τ. If the only sources of discrepancy between members of the set of equations were observational errors, it would be appropriate to derive by least squares a single equation connecting the stated quantities. An important single equation of condition governing H', β, β', and ρ/ρ' alone would thus be provided if separate means should be available for estimating A.

By a slightly modified procedure, it is possible to avoid introducing A. In early crustal investigations this was done in the following way. Let a set of numerical values of (β, β', ρ/ρ') based on bodily-wave and other evidence be postulated. Then (13.6) reduces to the form

$$\phi_5(\kappa H', v) = 0. \qquad (13.8)$$

Numerical values of $d(\kappa H'v)/d(\kappa H')$, and therefore of $d(\kappa v)/d\kappa$, can be derived from (13.8) in terms of values of v. Then, using (13.4), v and κ can be directly connected, and a set of values of H' derived using (13.8) again, each value corresponding to a particular observational value of the period $2\pi/\kappa v$. (In B, §5.3.2, the procedure is described in more concrete terms.) If the discordances among the derived values of H' turn out to be small, the mean gives a useful estimate of H'.

The simple procedures described above ignore various complications. For example, there may be departures from one-one correspondence between τ and V in (13.2); this will be the case if the observations are not restricted to relate to a single branch of the

[Refs. on p. 283]

tangent function in (8.42). In practice, some of the observations may relate to different branches and there may be a problem of deciding which branch a pair of observations of (τ, V) is associated with. This involves consideration of normal modes (§13.1.3).

The equations (13.6), (13.7) and (13.8) are of course reliable only to the extent that the structure being investigated turns out to have been reliably represented in terms of a Love-type model. Departures from the postulated model contribute to discrepancies between members of (13.7) [or between the derived values of H' when the method involving (13.8) is used], additionally to the observational errors. Thus the serviceability of the simple Love model is related to the degree of concordance between members of (13.7). Substantial discordance indicates that the Love model is seriously inadequate. The equations (13.7) can then still have some limited value, e.g. as a contribution to making broad comparisons between different geographical regions. Also, details of the discordance, coupled with an inspection of the theoretical and observational dispersion curves, may point the way towards improving the postulated type of structure.

Thus, altogether, analyses based on the simple Love model have the potentiality of supplying information, which may be general or detailed, on the geometrical character as well as the physical properties of the structure being investigated. The Love model gave useful first approximations in early stages of crustal investigation when information was more limited than now. However, the fullest use of modern surface-wave observations requires extended theory discussed in §13.2 for more complicated models.

13.1.3 *Modes of surface waves*

As a group of surface waves advances, the vibrations of particles of the medium will in general be the resultant of a series of normal modes of vibration, including a fundamental mode and overtones. The case of simple Love waves will be used again to provide a ready illustration of essential features of the normal modes.

It is convenient for this purpose to write the Love wave equation (8.41) as

$$\kappa s' H' = \tan^{-1}(\mu s / \mu' \iota s'), \qquad (13.9)$$

where $s^2 = (v/\beta)^2 - 1$ and $s'^2 = (v/\beta')^2 - 1$. The fundamental mode is associated with the principal branch of the tangent function in

[*Refs. on p. 283*]

(13.9); and the higher modes, in order, with successive branches. The dispersion details are in general different for the different branches.

For the fundamental and successive modes, (13.9) shows that the wave lengths λ lie in the ranges $\lambda > \lambda_1$, $\lambda_1 > \lambda > \lambda_3$, $\lambda_3 > \lambda > \lambda_5$, ... , respectively, where $(\lambda/s')_{2n+1} = 4H'/(2n + 1)$. Inspection of the first cosine factor in (8.43) shows that the modes have 0, 1, 2, ... (horizontal) nodal planes, respectively, inside the layer L'. (Nodal surfaces are defined in §3.6.)

Analogous series of surface-wave modes, likewise involving nodal surfaces, arise with structures less simple than the Love model. There is a close relation with modes of Earth oscillation in general (see e.g. §13.4 and Chapter 14). The problem of ensuring that observations of (τ, V) are associated with the correct modes is assisted by energy considerations. It is usually the case that much of the energy in surface waves goes into the fundamental mode. But difficulties sometimes arise in identifying particular modes and contribute to uncertainties in inferences from the observational data. The uncertainties are minimized in practice by bringing to bear trial-and-error experience.

Various processes of separating higher modes from the fundamental in complicated seismic surface-wave records, and of applying the results to determine features of the Earth's structure, are illustrated in papers of Oliver and Ewing (1957), Kovach and Anderson (1963), Crampin and Båth (1965), Kovach (1965), and Knopoff, Schwab, and Kausel (1973).

13.1.4 *Stationary group velocities*

The group velocity V, considered as a function of κ, may pass through a stationary value in circumstances that frequently occur in the observational context. [The asymptotic approximation (8.44) then has to be supplanted – see Jeffreys (1925) and B, §3.3.5.4.] The amplitudes of a group of waves for which V lies close to a stationary value are in general large compared with those of other groups, the relative size increasing with distance from the source.

This property featured in an early endeavour by Stoneley (1937) to provide evidence from surface waves on the 20° discontinuity (§10.3.2). He noted that, because of the sizeable depth involved, only the largest of the surface waves affected by the discontinuity would, with the recording instruments then available, be likely to

[*Refs. on p. 283*]

be observable. He assumed a simple Love-type structure with an outer layer thickness of order 400 km and, taking plausible values of ρ, α, and β, calculated theoretical minimum group velocities and corresponding periods of SH and P-SV waves. The calculation is interesting as the first to concern itself with what were later called mantle waves (see Ewing and Press, 1954).

Further detail on stationary group velocities is given in a review paper on surface waves by Ewing and Press (1956).

13.2 More complex model structures

More complex model structures include extensions to cases of two and more uniform layers L', L'', ... , separated by horizontal boundaries and resting on a uniform substratum L of unlimited depth, the layers being in welded contact when solid. Let unprimed, once primed, twice primed, ... symbols relate to properties of $L, L', L'', ...$, respectively. For either SH or P-SV waves, analysis extended from that in §§ 8.5.2–8.5.5 yields again a phase-velocity equation which is an extension of equations like (13.6) and (13.9). This equation involves the phase velocity v, the wave number κ, the thicknesses H', H'', ... , the densities ρ, ρ', ρ'', ... , the S velocities β, β', β'', ... , and (for P-SV waves) the P velocities α, α', α'', ... (For some examples, see B, § 5.5.)

The essential procedures described in § 13.1.1 may be applied to the new phase-velocity equation, resulting again in a set of equations, analogous to (13.7), from which a single equation may be derived by least squares, connecting all the unknowns (now more numerous than in the simple Love case). Again the yielded equation falls well short of determining the structure uniquely and has to be regarded as auxiliary or complementary to other information.

Other extensions include such modifications as allowance for variation of density inside a layer [see e.g. Jeffreys (1928), Stoneley (1934)]. Allowance for velocity variations inside a layer may be conveniently made using the form $v = ar^b$ (see Bullen 1945, and B, §§ 7.3.1, 7.5.4). Ewing, Jardetzky, and Press (1957) have collected together details for a large variety of cases with a limited number of surface layers.

With normal desk computing facilities, it proved practicable to work out the main characteristics of surface-wave transmission for models with up to three uniform surface layers. Various mathematical

[Refs. on p. 283]

devices, for example Rayleigh's principle, were brought to bear, especially by Jeffreys (1935), to assist in the task. The early results were applied principally to providing information on the structure and properties of the Earth's crust and properties immediately below.

With the advent of electronic computers, it became practicable to analyse models with an indefinitely large number of layers, covering a range of depth extending well down into the mantle. The capacity of surface-wave observations to provide information on mantle structure then became greatly increased, notably at the hands of Ewing, Oliver, Press, and Dorman.

As with many problems treated by electronic computing methods, the most effective procedure is often the reverse of that outlined in §13.1.2.1. Instead of starting from a set of observational data and applying equations of the type (13.7) directly to derive equations of condition on H, ρ, α, β, etc., it is now usual to start by specifying a model in full numerical detail, including the numerical distributions of ρ, α, and β. For a specified model, a relation of the type (13.6) thus reduces to being a relation between κ and v alone (or between v and τ alone, etc.). For a complex structure, a computer has to be used in deriving this relation, which is usually in the form of a table. Through the group-velocity relation (13.3), V and τ may also be connected. In this way a dispersion table or curve is derived for the model. Various mathematical devices, e.g. a matrix iteration method due to Thomson (1950) and adapted by Haskell (1953), may be brought to bear in arriving at the table. (See also Alterman, Jarosch and Pekeris, 1961.)

The procedure is usually repeated for each of a large set of specified models. A feature of electronic computing is the ease with which numerical tables can be derived for a large set of models, once the programme for a first model has been worked out. From the set of model dispersion curves thus derived, a selection is made of those that go nearest to fitting the observational curve [corresponding e.g. to (13.2)]. In this way a limited set of preferred models is arrived at. As already indicated, the process does not usually yield anything like a uniquely determined model, but it can narrow the issue by ruling out large classes of trial models (see §15.4). To the extent that the surviving models may have closely common features, a measure of constraint is thus placed on the allowable distributions of density etc. over the range of depth involved.

[Refs. on p. 283]

13.3 Direct observation of phase velocities

In §13.1, it was shown how inferences on the Earth's outer structure can be drawn from measurements on seismograms (a single seismogram may suffice) which give directly the group velocities V of trains of surface waves of periods τ. The corresponding phase velocities v may be derived through (13.3), but with the complication that an initially unknown constant A is involved. As seen in §13.1.2.1, a modified procedure may avoid the necessity of introducing A. But there are sometimes advantages in employing techniques in which v is observationally estimated instead of V. These techniques, in addition to by-passing A, have an advantage in that the observed v relate to local structures below limited geographical areas where the observations are taken, whereas the observed V are generally average values over the whole (usually considerable) distance from the source. Various techniques, starting from work of Valle (1949), have been developed to estimate v more or less directly. In contrast to the case for V, direct estimation of v requires seismograms to be read at more than a single station.

A technique applied by Press (1956) uses records of surface waves traced on seismograms of identical type at separate members of an array station. From three or more of the seismograms, the direction of advance of the waves can be inferred (see e.g. Bullen, 1954, p. 122). A set of crests (or troughs or other specified points), one crest on each seismogram and all arising from the same individual ground wave, is then identified and the period τ noted. The phase velocity v of that wave is determined from measurements of the differences between the arrival times of the crest (or other point) at the different member stations and knowledge of the station locations. A series of such determinations taking sets of crests for different ground waves gives a series of values of v in terms of τ. Members of a particular set of crests can usually be readily identified when the stations are fairly close. Supplementary techniques (Brune, Nafe, and Oliver, 1969) assist in identifying the crests when the stations are well separated. The determinations relate to the average structure below the area encompassed by the stations.

Brune and Dorman (1963) introduced a two-station technique in which phase velocities are measured at pairs of stations for earthquakes with epicentres lying close to the great circles through the station pairs. They used the technique to infer details of the mantle structure below the Canadian shield. Knopoff, Mueller, and Pilant (1966) applied an extension of the technique in investigating structure below the Alps. Knopoff (1972) stated that the two-station is superior to the three-station technique in that it minimizes errors arising from phase shifts in the presence of significant lateral inhomogeneity.

Satô (1958) and others — see e.g. Press (1966) — evolved techniques applicable to surface waves which are sufficiently unattenuated (and therefore sufficiently long and strong) to be satisfactorily recorded at angular distances exceeding 2π from an earthquake epicentre. On a single seismogram, a train of such waves is measured during successive passages through the site of the recording station at intervals corresponding to travel on great-circle paths right round the Earth. Let Δ be the angular distance of the station from the epicentre. Then the train is observed at distances $(\Delta + 2n\pi)$, where $n = 0, 1, 2, \ldots$. [A companion train is observed at distances $2(n + 1)\pi - \Delta$.] The observed periods of these mantle waves may reach several hundred seconds. In applying the techniques, allowance has to be made for a phase shift of $\frac{1}{2}\pi$ at each passage of the train through the anticentre and source (Brune, Nafe, and Alsop, 1961; see also §13.4). (The *anticentre* is diametrically opposite the epicentre.)

Other techniques that have been contrived depend on the Fourier analysis of seismograms and cross-spectral analysis. (See e.g. Toksök and Ben-Menahem, 1963.)

13.4 Allowance for Earth's curvature and gravity

With increasing periods and wave lengths, seismic surface waves become increasingly affected by the Earth's curvature and gravity [see e.g. the terms in U in (8.4,5), and also §8.3.3]. When the periods exceed the order of a minute, these effects may need to be taken into account. (The term *sphericity* is sometimes used for curvature in the present context, the Earth models considered being usually spherically symmetrical; some geophysicists love coining new terms.)

The first-approximation theory of free Earth oscillations also relates to spherically symmetrical Earth models and takes account of gravity terms. Results from that theory can be used as follows to supply corrections to results on surface waves derived for 'flat'

[*Refs. on p. 283*]

(horizontally layered) Earth models; [The theory of free Earth oscillations is treated in Chapter 14. Here, only certain broad considerations are needed; the corrections can of course be derived through pure surface-wave theory — see e.g. Satô (1959) — but not quite so readily.]

Let (r, ϕ, λ) be spherical polar coordinates (defined as in Chapter 3) of a point P inside a spherically symmetrical Earth model of radius a, taken so that $\phi = \tfrac{1}{2}\pi$ at a source generating free Earth oscillations. For a particular mode of oscillation let u be a component of displacement at P. For present purposes it is sufficient to represent u by the form $R(r)F(\phi, t)$, where [see e.g. the expression for u in the first equation of (14.6); see also Alterman, Jarosch, and Pekeris (1961)]

$$F = P_n(\sin \phi) \exp(\iota \gamma t). \tag{13.10}$$

In (13.10), P_n is a Legendre polynomial and $\gamma = 2\pi/\tau = \kappa v$. For large n and locations not too near the source or its anticentre, (3.12) and (13.10) then give u approximately proportional to

$$\exp\{\iota[\kappa vt - (n+\tfrac{1}{2})\phi + \tfrac{1}{2}n\pi]\} + \exp\{\iota[\kappa vt + (n+\tfrac{1}{2})\phi - \tfrac{1}{2}n\pi]\}. \tag{13.11}$$

which corresponds to the superposition of a pair of trains of surface waves, both of period $2\pi/\kappa v$ and wave length $2\pi/\kappa$, travelling with phase velocities v in opposite directions.

An incidental result, widely used for a variety of purposes, is

$$(n+\tfrac{1}{2})/a = \kappa = 2\pi/v\tau. \tag{13.12}$$

The result (13.12) is seen on comparing the first term of (13.11) with (8.44) and noting that x in (8.44) corresponds to $a\phi$. An equivalent result was derived by Jeans (1923) in a slightly more limited context.

In the neighbourhoods of $\phi = \pm\tfrac{1}{2}\pi$, the asymptotic approximation (3.12) for $P_n(\sin \phi)$ breaks down. One effect associated with the consequent inapplicability of (13.11) near $\phi = \pm\tfrac{1}{2}\pi$ is the polar phase shift mentioned in §13.3.

The free Earth oscillation theory yields relations between n and γ for assigned spherical Earth models. Using these relations and applying (13.12), theoretical dispersion curves which take full account of curvature and gravity can then be drawn for surface waves which correspond to either term in (13.11). Comparison with curves for flat-Earth models then yields the required corrections.

[Refs. on p. 283]

A comparison made by Alterman, Jarosch, and Pekeris (1961) showed that, for fundamental modes of *P-SV* surface waves, the flat-Earth theory gives accuracy within 1 per cent for v when $\tau < 50$ s and for V when $\tau < 250$ s; also that, for $\tau < 250$ s, the effects of curvature are larger than those of gravity. Bolt and Dorman (1961) estimated that the combined effect of curvature and gravity increases v by about $2\frac{1}{2}$ per cent at $\tau = 150$ s and 5 per cent at 300 s. A result of some generality is that deviations from the flat-Earth theory effect v considerably more than V.

Work of Jobert (1960) indicated that curvature corrections for *SH* surface waves are on the whole somewhat greater than for *P-SV* waves. Kovach and Anderson (1962) found that the effects on *SH* surface waves of deviations from the flat-Earth theory are very sensitive to such assumptions as the extent to which β may decrease with z below the crust. They also found the curvature correction to be important for periods as small as 20 s in some higher-mode *SH* surface waves. Anderson (1965) investigated the curvature correction for *SH* waves in further detail and emphasized the sensitivity of the corrections to the assumed model structure. For a model containing a low *S* velocity zone, there exists a range of periods of the fundamental *SH* surface waves such that the waves are largely confined to this zone and therefore travel round inside a sphere of radius less than that of the Earth.

Alterman, Jarosch, and Pekeris (1961) derived a specific 'Earth-flattening approximation' to take account of curvature effects. The process involves transforming a spherical Earth model into a model with a plane surface and transforming rays that are straight in the spherical model into curved rays. In effect, a perturbation term which varies linearly with the depth is superposed on the (*P* or *S*) velocity distribution. For fundamental modes of *P-SV* surface waves, the flattening approximation gives both v and V within 1 per cent accuracy when $\tau < 300$ s, approx.; allowance for gravity does not alter this result. For $\tau > 300$ s, recourse must be had to the free Earth oscillation theory. Effects of the presence of the Earth's core start to show when $\tau > 400$ s.

13.5 Evidence on crustal structure

Some specific surface-wave investigations of the structure of the Earth will now be described.

[*Refs. on p. 283*]

Soon after 1920, Tams, Angenheister, and Gutenberg showed that surface seismic waves travel faster across the Pacific Ocean than across continents (see B, §12.6). Applying single-layer theory (with horizontal boundaries) to derive dispersion curves, Stoneley (1935) showed from *SH* observations that the Pacific crust is markedly thinner than the Eurasian crust. This result was confirmed from *P-SV* observations (Bullen, 1939) using dispersion curves of Jeffreys (1935). With the addition of results for other geographical areas (see e.g. Wilson, 1940), it became well established that oceanic crusts are generally thinner than continental. The surface-wave periods used up to this time were mainly in the 15—60 s range.

Before 1940, extensions to two- and three-layer models had been applied by Stoneley and a start made on investigating effects of non-uniformity inside layers. The combination of evidence from bodily and surface waves at this time led to the two-layer representation depicted in §§10.3.1 and 10.4.1 for the average continental crust. Since detailed evidence on bodily waves inside the oceanic crust was then extremely meagre, a continental type of crust was taken as the global standard for the region *A* (§10.3.1) and used for the original *A*-type Earth models.

After 1940, surface-wave calculations were greatly extended, as described in §13.2. Among other things, Ewing and Press (1952) gave a satisfactory account of effects of oceans on surface-wave dispersion.

Considerable evidence has been amassed on differences between surface wave records for waves whose paths are in different geographical areas. For example, Santô (1965, 1966 – these papers include references to earlier work) provisionally classified *P-SV* dispersion curves over the 20—40 s period range into seven types – deep oceanic, oceanic, oceanic transitional, continental transitional, continental, mountainous, and high mountainous; he found, moreover, that even this classification is on the simple side. [See also de Lisle (1941), Bath (1959), Kovach (1965). For a general indication of the geographical differences, see B, Chapter 12.] The differences have made for considerable difficulties and uncertainties in interpreting surface-wave observations in fine detail. Well-determined results for the crust are now available for many limited geographical areas. But a comprehensive picture of the crust and its environs, including full details on the variations of crustal thickness and other factors relating to the crustal densities has not yet been

[*Refs. on p. 283*]

obtained.

In the primary task of estimating the best spherically symmetrical representation of the density distribution below the crust, the finer details for the crust are unimportant, an average global representation being the first requirement. The available evidence on the crustal differences has of course to be taken into account in estimating a suitable average.

The two-layer crustal model of §10.4.1, mentioned above, continues to serve usefully for some continental shield areas. Although this model has long been known to be inadequate for the crust as a whole, it was nevertheless used for many years, partly for reasons given in §10.3.1 and partly because of the principle that models used as standards (even when known to be a little outmoded) should not be replaced too frequently. (Much modern work has produced unnecessary confusion through disregard of this principle.) The stage has now been reached, however, where this early crustal model has had to be superseded in global representations. Not all continental areas show clear-cut evidence of two principal crustal layers, and even for those that do, the depth of the boundary between the layers is often not precisely determined. Consequently, in contexts involving the whole Earth, there has latterly been some reversion towards the use of uniform single-layer crustal models. A commonly used single-layer continental model has $H' = 33$ km, $\rho' = 2\cdot84$ g/cm^3, $\alpha' = 6\cdot30$, and $\beta' = 3\cdot55$ km/s.

As mentioned in §10.2, the Mohorovičić discontinuity is now known (from surface-wave and other evidence) to be generally 10 km or less below the floors of the main oceans. The ocean waters are of course less dense than the rocky crustal layers of continents. On the other hand, the sub-oceanic crust is mostly denser than the average continental crust, and evidence from gravity observations suggests that the immediate sub-crustal materials are also a little denser under oceans than continents (see e.g. Heiskanen and Vening Meinesz, 1958, p. 8).

A single-layer model for the global region A, used by a number of investigators, has the above values of ρ', α' and β', and $H' = 15$ km. This model, which takes account of the overall evidence, including the available surface-wave data for different geographical areas and also free Earth oscillation evidence (§14.7), probably gives about as good a representation of the average crust as can be inferred from present evidence (see §16.4).

[Refs. on p. 283]

In the secondary task of representing deviations from the globe-wide average, surface-wave dispersion data are specially pertinent. Brune's summary (1969) of results for crustal structures in different geographical areas has been given in §12.1.1.

13.6 Evidence on mantle structure

When the period range in seismic surface-wave observations became extended beyond 75 s, significant new information began to be supplied on the structure of the mantle. From bodily-wave observations alone, it is difficult (see §§9.3.5, 9.3.6, 9.6.1) to estimate the detailed variation of α and β inside regions where α or β decreases with z sufficiently rapidly to violate the condition (9.17). Negative velocity gradients cause less difficulty with surface-wave observations, which can thus help towards overcoming an important limitation of bodily-wave evidence.

In practice, surface-wave observations have contributed useful information on the S velocity β, rather than on the P velocity α. This is partly because (see e.g. §12.1.1) the observed violations of (9.17) appear to be much more marked and more widespread for β than for α inside the upper mantle, partly because varying an assumed distribution of α has less effect on the calculated surface-wave dispersion than does varying β. (At the same time the effect of varying α cannot be entirely neglected; even though the effect is small, disregard of it in applying surface-wave dispersion data has sometimes led to unduly low assessments of uncertainties in inferred values of β and ρ.)

Implications on density will be considered in §13.7.

13.6.1 *Low-velocity zone in outer mantle*

Reference has already been made in §§10.3.2 and 12.1.1 to the evidence of Gutenberg, Lehmann, and others for a low-velocity zone inside the outermost 200 km of the mantle.

A series of (V, τ) dispersion curves for $1 < \tau < 8$ min obtained by Ewing and Press (1954) for *P-SV* mantle waves was made the basis of an important test on this point by Dorman, Ewing, and Oliver (1960). Computations for a variety of Earth models with differing β distributions showed that the presence of a low S velocity zone inside the upper mantle would have a marked effect on the shape

[*Refs. on p. 283*]

of the (V, τ) curves for $75 < \tau < 225$ s. Using the procedures described in the latter part of §13.2, Dorman and his colleagues concluded that the (V, τ) observations of Ewing and Press agree reasonably well with either Gutenberg's (1954) β distribution in the outer mantle or with the Lehmann model (§12.1.1), both of which have a low-velocity zone for β; but not with the Jeffreys (1939) model, which lacks a low-velocity zone. Their results strongly suggested that a low S velocity zone does exist, at least in some geographical regions. Numerous other investigators of surface-wave observations have obtained results pointing the same way, and the conclusion is now well established. (For some numerical details, see below.)

As stated, changes in assumed velocity distributions in the mantle are found to affect the shapes of theoretical dispersion curves much less for α than for β. For this reason, surface-wave observations have not so far provided useful evidence on the occurrence of a low P velocity zone.

13.6.2 *The outermost 400 km of the mantle*

In view of the still limited bodily-wave evidence on the distribution of β below the oceans, Dorman and his colleagues proceeded independently of detailed bodily-wave observations to evolve, by successive approximation a sub-oceanic model structure which would fit relevant (V, τ) observations of P-SV surface waves within the observational scatter. Using flat-Earth theory, they arrived at their '8099 model' which has the following values of β (in km/s) in terms of the depth z: $11 < z < 60$ km, 4.61; $60 < z < 220$ km, 4.30; $220 < z < 320$ km, 4.60; $320 < z < 410$ km, 4.80; $410 < z < 500$ km, 5.19. This model has a low-velocity zone between 60 and 220 km depth (cf. Lehmann's model, §12.1.1) and, while by no means final, formed the basis of a number of later investigations.

Dorman also inferred that the high-velocity zone immediately below the crust is significantly thinner under oceans than under continents. Aki and Press (1961) showed that the evidence used for this purpose could be alternatively interpreted as indicating that β is slightly less in the low-velocity zone under oceans than under continents. Aki and Press also found that the 8099 model, while compatible with observations for paths across the Pacific ocean, is

[*Refs. on p. 283*]

somewhat less satisfactory for Atlantic-Indian ocean paths: a model in which β was reduced by 0·1-0·2 km/s in the outermost 50 km of the mantle gave better agreement below the Atlantic-Indian ocean.

The results from group-velocity observations of P-SV waves were soon supplemented by results from phase-velocity measurements, using the methods referred to in §13.3. (See e.g. Ewing, Brune, and Kuo, 1962). For $130 < \tau < 400$ s, Brune (1961, 1962) inferred that the phase velocity v is fairly independent of the geographical region.

Companion investigations involving SH surface waves made by Sykes, Landisman, and Satô (1962), Anderson and Toksöz (1963), and others, yielded results in moderate but not total agreement with Dorman's results for P-SV waves. The principal results of Anderson and Toksöz were embodied in a model called CIT 11. The variation of β in this model is approximately: $11 < z < 90$ km, β falls fairly steadily from 4·60 to 4·34; $90 < z < 160$ km, $\beta = 4.34$; $160 < z < 320$ km, $\beta = 4.50$; $360 < z < 450$ km, β rises to 5·40; $450 < z < 650$ km, $\beta = 5.40$. The model also has a low-velocity zone, but the diminution of β between the crust and this zone is more gradual than in the 8099 model.

Press (1970a, 1970b) used a Monte Carlo technique, taking account (among other things) of various analyses of surface-wave dispersion data, to derive a further series of models giving β distributions for the sub-oceanic mantle. All these models contained a low-velocity zone for S. For the model labelled 5.08, corresponding values of the depth z (in km) and β in (km/s) are: 10, 4·52; 71, 4·59; 146, 4·22; 221, 4·23; 296, 4·72; 371, 4·88; 421, 5·25; 621, 5·43. This model exhibits some of the main trends of the whole set and is selected for mention here because some other investigators (e.g. Kanamori, 1969) have used it as the basis of further calculations. Press's procedures and results will be further discussed in §15.3. See also §15.4.3.

A prominent property of the P-SV dispersion curve is a minimum group velocity equal to about 3·55 km/s near $\tau = 225$ s. Takeuchi, Press, and Kobayashi (1959) showed that the minimum is probably associated with a sharp increase in $d\beta/dz$ at a depth which could correspond to the 20° discontinuity (§10.3.2) or related feature. (The group velocity curve for observed SH waves shows no corresponding minimum.)

[Refs. on p. 283]

13.6.3 Longer-period surface waves

When the periods of surface seismic waves exceed 300 s, the formally calculated wave lengths are sizeable fractions of the Earth's radius. For example, for *P-SV* waves of periods 300, 400, and 700 s, the wave lengths are about 1600, 2400, and 4800 km. It therefore becomes to some extent a matter of convention as to whether the observations of the longer surface waves are treated in terms of an extension of the type of theory considered in the present chapter or in terms of the free Earth oscillation theory considered in Chapter 14. For many purposes it is indeed convenient and even preferable to use the latter theory. The following few comments relate to surface-wave observations as such.

Phase and group velocities have been measured for periods up to about 1000 s in the case of *P-SV* surface waves, and to about 700 s with *SH* surface waves. These periods are sufficiently long for the waves to be significantly affected by details of structure well below the outer mantle. For the fundamental mode of *P-SV* surface waves, the observed v rises steadily from 4·0 km/s at $\tau = 60$ s to a maximum of about 6·65 km/s at $\tau = 700$ s, and then falls. After passing through the minimum near $\tau = 225$ s (§13.6.2), the corresponding V rises steadily to a maximum of about 7·8 km/s near $\tau = 960$ s. These maxima for both v and V are possible consequences of the drop in α at the mantle-core boundary. For the fundamental mode of *SH* waves, representative results are: at $\tau = 100$ s, $v \approx 4·55$ and $V \approx 4·3$ km/s; at $\tau = 600$ s, $v \approx 6·1$ and $V \approx 5·0$ km/s; unlike the case for *P-SV* waves, there are no observed maxima in v and V. These results for *SH* waves are compatible with the presence of fluidity in the outer core.

The longer-period surface-wave observations are thus seen to contain evidence bearing on the Earth's deeper structure, including the mantle and even the outer core. Below 400 km depth, however, more precise results have been yielded in general through analyses using free Earth oscillation theory than through analyses using surface-wave theory. [But see also the reference to Dziewonski and Landisman (1970) in §14.5.4.]

13.6.4 Higher-mode surface waves

Higher-mode surface waves are important in two ways.

First, their presence adds to observational uncertainties in

[*Refs. on p. 283*]

measuring the periods of fundamental modes and complicates the task of identifying modes. Knopoff and others (see Knopoff, 1972) have held that, in drawing detailed inferences on mantle structure through the use of surface-wave dispersion data, higher-mode inference increases the uncertainties so substantially in the case of *SH* observations that, in general, efforts should for the present be principally concentrated on *P-SV* observations. (See also Thatcher and Brune, 1969.)

Secondly, theoretical work shows that dispersion curves for higher-mode surface waves have the potentiality of discriminating fairly finely between some particular features of outer mantle structure. While progress has been made in applying higher-mode observations to questions of crustal structure, the practical difficulty of disentangling the various modes has made for slow progress in applications to the mantle. Alterman (1969) has usefully summarized the possibilities.

13.6.5 *Lateral differences*

Reference was made in §13.5 to evidence from surface-wave data on the marked differences between crustal structures in different geographical areas. Surface-wave data, both *P-SV* and *SH*, also supplement evidence from seismic bodily-wave data (see §12.1.1) that the differences extend well below the crust. The overall evidence is now sufficient to show that the lateral differences persist, significantly but diminishingly, to depths exceeding 400 km.

Dispersion data for a variety of mixed paths have been gathered by many observers, including Brune, Nafe, and Alsop (1961), Ben-Manahem (1965), Toksöz and Anderson (1966), Aki (1966), and Dziewonski and Landisman (1970). Various investigators have assembled selections of the available data and sought to separate out details for pure paths from the mixed data. An example is provided in the work of Kanamori (1969) who treated the mixed paths as compounded of oceanic, continental shield, and 'actively tectonic' portions; the subscripts O, C, and T will apply to these three types of path. On this simple basis, Kanamori arrived at the phase velocities v_O, v_C, and v_T shown in Table 13.1 for pure paths. His assessed standard errors range from 0·01 to 0·08 km/s, but are less than 0·02 km/s for oceanic paths when $\tau \geqslant 175$ s. For both *P-SV* and *SH* waves, the table gives $|v_O - v_C|$ as fairly small for

TABLE 13.1
Kanamori's pure-path phase velocities (in km/s)

Period	P-SV waves			SH waves		
	v_O	v_C	v_T	v_O	v_C	v_T
125 s				4·75	4·73	4·53
150	4·32	4·27	4·21	4·79	4·83	4·67
175	4·44	4·45	4·36	4·86	4·91	4·74
200	4·59	4·60	4·50	4·93	4·98	4·83
225	4·74	4·75	4·70	4·99	5·06	4·92
250	4·93	4·90	4·89	5·07	5·13	4·99
275	5·13	5·06	5·04	5·16	5·20	5·05
300	5·30	5·30	5·25	5·24	5·28	5·14
325				5·32	5·37	5·25

most τ; the differences $(v_O - v_T)$ and $(v_C - v_T)$ are positive for all τ and significantly greater than $|v_O - v_C|$. Kanamori's analysis would thus indicate that the largest lateral deviations from the average outer mantle structure occur (as might be expected) in tectonically active regions. Taking account of the fact that the oceans occupy some 80 per cent of the Earth's surface area, and tectonically active regions only a small fraction of the remaining area, the analysis further suggests that the most suitable structure to take between depths of (say) 60 and 400 km in an average global Earth model would be fairly close to the corresponding sub-oceanic structure.

Various investigators have used results such as those in Table 13.1 extensively in the process of inferring distributions of β and ρ in the outer mantle. It needs to be noted that, because of a number of simplifications generally made, the uncertainties in the values of v used for this purpose are likely to be substantially greater than the standard errors formally derived. For example, Kanamori's classification of pure paths is much simpler than those suggested by Brune (§12.1.1) and Santô (§13.5); Knopoff (1972) preferred a classification into at least five pure paths. Other complications include the effects of the presence of higher-mode surface waves (see §13.6.4). Further comments on lateral variations will be made in §15.7.

13.7 Surface waves and density variation

The findings on β below the crust are immediately relevant to the

[Refs. on p. 283]

problem of the Earth's density variation in that they contribute usefully to knowledge of the internal layering. Further, the strong indication that $d\beta/dz$ is negative inside part of the outer mantle is evidence that a sizeable super-adiabatic temperature gradient exists. By (11.43,44), this in turn implies that the coefficient η (§11.5.2) is here significantly less than unity unless there should happen to be significant changes of composition or phase acting to offset the temperature effect. As will be seen (§14.7.2), the inference that $\eta < 1$ inside part at least of the region B (as defined in §10.3.2) is now well supported from other evidence.

As mentioned earlier, surface-wave data have the potentiality of supplying evidence on ρ in other combinations than those, μ/ρ and k/ρ, which occur in the expressions for β and α. Several investigators have inferred by this route that there is a density 'inversion' inside the upper mantle, i.e. that there exists a range of depth inside which η is actually negative. For example, Anderson's model CIT 11 has $\rho \approx 3.55$ g/cm^3 for $11 < z < 50$ km; ρ then falls to a minimum of 3.39 g/cm^3 at $z = 140$ km before starting to rise again. (See also Pekeris, 1966.) The model distributions of Press referred to in §13.6.2 also showed a density inversion. For example, corresponding values of z (in km) and ρ (in g/cm^3) in his model 5.08 are: 10, 3·35; 71, 3·59; 146, 3·52; 221, 3·53; 296, 3·42; 371, 3·26; 421, 3·55; 621, 3·92. This evidence, which will be further discussed in §15.3.3, makes it quite possible, but by no means certain, that the density near the top of the mantle may indeed exceed the earlier estimate 3.3 g/cm^3 of §10.4.3 by about 0·2, and then decline with increasing z for some distance. Should the possibility be later substantiated, some small adjustments would be required to recently determined Earth density models. The adjustments would be expected to be of order 0·2 g/cm^3 inside the region B but only minor below B. The adjustments would be fairly small since recent models take account of the already well-established result that $\eta < 1$ inside much of B (see §14.7).

13.8 Further remarks

13.8.1

There is a considerable overlap in the analyses of seismic surface waves and of free Earth oscillation data (see e.g. §13.4). In the practical procedures, free Earth oscillation aspects become increasingly important as the periods of the surface waves increase. Thus some of the detail to be given in Chapter 14 includes matter bearing additionally on surface waves.

13.8.2

Quite apart from uncertainties of observation and interpretation, and complications connected with lateral variations, there are problems connected with lack of uniqueness (see e.g. §13.2) in inferring Earth structure from surface-wave data. Reference will be made in §15.4 to some general questions of non-uniqueness in inferring distributions of α and β from seismic data.

The literature on seismic surface waves, both theoretical and observational, is vast. This applies in particular to mantle waves, and the papers cited in the foregoing sections are only a selection of some of the more important ones. Useful reviews, listing many other significant papers, have been prepared by Bolt (1963), Kovach (1965), Anderson (1965, 1967), Press (1966), and Caloi (1967).

REFERENCES

Aki, K. (1966). Generation and propagation of G waves from the Niigata earthquake of June 16, 1964. Part 1: A statistical analysis *Bull. Earthq. Res. Inst.* (Tokyo), **44**, 23–72.

Aki, K. and Press, F. (1961). Upper mantle structure under oceans and continents from Rayleigh waves. *Geophys. J., R. Astr. Soc.*, **5**, 292–305.

Alterman, Z. (1969). Higher-mode surface waves. In *The Earth's Crust and Upper Mantle*, American Geophysical Union, Washington, pp. 265–272.

Alterman, Z. Jarosch, H. and Pekeris, C.L. (1961). Propagation of Rayleigh waves in the Earth. *Geophys. J., R. Astr. Soc.*, **4**, 219–241.

Anderson, D.L. (1965). Recent evidence concerning the structure and composition of the Earth's mantle. *Phys. and Chem. of Earth*, **6**, 1–131.

Anderson, D.L. (1967). Latest information from seismic observations. In *The Earth's Mantle*, 355–420, Academic Press, London.

Anderson, D.L. and Toksöz, M.N. (1963). Surface waves on a sphere and upper mantle structure for Love waves. *J. Geophys. Res.*, **68**, 3483–3500.

Båth, M. (1959). Seismic surface wave dispersion – a world-wide survey. *Geofis. pura appl.*, **43**, 131–147.

Ben-Menahem, A. (1965). Observed attenuation and Q values of seismic surface waves in the upper mantle. *J. Geophys. Res.*, **70**, 4641–4651.

Bolt, B.A. (1963). Recent information on the Earth's interior from studies of mantle waves and eigenvibrations. *Phys. and Chem. of Earth*, **5**, 55—119.

Bolt, B.A. and Dorman, J. (1961). Phase and group velocities of Rayleigh waves in a spherical gravitating Earth. *J. Geophys. Res.*, **66**, 2965—2981.

Brune, J.N. (1961). Radiation pattern of Rayleigh waves from the southeast Alaska earthquake of July 10, 1958. In *A Symposium on Earthquake Mechanism*, 1—11, Dominion Observatory, Ottawa.

Brune, J.N. (1962). Correction of initial phase measurements for the southeast Alaska earthquake of July 10, 1958, and for certain nuclear explosions. *J. Geophys. Res.*, **67**, 3643—3644.

Brune, J.N. (1969). Surface waves and crustal structure. In *The Earth's Crust and Upper Mantle*, American Geophysical Union, Washington, pp. 230—242.

Brune, J.N. and Dorman, J. (1963). Seismic waves and Earth structure in the Canadian Shield. *Bull. Seismol. Soc. Amer.*, **53**, 167—209.

Brune, J.N., Nafe, J.E. and Alsop, L.E. (1961). The polar phase shift of surface waves on a sphere. *Bull. Seismol. Soc. Amer.*, **51**, 247—258.

Brune, J.N., Nafe, J.E. and Oliver, J. (1960). A simplified method for the analysis and synthesis of dispersed wave trains. *J. Geophys. Res.*, **65**, 287—304.

Bullen, K.E. (1939). On Rayleigh waves across the Pacific Ocean. *Mon. Not. R. Astr. Soc., Geophys. Suppl.*, **4**, 579—582.

Bullen, K.E. (1945). Features of the travel-time curves of seismic rays. *Mon. Not. R. Astr. Soc., Geophys. Suppl.*, **5**, 91—98.

Bullen, K.E. (1954). *Seismology*. Methuen, London.

Caloi, P. (1967). On the upper mantle. *Advances in Geophysics*, **12**, 79—210.

Crampin, S. and Båth, M. (1965). Higher modes of seismic surface waves: mode separation. *Geophys. J., R. Astr. Soc.*, **10**, 81—92.

de Lisle, J.F. (1941). On the dispersion of Rayleigh waves from the North Pacific earthquake of November 10, 1938. *Bull. Seismol. Soc. Amer.*, **31**, 303—307.

Dorman, J., Ewing, M. and Oliver, J. (1960). Study of shear-velocity distribution in the upper mantle by mantle Rayleigh waves. *Bull. Seismol. Soc. Amer.*, **50**, 87—115.

Dziewonski, A. and Landisman, M. (1970). Great circle Rayleigh and Love wave dispersion from 100 to 900 seconds. *Geophys. J., R. Astr. Soc.*, **19**, 37—91.

Ewing, M., Brune, J.N. and Kuo, J. (1962). Surface wave studies of the Pacific crust and mantle. In *Crust of Pacific Basin*, American Geophysical Union, Washington, pp. 30—40.

Ewing, W.M., Jardetzky, W.S. and Press, F. (1957). *Elastic Waves in Layered Media*. McGraw-Hill, New York.

Ewing, M. and Press, F. (1952). Crustal structure and surface wave dispersion, Part II, *Bull. Seismol. Soc. Amer.*, **42**, 315—325.

Ewing, M. and Press, F. (1954). An investigation of mantle Rayleigh waves. *Bull. Seismol. Soc. Amer.*, **44**, 121—147.

REFERENCES

Ewing, M. and Press, F. (1956). Surface waves and guided waves. *Encyclopedia of Physics*, **47**, 119–139, Springer-Verlag, Berlin.

Gutenberg, B. (1954). Low velocity layers in the Earth's mantle. *Bull. Geol. Soc. Amer.*, **65**, 337–348.

Haskell, N.A. (1953). The dispersion of surface waves in multilayered media. *Bull. Seismol. Soc. Amer.*, **43**, 17–34.

Heiskanen, W.A. and Vening Meinesz, F.A. (1958). *The Earth and its Gravity Field*, McGraw-Hill, New York.

Jeans, J.H. (1923). The propagation of earthquake waves. *Proc. R. Soc. A*, **102**, 554–574.

Jeffreys, H. (1925). On the surface waves of earthquakes. *Mon. Not. R. Astr. Soc., Geophys. Suppl.*, **1**, 282–292.

Jeffreys, H. (1928). The effect on Love waves of heterogeneity in the lower layer. *Mon. Not. R. Astr. Soc., Geophys. Suppl.*, **1**, 101–111.

Jeffreys, H. (1935). The surface waves of earthquakes. *Mon. Not. R. Astr. Soc., Geophys. Suppl.*, **3**, 253–261.

Jobert, N. (1960). Calcul de la dispersion des ondes de Love de grande période à la surface de la Terre. *C.R. Acad. Sci. (Paris)*, **250**, 3693–3695.

Kanamori, H. (1969). Velocity and Q of mantle waves. *Phys. Earth Planet. Interiors*, **2**, 259–275.

Knopoff, L. (1972). Observation and inversion of surface wave dispersion. (Preprint privately communicated.)

Knopoff, L., Mueller, S. and Pilant, W.L. (1966). Structure of the crust and upper mantle in the Alps from the phase velocity of Rayleigh waves. *Bull. Seismol. Soc. Amer.*, **56**, 1009–1044.

Knopoff, L., Schwab, F. and Kausel, E. (1973). Interpretation of L_g. *Geophys. J., Roy. Astr. Soc.*, **33**, 389–404.

Kovach, R.L. (1965). Seismic surface waves: some observations and recent developments. *Phys. and Chem. of Earth*, **6**, 251–314.

Kovach, R.L. and Anderson, D.L. (1962). Long period Love waves in a heterogeneous spherical Earth. *J. Geophys. Res.*, **67**, 5243–5255.

Kovach, R.L. and Anderson, D.L. (1963). Higher mode surface waves and their bearing on the structure of the Earth's mantle. *Bull. Seismol. Soc. Amer.*, **54**, 161–182.

Oliver, J. and Ewing, M. (1957). Higher modes of continental Rayleigh waves. *Bull. Seismol. Soc. Amer.*, **47**, 187–204.

Pekeris, C.L. (1966). The internal constitution of the Earth. *Geophys. J., R. Astr. Soc.*, **11**, 85–132.

Press, F. (1956). Determination of crustal structure from phase velocity of Rayleigh waves. Part 1: Southern California. *Bull. Geol. Soc. Amer.*, **67**, 1647–1658.

Press, F. (1966). Seismological information and advances. In *Advances in Earth Sciences* M.I.T. Press, Boston, pp. 247–286.

Press, F. (1970a). Earth models consistent with geophysical data. *Phys. Earth Planet. Interiors*, 3, 3–22.

Press. F. (1970b). Regionalized Earth models. *J. Geophys. Res.*, 75, 6575–6581.

Santô, T. (1965–6). Lateral variation of Rayleigh wave dispersion character. Parts I–III. *Pure and Appl. Geophys.*, 62, 49–66 and 67–80; 63, 40–59.

Satô, Y. (1958). Attenuation, dispersion and the wave guide of the G waves. *Bull. Seismol. Soc. Amer.*, 48, 231–251.

Satô, Y. (1959). Numerical integration of the equation of motion for surface waves in a medium with arbitrary variation of material constants. *Bull. Seismol. Soc. Amer.*, 49, 57–77.

Stoneley, R. (1925). Dispersion of seismic waves. *Mon. Not. R. Astr. Soc., Geophys. Suppl.*, 1, 280–282.

Stoneley, R. (1934). The transmission of Rayleigh waves in a heterogeneous medium. *Mon. Not. R. Astr. Soc., Geophys. Suppl.*, 3, 222–232.

Stoneley, R. (1935). On the apparent velocities of earthquake waves over the surface of the Earth. *Mon. Not. R. Astr. Soc., Geophys. Suppl.*, 3, 262–271.

Stoneley, R. (1937). Surface waves associated with the 20° discontinuity. *Mon. Not. R. Astr. Soc., Geophys. Suppl.*, 4, 39–43.

Sykes, L., Landisman, M. and Satô, Y. (1962). Mantle shear wave velocities determined for oceanic Love and Rayleigh wave dispersion. *J. Geophys. Res.*, 67, 5257–5271.

Takeuchi, H., Press, F. and Kobayashi, N. (1959). Rayleigh wave evidence for the low velocity zone in the mantle. *Bull. Seismol. Soc. Amer.*, 49, 355–364.

Thatcher, W. and Brune, J.N. (1969). Higher mode interference and observed anomalous apparent Love wave phase velocities. *J. Geophys. Res.*, 74, 6603–6611.

Thomson, W. (1950). Transmission of elastic waves through a stratified medium. *J. Appl. Phys.*, 21, 89–93.

Toksöz, M.N. and Anderson, D.L. (1966). Phase velocities of long-period surface waves and the structure of the upper mantle: 1. Great-circle Love and Rayleigh wave data. *J. Geophys. Res.*, 71, 1649–1658.

Toksöz, M.N. and Ben-Menahem, A. (1963). Velocities of mantle Love and Rayleigh waves over multiple paths. *Bull. Seismol. Soc. Amer.*, 53, 741–764.

Valle, P.E. (1949). Sulla Misusa della Velocita di Gruppo delle Onde Sismiche Superficiali. *Ann. Geophys.*, 2, 370–376.

Wilson, James T. (1940). The Love waves of the South Atlantic earthquake of August 28, 1933. *Bull. Seismol. Soc. Amer.*, 30, 273–301.

CHAPTER 14

Evidence from free Earth oscillations

Following the Chilean earthquake of 1960 May 22, important additional evidence on the distributions of the density ρ, incompressibility k, and rigidity μ in the interior of the Earth arose from recordings of free Earth oscillations with periods up to the order of one hour. A few hundred separate modes of free Earth oscillations have now been recorded and any acceptable Earth model has to be compatible with the better determined observational periods of these modes.

The present chapter starts with some general remarks on free oscillations of a dynamical system and outlines the theory whereby the free oscillation periods can be calculated for a body like the Earth.

In applying the theory of free Earth oscillations to the problem of the Earth's density variation etc., an appropriate procedure is to start with an Earth model, e.g. Model A'' (§ 10.7.2), which incorporates evidence from other sources, including evidence from records of seismic bodily waves, the revised estimate of the moment of inertia of the Earth, etc. The free oscillation periods for this model are then calculated and the residuals determined against the free oscillation periods recorded on seismograms and related records of ground motion or strain. By successive approximation, revised models are then derived, each successive model being designed with a view to reducing the residuals until they lie within suitably assessed uncertainties of the observational periods. At the same time, the effort is made to keep to models whose properties remain compatible with the evidence from other sources. There is an intimate connection between free oscillations and surface (including mantle) waves for the longer observed surface-wave periods (§§ 13.4, 13.6.3), and the application of free Earth oscillation data to determining outer mantle

[*Refs. on p. 319*]

FREE EARTH OSCILLATIONS [14-

structure merges with that of surface-wave data (and involves similar complications connected with geographical variations).

As in the case of the application of surface-wave observations, Earth models arrived at by the above procedure are by no means unique. But the addition of evidence from records of free Earth oscillations in this way has led to significant amendments to all the models discussed in earlier chapters. In the present chapter, emphasis is placed on certain amendments to earlier models which appear to be demanded by the oscillation data. More comprehensively based models which incorporate the overall evidence discussed in Chapters 10—14 are described in Chapter 16.

The present chapter includes tables giving most of the better determined observational oscillation periods, and discusses a selection of Earth models designed to fit the period data. Reference is made to the model HB_1 which follows the principles of scientific inference outlined in §6.1, especially in respect of economy in the introduction of new parameters. Important new evidence is given on the solidity of the Earth's inner core.

14.1 Free oscillations of a dynamical system

14.1.1 *General features*

Consider a non-rigid dynamical system of N degrees of freedom, subject to internal constraints which do not cause too rapid a dissipation of mechanical energy. When mechanical energy is imparted to such a system and the source of energy is then removed, the system in general vibrates (oscillates) freely for a time. The vibratory motion is commonly regarded as the resultant of component vibrations each of which is characterized by a relatively simple mathematical form. The number of component vibrations depends on N and is indefinitely large when N is indefinitely large.

For some systems, the mathematical forms of the component vibrations include more than one distinct class (e.g. torsional and spheroidal oscillations below). In a wide variety of vibrating systems, the form for any one class includes one or more integer parameters, $l, m, n, ...$, say. A particular normal mode of oscillations of a given class is specified by the numerical values assigned to $l, m, n, ... $. One of these parameters, l say (see §3.6), is singled out to indicate the 'fundamental' or 'gravest' mode of oscillation and the 'overtones' for

[Refs. on p. 319]

a series of modes in which each of the other parameters $m, n, \ldots$ has an assigned value. The lowest value of l (commonly taken as zero, but sometimes otherwise) gives the fundamental mode of the series, and successive values of l give the overtones. In general (but sometimes with exceptions at small values of l) the periods of these modes form a decreasing sequence as l increases. The roles played by $(m, n, \ldots)$, as well as l, will be displayed in the detail to follow. The entire set of periods of free oscillations constitutes the *spectrum* of the system. For details relating to some particular vibrating mechanical systems, see **B**, Chapter 3.

14.1.2 *The Earth as an oscillating system*

The Earth is a dynamical system in which N is indefinitely large and which therefore has an indefinitely large number of modes of oscillation. Before 1960, the motions generated in the Earth through the release of mechanical energy at earthquake foci were mostly investigated in terms of the theory of bodily and surface waves considered in earlier chapters. While bodily and surface waves are intimately connected with the Earth's free normal modes of vibration (see e.g. §§ 13.4, 13.6.4, and **B**, § 3.2.5), the emphases in the two approaches are different. The usual theory of seismic wave transmission looks at the associated Earth motions as travelling disturbances which affect only a small part of the Earth at any given instant, whereas the theory of free Earth oscillations looks at the motions as normal modes of vibration affecting the whole Earth simultaneously. (For many particular modes, the amplitudes may, however, be nearly zero throughout all but a small part of the Earth's volume.) The wave theory approach is usually appropriate when the periods are comparatively short but, as already printed out in Chapter 13, there is a considerable range of periods for which the surface-wave and oscillation theory approaches overlap.

When modern instrumental developments at the hands of Benioff (1935), Ewing (1961), Slichter (see Ness, Harrison, and Slichter, 1961) and others led to an extension in the reliable recording of Earth motions to periods much greater than a minute, it became feasible to bring to bear theory on the Earth's free fundamental oscillations and overtones. The names of Lamb (1882), Love (1911), Jeans (1923), Matumoto and Satô (1954), Jobert (1956), Alterman, Jarosch, and Pekeris (1959), Takeuchi (1959), Backus and Gilbert

[*Refs. on p. 319*]

(1961), and Slichter (1961) are prominent in the theoretical development.

Observations of the Earth's vibration spectrum now extend to periods of one hour and greater, and so cover the whole range of free Earth oscillations. (The observed spectrum actually goes much beyond periods of one hour to periods of interest in tidal observations —see Jeffreys, 1970.) Each oscillation period measurement provides in effect one new equation of condition which the distributions of ρ, k, and μ must satisfy. Altogether, the observed oscillation periods constitute a body of important new evidence on the values of (ρ, k, μ) inside the Earth. As stated earlier, the evidence does not by itself determine (ρ, k, μ) uniquely, but it usefully supplements the evidence from bodily waves and complements the evidence from from surface waves.

14.2 Approach to the theory of Earth oscillations

14.2.1 Displacement in a normal mode

As in Chapter 3 and §13.4, we shall use spherical polar coordinates consisting of the radius vector r, latitude ϕ, and longitude λ, and sometimes write $\mu = \sin \phi$. The corresponding cartesian coordinates x_i are

$$x_1 = r \cos \phi \cos \lambda, \quad x_2 = r \cos \phi \sin \lambda, \quad x_3 = r \sin \phi. \tag{14.1}$$

Let $P_n^m(\mu)$ be an Associated Legendre Function of the First Kind (§3.4). In §3.6, it was shown that

$$y = r^n P_n^m(\mu) \cos m\lambda, \tag{14.2}$$

where m and n are positive integers such that $m \leqslant n$, is a solution of Laplace's equation $\nabla^2 y = 0$.

Expressions related to $\nabla^2 y$, and acceleration terms of the type $\partial^2 y / \partial t^2$, appear in differential equations governing the free oscillations of a dynamical system. Although the equations for free Earth oscillations are substantially more complicated than, say, the classical wave equation $\nabla^2 y = v^{-2} \partial^2 y / \partial t^2$, it has been found appropriate to seek solutions that are related to the form (14.2). [The procedure is analogous to that in simpler wave problems where forms of the type (8.20) are substituted into equations of motion to obtain particular solutions.] The procedure is justified, as a trial and error procedure, by the degree of success with which it yields a class of particular solutions. Trial forms based on (14.2) derive special value through the readiness with which they can be physically interpreted and the extent to which complicated observed motions can be described in terms of combinations of them.

In the case of Earth oscillations, trial and error led to the assuming of forms for displacement components of the type (3.32), namely,

$$_l R_n^m (r) F_n^m (\mu) \cos m\lambda \exp(\iota \gamma t), \qquad (14.3)$$

where $_l R_n^m$ is a function of r, $F_n^m(\mu)$ is identical with or closely related to $P_n^m (\mu)$, and $2\pi/\gamma$ is the period τ of the oscillation ($\gamma/2\pi$ is the frequency). Both γ and R depend in general on l, m, and n; F depends on m and n. The details in §§3.4 and 3.6 on nodal surfaces are relevant to the form (14.3).

14.2.2 *Effects of axial rotation and ellipticity*

For spherically symmetrical non-rotating Earth models, the free oscillation periods are found to be independent of m, so that the system is degenerate (§3.6) with respect to m. The degeneracy is removed when the model is given a non-zero angular speed Ω about its axis. [There is a close analogy with the Zeeman effect of the splitting of lines of atomic spectra by a magnetic field, as pointed out by MacDonald and Ness (1961) and Pekeris, Alterman, and Zarosch (1961).]

In this connection, it is convenient to put the cosine factor in (14.3) in exponential form and to consider a pair of trial forms, namely,

$$_l R_n^m \; F_n^m (\mu) \exp\{\iota\gamma(t - m\gamma^{-1}\lambda)\} \qquad (14.4)$$

and a companion form in which the sign before m is changed. In general, each of these forms gives a travelling wave with wave front on a diametral plane which passes through the polar axis and rotates about the axis with angular speed γ/m: west-east with (14.4), and east-west with the companion wave. For assigned n and l, there are $(2n + 1)$ modes of oscillation, namely, the mode with $m = 0$ and $2m$ modes in which m takes the values 1, 2, ..., n for each of (14.4) and its companion; the frequencies $\gamma/2\pi$ are approximately equally spaced in this set of modes.

Since the periods of the free Earth oscillations do not exceed the

[Refs. on p. 319]

order of an hour, they are fairly small compared with the period of the Earth's axial rotation. The effect of axial rotation is therefore fairly slight and is neglected in the first (and main) approximation. For any assumed values of m, the west-east and east-west waves as given by (14.4) and its companion combine in this case to give standing waves in which the frequency is independent of the value taken for m.

An effect of axial rotation is therefore to produce 'split modes', involving the replacement of a single frequency in the non-rotating case by a set of frequencies. Backus and Gilbert (1961) showed that the presence of axial rotation causes the nodal pattern to drift westward around the Earth's axis with angular speed ω equal to the frequency difference between successive split modes; the ratio ω/Ω, where Ω is the Earth's axial angular speed, has been called the 'splitting parameter'.

As seen in Chapter 5, the axial rotation causes surfaces of constant density in the Earth to be elliptical, and this effect contributes further to the splitting of the modes. For detail on the splitting due to both axial rotation and the consequent ellipticity effect, reference may be made to Pekeris, Alterman, and Jarosch (1961), Backus and Gilbert (1961), Usami and Satô (1962), Caputo (1963), Slichter (1967), and Dahlen (1968, 1969).

14.2.3 *Effects of further departures from spherical symmetry*

Reference has been made in §13.6.5 to lateral differences other than those associated with ellipticity, extending down to several hundred kilometres depth inside the Earth's mantle. For the lowest values of n, the fundamental free oscillation periods depend predominantly on the deeper structure of the Earth, and these lateral differences have negligible effect.

As n increases, the periods become increasingly influenced, however, by the structure inside a diminishing range of depth below the Earth's surface. When n exceeds a certain value, n' say (n' is not in the circumstances closely determined — see §14.5.2 for further detail), the periods depend principally on structure in the outer mantle and lateral differences cease to be negligible. The modes of oscillation for which $n > n'$ have periods inside the range for which surface-wave theory is commonly applied; and the same difficulties arise in applying the oscillation results for these periods as in the case of applying surface-wave data (see §13.6.5 again).

Thus the free Earth oscillation data are most serviceable for those fundamental modes for which $n < n'$. A further practical point is that the intervals between the periods of successive modes in general decrease as n increases. Hence errors due to misidentifications of modes (see §14.5.1) are in general most likely to occur with those modes which have the potentiality of providing information on the outer mantle.

14.2.4 *Torsional and spheroidal oscillations*

Work of Lamb (1882) and Hoskins (1920) — see Stoneley (1961) — distinguished two broad classes of free oscillations for a spherically symmetric Earth model, both classes being of the type (14.3). The components of displacement at the point (r, ϕ, λ) of the model in the directions of r, ϕ, λ increasing will be denoted as (u, v, w).

The oscillations of the first class are defined by

$$\left. \begin{array}{l} u = 0, \\ v = -(\cos\phi)^{-1} V(r)(\partial S_n^m/\partial\lambda)\exp(\iota\gamma t), \\ w = V(r)(\partial S_n^m/\partial\phi)\exp(\iota\gamma t), \end{array} \right\} \quad (14.5)$$

where $S_n^m(\mu, \lambda)$ is a spherical harmonic of **degree** n and order m and may be taken as $P_n^m(\mu)\cos m\lambda$; the factor $V(r)$ depends on l, m, and n, where l is the parameter which indicates fundamental and overtone oscillations (§14.1.1). A characteristic of the oscillations (14.5) is that the dilatation Δ is always zero everywhere, for which reason they are commonly called *torsional oscillations*. They have also been called toroidal oscillations because the displacements are confined to the surfaces of concentric spheres. Because $\Delta = 0$, torsional oscillations cause no disturbance of density or the gravitation field. For higher values of n, torsional oscillations are intimately related to *SH* surface waves.

The oscillations of the second class are defined by

$$\left. \begin{array}{l} u = U(r) S_n^m(\mu, \lambda)\exp(\iota\gamma t), \\ v = V(r)(\partial S_n^m/\partial\phi)\exp(\iota\gamma t), \\ w = V(r)(\cos\phi)^{-1}(\partial S_n^m/\partial\lambda)\exp(\iota\gamma t), \end{array} \right\} \quad (14.6)$$

where S_n^m is as before, and U and V depend on l, m, and n. A characteristic of these oscillations, called *spheroidal oscillations*, is that

[*Refs. on p. 319*]

the radial component of curl u, where u is the displacement, is everywhere zero. For higher values of n, spheroidal oscillations are intimately related to P-SV surface waves.

Torsional and spheroidal oscillations are commonly denoted as $_l T_n^m$ and $_l S_n^m$, respectively. In the case of degeneracy with respect to m, the notation $_l T_n$, $_l S_n$ is used; n is then commonly referred to as the order number and l, which is connected with the numbers of zeroes in U and V, as the overtone number. For given n, the mode $_0 T_n$ is the fundamental, and $_1 T_n$, $_2 T_n$, ... are the overtones; and similarly with $_0 S_n$, $_1 S_n$, $_2 S_n$ ... . [The symbol S in $_l S_n^m$ and $_l S_n$ indicates spheroidal oscillations and is not the spherical harmonic S which appears in equations (14.5) and (14.6).]

In the particular case of those spheroidal oscillations for which $n = 0$, the oscillations are purely radial and are expressed by

$$u = U(r)\exp(\iota\gamma t), \quad v = w = 0. \tag{14.7}$$

Certain particular modes may be unobservable. For example, $_l T_n^m$ would represent an unobservable rigid-body rotation about the polar axis when $l = m = 0$ and $n = 1$; the modes $_n T_0^0$ have $u = v = w = 0$.

14.3 Equations of motion of an oscillating Earth model

We now consider the derivation of equations of motion for a spherically symmetrical Earth model in a form suitable for numerical application to the free Earth oscillation problems. Both spherical polar coordinates (e.g. Alterman, Jarosch, and Pekeris, 1959) and cartesian coordinates (e.g. Jeffreys and Vicente, 1966) have been used, and have certain relative advantages and disadvantages in the practical applications. Here, spherical polars will be used. The derivation is heavy and will be given only in outline. Complementary detail may be found in accounts by, for example, Pekeris, Alterman, and Jarosch (1958, 1959), Carr (1961, 1963), Slichter (1967), Coulomb and Jobert (1973), and in B, Chapter 14. [Note that in some of these references the spherical polar coordinates are taken as (r, θ, ϕ), where θ and ϕ correspond to $(\frac{1}{2}\pi - \phi)$ and λ, respectively.]

For the purposes of the first (and main) approximation, a spherically symmetric non-rotating Earth model will be taken, composed of perfectly elastic, isotropic materials. Let ρ, k, μ, α, β, and g be the density, incompressibility, rigidity, P and S velocities,

[Refs. on p. 319]

and gravitational intensity at distance r from the centre, and let G be the constant of gravitation. (The symbol μ used above to denote $\sin\phi$ will from now on denote only rigidity.) To sufficient accuracy, the symbols ρ, k, μ appearing below may be regarded as relating to the initial undisturbed configuration.

A state of hydrostatic stress is assumed in the equilibrium configuration. Thus the initial stress is specifiable in terms of a pressure distribution $p(r)$, and $dp/dz = -g\rho$.

The first step towards deriving the desired equations of motion is to convert, by routine methods, the basic cartesian equations of motion (8.1) into spherical polar equivalents. Apart from the slight difference of notation mentioned above, the resulting equations are identical with the equations (4) and (5) given in B (§14.1.1). (In the first of these equations in B, a minus sign needs to be inserted before ρg_0, and $\cot\phi$ needs to be replaced by $\cot\theta$.)

The second step is to introduce spherical polar stress-strain relations, through which the equations of motion are expressed in terms of the strain components e_{rr}, etc. The stress-strain relations taken will be the spherical polar equivalents of the cartesian relations (8.2). It is convenient to make auxiliary use of the dilatation Δ in incorporating the effects of initial hydrostatic stress in the stress-strain relations, as in B, §14.1.1, (3). The e_{rr}, etc., and Δ have to be expressed in terms of the displacement components u, v, w [see B, §14.1.1, (1) and (2)]; $u, v,$ and w are taken as zero in the initial state. It is also convenient to introduce the gravitational potential ψ associated with the disturbance in g consequent upon the density disturbance caused by the oscillations.

The following equations are yielded:

$$\rho \frac{\partial^2 u}{\partial t^2} = \rho g \Delta + \rho \frac{\partial \psi}{\partial r} - \rho \frac{\partial (gu)}{\partial r} + \frac{\partial}{\partial r}\left\{(k - \tfrac{2}{3}\mu)\Delta + 2\mu \frac{\partial u}{\partial r}\right\}$$

$$+ \frac{2}{r}\frac{\partial(\mu e_{r\phi})}{\partial \phi} + \frac{2}{r\cos\phi}\frac{\partial(\mu e_{r\lambda})}{\partial \lambda}$$

$$+ \frac{2\mu}{r}(2e_{rr} - e_{\phi\phi} - e_{\lambda\lambda} - e_{r\phi}\tan\phi), \tag{14.8}$$

$$\rho \frac{\partial^2 v}{\partial t^2} = \frac{\rho}{r} \frac{\partial \psi}{\partial \phi} + \frac{\partial}{\partial r}(2\mu e_{r\phi}) + \frac{1}{r}\frac{\partial}{\partial \phi}\{-g\rho u + (k - \tfrac{2}{3}\mu)\Delta + 2\mu e_{\phi\phi}\}$$

$$+ \frac{2}{r \cos \phi} \frac{\partial(\mu e_{\phi\lambda})}{\partial \lambda} - \frac{2\mu \tan \phi}{r^2}\left(\frac{\partial v}{\partial \phi} + v \tan \phi - \frac{\partial w}{\partial \lambda} \sec \phi\right)$$

$$+ \frac{6\mu e_{r\phi}}{r}, \tag{14.9}$$

$$\rho \frac{\partial^2 w}{\partial t^2} = \frac{\rho}{r \cos \phi} \frac{\partial \psi}{\partial \lambda} + \frac{\partial}{\partial r}(2\mu e_{r\lambda}) + \frac{2}{r}\frac{\partial(\mu e_{\phi\lambda})}{\partial \phi} + \frac{6\mu e_{r\lambda}}{r}$$

$$+ \frac{1}{r \cos \phi}\frac{\partial}{\partial \lambda}\{-g\rho u + (k - \tfrac{2}{3}\mu)\Delta + 2\mu e_{\lambda\lambda}\}$$

$$- \frac{4\mu}{r} e_{\phi\lambda} \tan \phi, \tag{14.10}$$

$$\nabla^2 \psi = 4\pi G\{\rho\Delta + u \mathrm{d}\rho/\mathrm{d}r\}. \tag{14.11}$$

After replacement of the strain components e_{rr}, etc. and Δ in terms of u, v, w, the result is a very complicated set of second-order differential equations, expressing u, v, w, and ψ in terms of r, ϕ, λ, t, and the properties $\rho(r)$, $k(r)$, and $\mu(r)$ of the Earth model. Adaptations of the equations are made to meet the requirements of particular classes of oscillations.

14.4 Solving the equations of motion

14.4.1 *Spheroidal oscillations*

Spheroidal oscillations as defined by (14.6) will be considered first. We continue to assume spherical symmetry and ignore rotation effects, so that it is sufficient to replace S_n^m in (14.6) by S_n. It can then be deduced from (14.6) that

$$\Delta = X(r) S_n \exp(\iota \gamma t), \tag{14.12}$$

where
$$X = \mathrm{d}U/\mathrm{d}r + \{2U - n(n+1)V\}r^{-1}. \tag{14.13}$$

Then, by (14.11),

$$\psi = P(r) S_n \exp(\iota \gamma t), \tag{14.14}$$

where $P(r)$ has to satisfy

$$d^2P/dr^2 + 2r^{-1}dP/dr - n(n+1)Pr^{-2} = 4\pi G(X\rho + Ud\rho/dr). \tag{14.15}$$

Through (14.12)–(14.15), the equations (14.8)–(14.11) can be reduced to a set of three second-order differential equations involving γ and the dependent variables U, V, and P. In the process followed by Alterman, Jarosch, and Pekeris, these equations are converted into a set of six first-order equations by introducing new dependent variables y_i ($i = 1, ..., 6$), where

$$\left.\begin{aligned} y_1 &= U, \quad y_2 = (k - \tfrac{2}{3}\mu)X + 2\mu dU/dr \quad y_3 = V, \quad y_5 = P, \\ y_4 &= \mu\{(U-V)r^{-1} + dV/dr\}, \quad y_6 = -4\pi G\rho U + dP/dr. \end{aligned}\right\} \tag{14.16}$$

The equations governing the y_i and γ are equivalent to the following:

$$\alpha^2 dy_1/dr = (\alpha^2 - 2\beta^2)\{-2y_1 + n(n+1)y_3\}r^{-1} + y_2\rho^{-1}, \tag{14.17}$$

$$\begin{aligned}\alpha^2 dy_2/dr =\ &\rho\{-\gamma^2\alpha^2 r^2 - 4\alpha^2 gr + 4\beta^2(3\alpha^2 - 4\beta^2)\}y_1 r^{-2} - 4\beta^2 y_2 r^{-1} \\ &+ n(n+1)\rho\{g\alpha^2 r - 2\beta^2(3\alpha^2 - 4\beta^2)\}y_3 r^{-2} \\ &+ n(n+1)\alpha^2 y_4 r^{-1} - \rho\alpha^2 y_6,\end{aligned} \tag{14.18}$$

$$dy_3/dr = (y_3 - y_1)r^{-1} + y_4/(\rho\beta^2), \tag{14.19}$$

$$\begin{aligned}\alpha^2 dy_4/dr =\ &\rho\{g\alpha^2 r - 2\beta^2(3\alpha^2 - 4\beta^2)\}y_1 r^{-2} - (\alpha^2 - 2\beta^2)y_2 r^{-1} \\ &+ \rho\{-\gamma^2\alpha^2 r^2 + 2\beta^2[2n(n+1)(\alpha^2-\beta^2) - \alpha^2]\}\ y_3 r^{-2} \\ &- \alpha^2(3y_4 + \rho y_5)r^{-1},\end{aligned} \tag{14.20}$$

$$dy_5/dr = 4\pi G\rho y_1 + y_6, \tag{14.21}$$

$$dy_6/dr = n(n+1)(-4\pi G\rho y_3 + y_5)r^{-2} - 2y_6 r^{-1}. \tag{14.22}$$

For fluid regions, $\beta = 0$, $y_2 = kX$, and $y_4 = 0$, and the equations simplify considerably.

Equations such as (14.17)–(14.22) can be solved for specific Earth models in which the distributions of (ρ, k, μ) or (equivalently) of (ρ, α, β) are prescribed functions of r; then g is also a given function of r by (10.5). In practice, the prescribed functions are usually in the form of numerical tables, as for example the distributions for A- and B-type Earth models. The equations are then solved subject to the boundary conditions which include the conditions at the outside surface and at each internal surface where ρ, k, or μ changes discontinuously. Requirements at $r = 0$ [see e.g. Carr (1961, 1963) for details] also have to be met. Solving (14.17)–(14.22) as a set of simultaneous equations yields expressions for the six y_i, and thence U, V, etc., in terms of r and γ. On applying the boundary conditions, an eigen-value equation for γ is derived. Solving this equation gives the periods $2\pi/\gamma$ for the various normal modes of spheroidal oscillations of the specified model, each particular mode and corresponding period being associated with a particular pair of values of (l, n). Other properties of the modes such as the spatial amplitude distributions may then be derived.

The equations of motion are readily solvable only for artificially simple Earth models, and then only for the simplest modes of oscillation. The solution for radial oscillations ($n = 0$) of an Earth model in which ρ, k, and μ are all constant was given by Love (1911). Study of this solution (see e.g. B, § 14.2.1) gives insight into the general process whereby the normal mode periods $2\pi/\gamma$ are derived in more difficult cases. The equations for complicated Earth models can be solved only with the help of electronic computers, usually by trial and error processes.

In respect of the problem of the Earth's density variation, the principal use so far made of free Earth oscillation data has been of the period results. Discrepancies between observational periods derived from measurements of instrumental records and theoretical periods calculated for particular Earth models point the way towards improved models.

As already indicated, the process of solving the above equations of motion, formidably complicated though it is, meets conditions in the Earth to a first approximation only. Corrections for such effects as axial rotation, departures from spherical symmetry, and damping have to be considered. The first approximation is sufficiently reliable, however, to be closely applicable to the problem of the Earth's density variation, and small corrections are left over to later stages. (See also § 14.5.2).

14.4.2 *Torsional oscillations*

The derivation of the solution for torsional oscillations, as defined by

(14.5), follows the same general principles as for spheroidal oscillations, but the equations are rather less complicated since (14.5) involve a single function $V(r)$ in place of the two functions $U(r)$ and $V(r)$ of (14.6); also $\Delta = 0$ and $\psi = 0$.

On substituting from (14.5) into (14.8)–(14.11), it is found that (14.8) and (14.11) are satisfied identically, while each of (14.9) and (14.10) gives

$$\mu(V'' + 2r^{-1}V') + \mu'(V' - r^{-1}V) + \{\gamma^2 \rho - n(n+1)\mu r^{-2}\}V = 0, \tag{14.23}$$

primes here denoting differentiations with respect to r.

Pekeris introduced new dependent variables y_1 and y_2, where

$$y_1 = V, \quad y_2 = \mu(V' - r^{-1}V), \tag{14.24}$$

enabling (14.23) to be reduced to a pair of simultaneous first-order equations, equivalent to

$$dy_1/dr = r^{-1}y_1 + (\rho\beta^2)^{-1}y_2, \tag{14.25}$$

$$dy_2/dr = (n-1)(n+2)\rho\beta^2 r^{-2}y_1 - \gamma^2 \rho y_1 - 3r^{-1}y_2. \tag{14.26}$$

[The variables y_i in (14.16) and (14.24) were selected with a view to avoid having to compute derivatives of ρ, k, and μ explicitly from the tables specifying the Earth models used.]

Although the pair of equations (14.25,26) is much simpler than the set (14.17)–(14.22), the solution in general still requires the services of an electronic computer.

A point of practical significance is that, unlike spheroidal oscillations, for which ψ does not vanish, torsional oscillations do not disturb the gravitational field. One consequence is that instruments designed to measure small fluctuations in gravity may record spheroidal, but not torsional, oscillations. This property was dramatically illustrated at an international meeting in Helsinki in August 1960 when representatives of four groups of observers (Benioff and Press, Slichter, Ewing, Bogert, and colleagues, 1961) independently presented observations that appeared to relate to free Earth oscillations excited by the Chilean earthquake earlier that year. The series of periods presented by the four groups were in fair agreement except that some periods were missing on the records of Slichter's group. A rapid scrutiny on the spot indicated that the missing periods were ones that would be expected to be associated with torsional

[Refs. on p. 319]

oscillations — that Slichter had recorded only spheroidal oscillations and the other three groups both spheroidal and torsional. The further revelation that Slichter's records had been taken on a LaCoste-Romberg tidal gravity meter banished any lingering doubts that free Earth oscillations had been authentically recorded.

The same property is of considerable assistance in the practical problem of identifying modes (§ 14.5.1). Serious difficulties arise in practice because of crowding of periods in some parts of the spectrum and also because of amplitude vagaries. Instruments which measure only the gravitational disturbances therefore assist in reducing the uncertainties of identification by discriminating between spheroidal and torsional modes. They also contribute to finer assessment of the spheroidal modes through the absence of torsional contamination.

14.4.3 *Other free Earth oscillations*

Alterman, Jarosch, and Pekeris (1959) drew attention to a class of spheroidal oscillations, called *core oscillations*, in which the theoretical motion is predominantly in the core. Some interest attaches to their result that, for models whose central density ρ_0 is sufficiently high, there exist some modes of core oscillations with periods exceeding an hour, whereas the greatest period of all the oscillations which have significant amplitudes inside the mantle is theoretically only 54 min. In particular, their calculations gave core oscillations with periods of about 76, 85, and 100 min in the case of the original Earth model B (§ 11.8.1), for which $\rho_0 \approx 18 \text{ g/cm}^3$. On the other hand, the greatest period did not exceed 54 min for models in which $\rho_0 \leqslant 13 \text{ g/cm}^3$. Thus evidence bearing on the density in the inner core might be provided if core oscillations could be observed; but the amplitudes would be extremely small in the mantle.

Some records have in fact suggested the occurrence of free Earth oscillation periods exceeding an hour, leading to further theoretical investigations. For example, Slichter (1961) examined the effect of an oscillation of the inner core relative to the remainder of the Earth in an endeavour to interpret a period of 86 min apparently recorded amid other Earth oscillations on a gravimeter. Making plausible simplified assumptions, he estimated that such an interpretation would require the presence of small but significant rigidity (of order $10^9 - 10^{10} \text{ N/m}^2$) in the outer core if the density of the inner core were not to exceed about 15 g/cm^3.

Atmospheric oscillations can be recorded by instruments which record free Earth oscillations, and there can sometimes be a problem of distinguishing these two types of oscillation on the records. The longer-period atmospheric oscillations exceed an hour and there is some likelihood that they probably account for observations so far made of periods between one and two hours.

14.5 Observational data

The first indisputable observations of free Earth oscillations were those presented at Helsinki by the four groups mentioned in §14.4.2, following the Chilean earthquake of 1960 May 22. Periods were presented for fundamental spheroidal modes for most values of n up to 42, for fundamental torsional modes up to $n = 11$, and for a small number of spheroidal overtones. (Comparative results of the four groups for the smaller n are shown in B, §14.5.2.) The Alaskan earthquake of 1964 March 28 excited a further large series of free Earth oscillations. By that time, various refinements had been added to both the observational techniques and the processes of inferring the periods, with the result that a large quantity of improved data soon became available. By far the greater part of the free Earth oscillation data so far utilized has come from analyses of records of those two great earthquakes.

By 1966, Pekeris had assembled a fairly reliable set of observational periods for fundamental spheroidal and torsional oscillations for the values of n given in §14.5.3. Derr (1969a) summarized the observational periods available by 1968, including data from several additional earthquakes. Dziewonski and Gilbert (1972), in a specially detailed analysis of 84 recordings of the 1964 Alaskan earthquake, considerably extended the observational period data.

The regular method of deriving the free Earth oscillation periods is to apply Fourier analysis and power-spectrum techniques (see e.g. Blackman and Tukey, 1958) to suitable records of the ground motion or changes of strain excited by large earthquakes. Spectral curves are derived which show, for example, the distribution of energy density with respect to the period or frequency in the recorded motions. The spectral peaks are attributed to specific normal modes of oscillation, and periods are correspondingly assigned to the modes.

The following subsections are concerned with numerical results on the periods, including questions of reliability and uncertainty in

[Refs. on p. 319]

the period assessments.

14.5.1 *Problems of identification*

A first problem in interpreting the records is to identify correctly the modes which correspond to particular peaks on the observational spectral curves. There is little difficulty in arriving at a good first approximation to the identification of many fundamental modes and some overtone modes, because of the availability of theoretically calculated periods for a variety of Earth models. In practice, the identifications are fairly straightforward for most fundamental modes for which n has values up to 40–50: for most of these modes, the spectral peaks are well indicated on good records and are not so close to other peaks as to be confused with or 'contaminated' by them.

Even so, there can be difficulties in some cases. For example, the period of the fundamental mode $_0S_{10}$ happens to be close to the overtone $_3S_2$, and of $_0S_{11}$ to $_2S_7$; the period of $_0S_{34}$ is nearly identical with those of $_4S_0$, $_4S_{12}$ and $_7S_5$ – see Tables 14.1 and 14.3. (Other cases of near coincidence of periods are apparent in tables 5 and 6 presented by Haddon and Bullen, 1969.) However, near-coincidence of a pair of periods does not always cause difficulty with the mode of larger amplitude: there may be theoretical evidence that the smaller mode has negligible amplitude compared with the larger. But each case of near coincidence has to be individually examined.

The problems of identification are further complicated by the presence of spurious peaks on the observational spectral curve, arising from accidental combinations of circumstances. Again, comparison with results for theoretical models assists toward discriminating between peaks which are and are not spurious, but there remain doubtful cases for which reliable criteria are not available.

Uncertainties of identification are still serious with a number of overtones for which n is small and with many modes, both fundamental and overtone, for which n is large. The process of identification is assisted in some cases by using evidence relating to nodal patterns and by comparing results from records taken on different types of instrument, e.g. gravimeters and other instruments (see §14.4.2). Examples of progress made in identifying modes in some of the difficult cases have been given (among others) by Haddon and Bullen (1969) and Dziewonski and Gilbert (1972, 1973).

14.5.2 Problems of precision

When modes have been correctly identified, there remains the problem of estimating the most probable values of the periods and assessing the uncertainties. Sources of uncertainty include the as yet not satisfactorily determined influence of damping on the periods (see § 12.4) and, especially with modes for which n is of the order of 50 or more, the presence of lateral variations of structure in the outer mantle. The uncertainties arising from lateral variations are similar to those already discussed in the case of surface waves (§ 13.6.5), and so are considerable for the larger n. Other uncertainties arise through interference between the ground movements associated with pairs of modes of nearly equal periods, and through rotational splitting (§ 14.2.2).

Somewhat independently of the foregoing sources of uncertainty, there are awkward questions of the significance to be attached to the results of applying harmonic analysis to the recorded data: underlying statistical problems are difficult and have not yet been solved to general satisfaction. Jeffreys (1967, 1970) has warned against accepting many of the period assessments, including formal assessments of uncertainty, at face value. Dziewonski and Gilbert (1972, 1973) and others have looked into the question of the resolving power of free Earth oscillation data in some detail and sought checks on the extent of influence of systematic errors.

On the overall evidence to date, much caution is still needed in attempting to set close bounds of uncertainty to the estimated oscillation periods. The difficulties are least with most fundamental modes for which n is not greater than 40 to 50. With these modes, it is likely that the uncertainties of the observational periods derived by some investigators do not greatly exceed the formally assessed standard errors. These particular period results can therefore supply important well-determined restrictions on the Earth's internal structure. But the difficulties are fairly serious with most fundamental modes for which $n > 50$ and with overtones in general: the real uncertainties may be decidedly greater than the apparent uncertainties. Nevertheless, when realistic allowance is made for uncertainties, the observational periods for these latter modes can sometimes contribute important information on the Earth's interior. An example is the additional evidence provided (§ 14.9) on the solidity of the Earth's inner core.

[Refs. on p. 319]

14.5.3 Observational periods of the main fundamental modes

The observational data assembled by Pekeris (1966) included periods of fundamental spheroidal oscillations for $2 \leq n \leq 48$ and of fundamental torsional oscillations for $2 \leq n \leq 44$, along with assessments of standard errors. The data were compiled mainly from work of Landisman, Satô, and Nafe (1965), Smith (1966), and Slichter, MacDonald, Caputo, and Hager (1965). Along with Slichter's determination (1966) of the period of $_0S_0$, these data constituted a useful (though not exclusive) body of evidence from free Earth oscillations through which earlier estimates of the Earth's density variation could be noticeably improved. (The mode $_0S_1$ is not sufficiently well observed to be useful, the displacement at the Earth's surface being nearly horizontal and relatively small. The theoretically estimated period is about 42 min. For further detail on $_0S_1$, see Pekeris, Alterman, and Jarosch, 1963.) A selection of the periods is shown in Table 14.1. The column headed 'Range of activity' gives, for spheroidal oscillations, rough estimates of the ranges of depth below the Earth's surface inside which the motion is significant. The uncertainties shown are, in all cases except $_0S_0$, the standard errors of the means as determined by Pekeris. For $_0S_0$, the period is Slichter's (later closely confirmed by Dziewonski and Gilbert, 1972), with the standard error of the mean as assessed by Dziewonski and Gilbert.

The periods in Table 14.1 have been used in Earth model calculations by a number of investigators and, in particular, in some of the calculations to be referred to later. Several attempts have been made to improve the estimates of the periods and to narrow the assessments of error. It is questionable, however, whether any significant improvements to the periods have yet been made [cf. the preferred values arrived at by Dziewonski and Gilbert (1972) using a much larger quantity of data] and also, having regard to the difficulties referred to in §14.5.2, whether the real uncertainties have been reduced. The periods in Table 14.1 therefore continue to be serviceable.

14.5.4 Observational periods of other fundamental modes

In the case of fundamental modes for which n is larger than the values shown in Table 14.1, the various complications mentioned in §14.5.2 become more serious, and it is questionable whether the

14.5] PERIODS OF FUNDAMENTAL MODES

Table 14.1

Observational (O) periods (in s) and residuals (O − C) against calculated periods (for Model HB_1) of fundamental free Earth oscillations; results for smaller n

	Spheroidal oscillations			Torsional oscillations	
n	(O)	(O − C)	Range of activity	(O)	(O − C)
0	1227·7 ± 0·1	−1·1	6370 km		
2	3233·1 ± 7·3	+6·2	5850	2642·5 ± 24·4	+13·2
3	2139·2 ± 4·3	+3·6	5500	1704·7 ± 4·9	+2·7
4	1546·0 ± 2·1	−1·2	5270	1305·1 ± 1·8	+1·4
5	1188·4 ± 1·5	−3·0	5090	1075·5 ± 1·2	−0·2
6	962·3 ± 0·9	−2·0	4630	926·5 ± 1·4	+0·7
7	809·1 ± 1·0	−3·4	4630	818·2 ± 1·3	−0·1
8	707·7 ± 0·4	−0·2	4630	735·0 ± 1·3	−1·7
9	634·0 ± 0·3	+0·1	4630	669·8 ± 1·5	−2·2
10	579·3 ± 0·8	−0·1	3940	618·9 ± 0·5	−0·3
11	536·8 ± 0·5	−0·3	3590	575·2 ± 0·4	+0·1
12	502·3 ± 0·2	−0·3	3590	536·6 ± 0·8	−1·0
13	473·2 ± 0·2	−0·3	3250	505·3 ± 0·7	+0·1
14	448·4 ± 0·3	0·0	3250	476·8 ± 0·4	0·0
15	426·3 ± 0·2	−0·1	3250	453·2 ± 0·9	+1·5
16	406·8 ± 0·3	−0·2	3250	430·1 ± 0·3	+0·7
17	389·3 ± 0·2	−0·4	2800	408·3 ± 0·4	−1·0
18	373·9 ± 0·2	−0·3	2800	390·7 ± 0·6	−0·4
19	361·5 ± 0·8	+1·4	2600	374·8 ± 0·4	+0·2
20	347·3 ± 0·3	−0·1	2600	359·7 ± 0·8	+0·3
21	335·8 ± 0·2	0·0	2600	346·1 ± 0·4	+0·6
22	324·8 ± 0·3	−0·4	2400	332·9 ± 0·6	+0·2
23	315·5 ± 0·1	+0·2	2400	321·4 ± 0·7	+0·6
24	306·3 ± 0·2	+0·1	2400	310·7 ± 0·6	+0·9
25	297·6 ± 0·1	−0·1	2200	298·3 ± 1·2	−1·2
26	289·9 ± 0·2	+0·2	2200	289·8 ± 0·3	−0·1
27	281·8 ± 0·2	−0·5	1800	281·4 ± 1·0	+0·6
28	275·2 ± 0·2	0·0	1800	272·7 ± 1·4	+0·3
29	268·4 ± 0·1	−0·1	1800	264·4 ± 1·5	0·0
30	262·1 ± 0·1	−0·1	1800	256·5 ± 0·8	−0·4
31	256·2 ± 0·2	0·0	1800	249·6 ± 0·7	−0·2
32	250·3 ± 0·2	−0·2	1800	243·4 ± 1·1	+0·3
33	245·0 ± 0·1	0·0	1800	235·5 ± 1·8	−1·2
34	239·8 ± 0·2	0·0	1800	233·2 ± 2·0	+2·5
35	234·9 ± 0·2	+0·1	1400	224·9 ± 1·3	−0·1

[*Refs.* on p. 319]

Table 14.1 (Continued)

n	Spheroidal oscillations (O)	(O − C)	Range of activity	Torsional oscillations (O)	(O − C)
36	229·9 ± 0·3	−0·1	1400 km	219·4 ± 0·9	−0·1
37	224·9 ± 0·5	−0·5	1400	213·3 ± 1·0	−1·0
38	219·8 ± 0·3	−1·1	1400	209·3 ± 0·8	−0·1
39	216·4 ± 0·5	−0·2	1400	203·2 ± 1·5	−1·4
40	212·3 ± 0·4	−0·2	1400	200·0 ± 1·3	−0·1
41	208·3 ± 0·4	−0·2	1400	196·3 ± 1·5	+0·5
42	204·7 ± 0·2	0·0	1400	190·7 ± 2·0	−0·9
43	200·8 ± 0·3	−0·1	1400	186·2 ± 2·0	−1·4
44	197·6 ± 0·1	+0·3	1400	182·6 ± 2·0	−1·2
45	194·0 ± 0·2	+0·1	1400		
46	191·2 ± 0·1	+0·7	1400		
47	187·4 ± 0·6	+0·2	1400		
48	184·3 ± 0·5	+0·2	1400		

results of regular spectral analysis of records for these modes can lead to improvements on the information derivable through surface-wave analysis.

A list of apparent observational periods, derived mainly through spectral analysis, was assembled by Derr (1969a) for many further fundamental modes including $_0S_n$ ($n = 49-97$) and $_0T_n$ ($n = 45-99$).

Dziewonski and Landisman (1970) produced another list, using techniques involving P-SV and SH surface-wave theory and both phase and group velocity observations. These authors sought to reduce observational uncertainties by applying an auto-correlation method developed by Landisman, Dziewonski, and Satô (1969). Group-velocity as well as phase-velocity information was used to help discriminate between different modes, resulting in some gain of resolution for the larger n. (For the smaller n, spectral analysis gives the better results.) The periods listed by these authors relate to two particular great-circle paths; each path goes through the epicentre of the exciting earthquake, the 1960 Chilean earthquake in one case, and the Kurile Islands earthquake of 1963 October 13 in the other. The paths cut approximately at right-angles, and differences between corresponding periods for the two paths were taken by the authors as a tentative index of uncertainties introduced by lateral variations. But since both paths are mixed, the interpretation involves the usual

[Refs. on p. 319]

difficulties encountered in surface-wave analysis, already referred to in Chapter 13, and again the uncertainties may be substantially greater than these differences.

Table 14.2 gives a list of observational periods for $_0S_n$ and $_0T_n$ for larger n, derived from the results of Derr, Dziewonski, and Landisman. Because of doubt about the size of uncertainties, these periods are given only to the nearest second; they are set down for the sake of general information rather than for detailed application.

Table 14.2

Observed periods (in s) of fundamental free Earth oscillations for larger n

n	Period of spheroidal oscillations	Period of torsional oscillations
45		180
50	178	164
55	165	151
60	154	139
65	143	129
70	134	121
75	127	113
80	119	107
85	113	101
90	108	96
95	102	92

14.5.5 Observational periods of overtones

The number of free Earth overtone oscillations, even when limited to those with periods τ greater than (say) 100 s is legion. A partial list prepared by Haddon and Bullen (1969) for $\tau > 178$ s (the period of $_0S_{50}$ — Table 14.2) included 155 spheroidal and 91 torsional overtones.

As late as 1969, only 16 of these overtones had been identified on three or more records (see Derr, 1969a). (Another 55 had been possibly identified but on only one or two records in each case.) By 1972, however, the number of observed overtone periods had been much increased.

Table 14.3 includes the full list of spheroidal overtone periods observationally derived by Dziewonski and Gilbert (1972). The

[Refs. on p. 319]

periods are means for an average of 18 records. The formally computed standard errors of the means are all less than 0·1 per cent. Dziewonski and Gilbert applied a number of tests in seeking to minimize the uncertainties, but the difficulties stated in § 14.5.2 make it still desirable to allow for uncertainties larger than those formally indicated. Entries in brackets in Table 14.3 are taken from Derr's list and are for overtones not observed by Dziewonski. The results are shown in considerable detail for several reasons: to illustrate features of the 'crowding' that can occur with some groups of periods; to provide background enabling some of the difficulties referred to in §§ 14.5.1 and 14.5.2 to be better appreciated; to indicate the order of accuracy now being sought in applications of free Earth oscillation data; and to provide an informative comparison with some theoretical results to be given later.

Table 14.4 contains similar results for torsional overtones. The number of observational periods available is much smaller than for spheroidal overtones, and the average number of records suitable for deriving the torsional overtone periods is smaller. The formally assessed standard errors of the means are, however, mostly still below 0·1 per cent.

Table 14.3

Observational (O) and calculated (HB_1) periods (in s) of free Earth spheroidal overtones; x = percentage difference (O − C)

Overtone	(O)	HB_1	x	Overtone	(O)	HB_1	x
$_1S_1$		2477·9		$_3S_2$		581·6	
$_1S_2$	1470·8	1468·2	+0·1	$_1S_8$	555·8	556·3	−0·1
$_1S_3$	1060·8	1063·3	−0·2	$_2S_7$	(537·1)	536·3	(+0·1)
$_2S_1$	1058·1	1060·0	−0·2	$_1S_9$	509·6	510·0	−0·1
$_2S_2$	904·2	915·1	−1·2	$_4S_1$	505·8	504·5	+0·3
$_1S_4$	852·7	852·0	+0·1	$_3S_3$	489·1	489·1	0·0
$_2S_3$	804·2	803·5	+0·1	$_2S_8$	488·0	488·0	0·0
$_1S_5$	730·6	729·2	+0·2	$_4S_2$	479·3	476·9	+0·5
$_2S_4$	724·9	725·3	−0·1	$_1S_{10}$	465·4	466·3	−0·2
$_3S_1$		707·2		$_4S_3$	460·8	459·3	+0·3
$_2S_5$	660·4	660·5	0·0	$_2S_9$	448·4	448·6	−0·1
$_1S_6$	657·6	656·7	+0·2	$_5S_1$		447·0	
$_1S_0$	613·6	607·4	+1·0	$_3S_4$	439·2	437·1	+0·5
$_1S_7$	603·9	604·2	0·0	$_1S_{11}$		426·8	
$_2S_6$	594·7	595·3	−0·1	$_4S_4$	420·1	421·1	−0·2

[*Refs. on p. 319*]

Table 14.3 (Continued)

Overtone	(O)	HB_1	x	Overtone	(O)	HB_1	x
$_2S_{10}$	415·7	416·0	−0·1	$_5S_6$		293·2	
$_3S_5$	415·1	413·2	+0·5	$_6S_4$	293·2	292·1	+0·4
$_5S_2$	397·4	399·1	−0·4	$_2S_{16}$	(293·3)	291·0	(+0·8)
$_2S_0$	398·5	394·0	+1·1	$_1S_{17}$	286·0	286·0	0·0
$_1S_{12}$	(396·7)	392·2	(+1·1)	$_3S_{13}$		284·2	
$_3S_6$	392·3	390·8	+0·4	$_4S_8$	283·6	283·0	+0·2
$_2S_{11}$	388·3	388·6	−0·1	$_7S_3$	281·4	282·1	+0·1
$_3S_7$	372·0	371·0	+0·3	$_5S_7$		281·1	
$_4S_5$	369·7	370·5	−0·2	$_1S_{18}$	(271·3)	274·2	(−1·1)
$_2S_{12}$	365·1	365·1	0·0	$_2S_{17}$		274·0	
$_1S_{13}$	(365·1)	362·2	(+0·8)	$_6S_5$	273·5	273·0	+0·2
$_5S_3$	353·5	354·5	−0·3	$_3S_{14}$		272·5	
$_3S_8$	354·6	353·5	+0·3	$_8S_1$	272·1	270·5	+0·6
$_6S_1$	348·4	346·6	+0·5	$_4S_9$	269·7	268·7	+0·4
$_2S_{13}$	344·9	344·6	+0·1	$_5S_8$		267·4	
$_3S_9$	339·1	337·7	+0·4	$_1S_{19}$	(263·2)	263·5	(−0·1)
$_1S_{14}$	337·0	336·6	+0·1	$_3S_{15}$		261·5	
$_4S_6$	332·1	332·5	−0·1	$_7S_4$		258·9	
$_2S_{14}$	326·3	326·3	0·0	$_2S_{18}$		258·7	
$_5S_4$		323·5		$_4S_{10}$	258·9	257·8	+0·4
$_3S_{10}$	323·8	323·1	+0·2	$_1S_{20}$	(252·9)	253·7	(−0·3)
$_1S_{15}$	316·1	315·5	+0·2	$_6S_6$		252·5	
$_7S_1$		311·5		$_5S_9$		252·2	
$_6S_2$		311·1		$_3S_{16}$		251·2	
$_3S_{11}$	310·8	309·4	+0·4	$_4S_{11}$		248·4	
$_2S_{15}$	309·2	308·9	+0·1	$_8S_2$	247·7	247·2	+0·2
$_7S_2$	310·1	308·5	+0·5	$_2S_{19}$		244·9	
$_5S_5$		305·7		$_1S_{21}$		244·7	
$_4S_7$	304·0	303·8	+0·1	$_3S_{17}$		241·5	
$_6S_3$		303·5		$_7S_5$	240·0	240·1	0·0
$_3S_0$	305·8	300·9	+1·6	$_4S_0$	243·6	239·9	+1·5
$_1S_{16}$	299·9	299·4	+0·2	$_4S_{12}$		239·9	
$_3S_{12}$		296·4					

14.5.6 *Expression of results in terms of phase velocity*

In the preceding subsections, results have been presented in terms of the period τ. In practice, the results are instead commonly presented in terms of v, where v is given by (13.12), namely

[*Refs. on p. 319*]

FREE EARTH OSCILLATIONS [14.5-

$$(n + \tfrac{1}{2})v\tau = 2\pi a, \qquad (14.27)$$

where a is the mean radius of the Earth. The entity v corresponds to phase velocity in contexts (e.g. §§ 13.5, 14.5.4) where the oscillations are looked at as combinations of travelling waves. When spectral analysis is being applied to the records, it is sufficient to regard v simply as an alternative parameter to τ in presenting the results; it is frequently convenient to use v instead of τ because v varies rather more evenly with n.

Table 14.4
Observational (O) and calculated (HB_1) periods (in s) of free Earth torsional overtones (x as in Table 14.3)

Overtone	(O)	HB_1	x	Overtone	(O)	HB_1	x
$_1T_1$		805.8		$_1T_{10}$	381.6	380.5	+0.3
$_1T_2$	756.6	754.6	+0.3	$_2T_7$	363.7	361.4	+0.6
$_1T_3$	695.2	692.3	+0.4	$_1T_{11}$		358.3	
$_1T_4$	630.0	628.4	+0.3	$_2T_8$	343.5	341.7	+0.5
$_1T_5$	(570.6)	569.2	(+0.2)	$_1T_{12}$	(338.1)	338.9	(−0.2)
$_1T_6$	519.1	517.5	+0.3	$_2T_9$		322.7	
$_1T_7$	(474.2)	473.7	(+0.1)	$_1T_{13}$		321.8	
$_2T_1$		454.0		$_3T_1$		311.0	
$_2T_2$		445.3		$_3T_2$		308.1	
$_1T_8$	438.5	437.1	+0.3	$_1T_{14}$		306.6	
$_2T_3$		432.9		$_2T_{10}$		304.9	
$_2T_4$	421.8	417.6	+1.0	$_3T_3$		303.9	
$_1T_9$		406.4		$_3T_4$		298.5	
$_2T_5$		400.1		$_1T_{15}$	(293.3)	292.9	(+0.1)
$_2T_6$		381.1		$_3T_5$		292.0	

Table 14.5 shows approximate corresponding values of τ and n for a selection of fundamental modes. Further details for $n \leqslant 50$ are given in Table 4 of Haddon and Bullen (1969).

14.6 Early inferences from free Earth oscillation data

It was pointed out in § 14.1.2 that each observational period for a mode of free Earth oscillations in effect contributes an equation of condition restricting the possible distributions of (ρ, k, μ) or (ρ, α, β) in the Earth's interior. The observed periods for the main fundamental

Table 14.5

Corresponding values of τ and v for fundamental free Earth oscillations

n	$\tau(_0S_n)$ (s)	$v(_0S_n)$ (km/s)	$\tau(_0T_n)$ (s)	$v(_0T_n)$ (km/s)
2	3227	4·96	2629	6·09
3	2136	5·36	1702	6·72
4	1547	5·75	1304	6·82
5	1191	6·11	1076	6·77
6	964	6·39	926	6·65
8	708	6·65	737	6·39
10	579	6·58	619	6·16
15	426	6·06	452	5·72
20	347	5·62	359	5·43
25	298	5·27	300	5·24
30	262	5·00	257	5·11
35	235	4·80	225	5·01
40	212	4·65	200	4·94
45	194	4·54	180	4·88
50	178	4·45	164	4·84
60	153	4·32	140	4·72
70	134	4·23	122	4·67
80	119	4·17	107	4·63
90	107	4·13	96	4·60
100	97	4·10	87	4·57

modes listed in Table 14.1 alone provide the equivalent of nearly 100 such equations. These periods constitute a large, but not the whole, part of the free Earth oscillation data so far successfully applied to the problem of the Earth's internal structure.

As previously stated, the data do not of themselves lead to a closely determined Earth model. (For some further detail on lack of uniqueness, see Backus and Gilbert, 1968, 1970; see also §15.4.3.) The principal ways in which they have been brought to bear are: (i) checking the reliabilities of previously existing Earth models; (ii) investigating changes to previous models that appear to be demanded by the oscillation data; (iii) drawing fairly sharp inferences on some specific features of the Earth's interior where this is possible; (iv) contributing towards the derivation of Earth models which take account of the overall evidence. The remainder of the chapter will be principally concerned with (i), (ii), and (iii); the point (iv) is

[Refs. on p. 319]

considered in Chapter 16. In §§ 14.6.1 and 14.6.2, some early inferences from oscillation data, chiefly relating to the points (i) and (iii), are discussed. The points (ii) and (iii) are involved in § 14.7, and (iii) again in § 14.9.

14.6.1 *Evidence on the reliability of the Earth models A and B*

Reference has been made to the free Earth oscillation periods derived from early analyses of records of the 1960 Chilean earthquake by the four groups of observers mentioned in § 14.4.2. These observational periods were compared with periods calculated for the original Earth models A and B, and for several variants (due to Birch, Bolt, Bullard, and others) based on the original models but incorporating assigned deviations from them. The comparison showed sufficiently good agreement with the original models to make it strongly evident both that free Earth oscillations had been authentically recorded and that models which are not fairly close to the A and B types are untenable. In particular, the early studies showed that the original A and B model distributions of ρ, k, and μ are reliable within the order of accuracy that had been previously estimated. (See **B**, § 14.5.2.)

With this agreement established, attention was turned to the problem of using oscillation data to improve the existing Earth models. The early studies pointed to ranges of depth in both the A and B models where the distributions, particularly of ρ and μ, seemed capable of some improvement. In the immediately following years, oscillation periods were calculated for a large variety of deviant models. In the scramble to obtain early results, a number of rather unsatisfactory hybrid models were used, for example, models which combined Model A or B density distributions with Gutenberg's distributions of α and β (§ 9.6.2).

Most of the more detailed inferences made during these early years were, however, largely nullified by the sizeable revision of the value of y (the Earth's moment of inertia coefficient) shown in 1963 to be required by artificial satellite data. The revision made it imperative to revise all pre-1963 Earth models independently of the free Earth oscillation evidence, resulting in models such as A'' (§ 10.7.2). This necessity led to a fresh start in the testing of models against the oscillation data, and the testing subsequently proceeded in a more orderly way.

[*Refs, on p. 319*]

14.6.2 Density gradient in the lower mantle

Landisman, Satô, and Nafe (1965) were the first to take account of the revised y in applying free Earth oscillation data to questions of the Earth's internal structure. Their oscillation data consisted principally of observations of fundamental spheroidal periods up to $n = 25$ and of fundamental torsional periods up to $n = 11$. They sought a density distribution which would be consistent with these data and various background data, including evidence on P and S velocities derived from analyses of bodily seismic waves.

They were unable to achieve consistency without departing considerably from previous model density distributions. In particular, their calculations led them to infer that $d\rho/dz \approx 0$ throughout an assumed abnormal zone lying between 1600 and 2800 km depth, and they arrived at a pair of preferred Earth models called M1 and M3 incorporating this feature. (The structures of M1 and M3 differed only in the core.) They suggested that the abnormal zone would be characterized by the presence of both a super-adiabatic temperature gradient of $2°K/km$ and an iron content decreasing with depth. Having regard to the extent of decrease of iron required, this explanation could not, however, be regarded as very plausible; in any case, strong arguments can be raised against the presence of a near-zero density gradient in the mantle. For example, since $\beta^2 = \mu/\rho$ and the variation of β is known to be close to normal inside most of the abnormal zone, taking $d\rho/dz = 0$ would give $d\mu/dz$ abnormally high, a result that is quite contrary to all other evidence for this part of the Earth.

The thoroughness of the analysis carried out by Landisman and his colleagues (the formal correctness of their calculations was later well confirmed in work of Bullen and Haddon, 1967a, 1967b) and the surprising conclusion they reached kindled renewed interest in the application of free Earth oscillation data to the Earth's interior. Attention was focused on some new characteristics of the oscillation data, and endeavours were made to find alternatives to the conclusion that $d\rho/dz \approx 0$ inside much of the lower mantle.

As part of their background data, Landisman and his colleagues had assumed for the core-radius R_c the value 3470 km, which is compatible with the estimate 3473 ± 3 km arrived at by Jeffreys in 1939. Bullen and Haddon (1967a) showed, however, that, by increasing the assumed core-radius by about 15-20 km and otherwise

[*Refs. on p. 319*]

adhering to the data and assumptions of Landisman, the coefficient η (§11.5.2) could be held equal to unity, and thus $d\rho/dz$ kept fully normal, throughout the region D'.

An incidental consequence was the direction of attention to the problem of reconciling different lines of evidence on R_c, already discussed in §12.4. This is one example of free Earth oscillation data contributing (though not in isolation) to knowledge of a specific feature of the Earth's internal structure. Other examples are indicated in §§14.7 and 14.9.

14.7 The model HB_1

14.7.1 Background

In seeking an alternative to Landisman's conclusion, Bullen and Haddon set themselves the task of estimating the modifications needed to existing Earth models that appeared to be demanded by the observational oscillation periods listed in Table 14.1. [Strictly, the word 'demand' is a shade too strong because of lack of uniqueness in the inversion problem (see §15.4). The word is used in practice when all of numerous trial-and-error results point the same way.]

An appropriate procedure is to start with a previously existing Earth model and to perturb it inside the previously assessed range of uncertainties until the oscillation data are fitted within their uncertainties. The starting point taken by Bullen and Haddon was Model A'' (§10.72), which had been constructed explicitly for this purpose. In spite of lack of uniqueness in the end product, and although the changes from A'' were mainly quite small, many perturbed models had to be generated and analysed before a satisfactory model (HB_1) was found in 1967. Wiggins (1968) later supplied tables, derived by perturbing the Lagrangian integral for free oscillations of spherically symmetrical Earth models, which greatly assist in the process of fitting the oscillation data.

The entries under 'Range of activity' in Table 14.1 show that the motion in the Earth's inner core in the modes $_0S_n$ is quite negligible for $n > 5$, and in the whole core for $n > 15$. All the modes $_0S_n$ are, however, influenced by the structure above the core. Consequently, the possibility of using the observational periods in Table 14.1 to provide improved knowledge of the core structure depends on the precision with which the mantle structure is determined. Since this precision proved to be not fine enough for the purpose, the oscillation data in question did not demand any change from the simple entirely fluid core taken in Model A''. (But evidence on oscillation overtones can lead to increased knowledge of core structure — see §14.9.) Also no changes were demanded from the P velocity distribution of Model A''.

On the other hand, the oscillation data of Table 14.1 did demand modifications in respect of ρ and β. The Model HB_1, designed to be the parametrically simplest spherically symmetrical model meeting the new demands, incorporated these modifications.

14.7.2 *Features of the model*

In constructing Model HB_1, the thickness of the crustal region A was reduced to 15 km from the continental value of 33 km taken in Model A''. The region B was subdivided into B' ($15 < z < 60$ km) and B'' ($60 < z < 350$ km), and C into C' ($350 < z < 650$ km) and C'' ($650 < z < 984$ km). None of the depths of the boundaries between these regions is closely determined by the oscillation data: a considerable variety of alternative models fit the data within the uncertainties. Nevertheless a more complex upper mantle structure than that of Model A'' was demanded; Model HB_1 has the minimum additional complexity needed.

For the region B, a substantial reduction was demanded in the coefficient η below the Williamson-Adams value of unity which holds in Model A''. The value taken in Model HB_1 was 0·45 but could have been less. A substantial reduction was also demanded in the average value of $d\beta/dz$ inside the region B. Subject to the not quite certain proviso that the globally average value of β is at least 4·6 km/s at the top of B, it was confirmed that β must fall inside B, in agreement with other evidence that a low β layer here exists. In Model HB_1, β was conventionally taken constant inside each of B' and B'', the values being 4·625 and 4·5 km/s.

For the region C, the principal changes from Model A'' were inside the upper subdivision C', where smaller gradients of ρ and β were indicated than in Model A'', Below C, the changes from Model A'' were, apart from the changes in R_c, of the nature of secondary adjustments.

[*Refs. on p. 319*]

Table 14.6

Densities, pressures, and P and S velocities in the upper part of Model HB_1

Region	Depth (km)	Pressure (10^{11} N/m²)	Density (g/cm³)	P velocity (km/s)	S velocity (km/s)
---	0	0·000	2·84	6·30	3·55
A	15	0·004	2·84	6·30	3·55
---	15	0·004	3·31	7·70	4·625
B'	60	0·019	3·33	7·83	4·625
---	60	0·019	3·33	7·83	4·50
B''	350	0·116	3·44	8·75	4·50
---	350	0·116	3·70	8·75	4·50
C'	650	0·234	4·15	10·48	5·80
---	650	0·234	4·20	10·48	5·90
C''	984	0·380	4·53	11·42	6·35

Table 14.6 gives a selection of values of (ρ, α, β) for Model HB_1 down to the bottom of the region C, covering the range of depth inside which (apart from the sizeable increase in R_c) the essential changes were made from Model A''. For more complete details, see Haddon and Bullen (1969). The free oscillation periods of the main fundamental modes of HB_1 are obtained from Table 14.1 by subtracting the entries under (O) and (O − C); the table shows the degree of fit attained.

At the time when Model HB_1 was constructed, observational overtone periods of free Earth oscillations were too meagre and uncertain to be used. A large number of overtone periods of HB_1 were, however, computed, a partial list being shown in Table 14.3. The considerable degree of agreement with overtone periods subsequently observed indicates an important role of the table in enabling specific overtones to be identified in spectral analyses of recorded free Earth oscillations. At the same time, there are a few significant discrepancies between the observational and calculated periods in Table 14.3. These discrepancies are referred to in § 14.9 and have an important implication.

A comparison of the degrees of fit of HB_1 and several other models with the observational oscillation data is shown in Figs. 1 and 2 of Bullen and Haddon (1967b).

[*Refs. on p. 319*]

14.7.3 Significance of Model HB_1

In view of misconceptions by some other writers (see Bullen and Haddon, 1973), it needs to be stressed that Model HB_1 was not intended to be in any sense a final Earth model. As stated, the central purpose in constructing it was to provide the simplest model which would incorporate the demands of a primary set of observational oscillation data and so serve as a guide to further progress. In this respect, HB_1 marks a specific stage in the evolution of Earth models. It is partly for this reason that the model has been here discussed at some length. The fact that it incorporates the principal changes *demanded* by the oscillation data used has the consequence, moreover, that the fundamental oscillation periods of well-determined models which incorporate additional detail (e.g. those which allow for a less simple core) will not be expected to differ very much from the HB_1 periods.

14.8 Other models using free Earth oscillation data

Among the more important models constructed around the same time as HB_1 is a set due to Pekeris (1966). The model of this set which goes closest to fitting the oscillation data, called M3(G—LSN), is broadly similar to the Landisman model M3; some other models of the set are of interest through showing density inversions (see § 13.7) inside the outermost 200 km. Other contemporaneous models of some interest include the model Q1 of Gilbert and Backus (1968), and the model DI—11 of Derr (1969b) which incorporated (among other things) his inferences in § 14.9 on the inner core.

The most recent models of importance include the models UTD124 (A', B') and B497 of Dziewonski and Gilbert (1972, 1973) and the model B1 of Jordan and Anderson (1974). These models all have core radii between 3482 and 3485 km and inner-core radii between 1215 and 1250 km. Further details of the two last models will be discussed in Chapter 16.

14.9 Oscillation evidence on solidity of inner core

As stated, the core of Model HB_1 was taken entirely fluid because the reliable oscillation evidence available by 1969 did not demand otherwise. The extensive additional observational data on overtone

[*Refs. on p. 319*]

periods provided by Dziewonski and Gilbert (1972) led, however, to additional demands. Inspection of Table 14.3 shows immediately that the percentage differences between the observational periods and those calculated for Model HB_1 are markedly greater for the modes $_2S_2$, $_1S_0$, $_2S_0$, $_3S_0$, and $_4S_0$ than for all other spheriodal overtones (excluding less reliably observed periods shown in brackets). Analysis shows that these differences (together with smaller differences with certain other modes) are just what are to be expected if the Earth's inner core is solid.

Derr (1969b) had inferred support for solidity in the inner core, for which he estimated β to be 2·2 km/s, using a far more limited set of overtone data than that shown in Table 14.3. There was, however, considerable uncertainty in his inference because of uncertainties in the identification of modes (several corrections had later to be made — see e.g. Dziewonski and Gilbert, 1972) and in the periods used. Further, his data simultaneously implied a density jump of 2 g/cm^3 at the base of the outer core, which has not been found necessary in later work. Haddon (see Bullen, 1972) inferred, also from more limited oscillation data than in Table 14.3, that it is strongly probable that significant rigidity exists somewhere inside the core, and on other grounds this rigidity must be located in the inner core.

The evidence from oscillation data for a solid inner core was strengthened by Dziewonski and Gilbert (1971, 1972, 1973) using their much more extensive and better-determined overtone periods. They found that their attempts to fit the data with fluid-core models results in quite implausible decreases in density at the inner-core boundary. They also explored, again with implausible consequences, such alternatives as a strongly super-adiabatic temperature gradient in the lower outer core. Their model B497 gave β as ranging from 3·7 at the top of the inner core to 3·5 km/s at the centre [cf. the estimate 3·9–2·9 km/s derived (§ 12.6.2) by Bullen and Haddon (1967c) on other grounds]. The model B1 of Jordan and Anderson gave a mean value of 3·5 km/s in the inner core. Dziewonski and Gilbert (1973) regarded their identifications of the oscillation overtones $_{10}S_2$ and $_{11}S_2$ with periods near 247 s as crucial evidence that the inner core is solid. [Haddon and Bullen (1969) had given 206 s for $_{10}S_2$ in Model HB_1, which assumes a fluid inner core.] It is desirable, however, that the new identifications be confirmed.

The foregoing discussions mention only a few of the legions of Earth models produced during the last few years of this computer age. Even from the small sample given, readers will doubtless be intrigued by the different tastes of authors in naming their brain-children.

REFERENCES

Alsop, L.E., Sutton, G.H. and Ewing, M. (1961). Free oscillations of the Earth observed on strain and pendulum seismographs. *J. Geophys. Res.*, 66, 631–641.

Alterman, Z., Jarosch, H. and Pekeris, C.L. (1959). Oscillations of the Earth. *Proc. Roy. Soc.*, A, 252, 80–95.

Backus, G. and Gilbert, F. (1961). The rotational splitting of the free oscillations of the Earth. *Proc. Nat. Acad. Sci. USA.*, 47, 362–371.

Backus, G. and Gilbert, F. (1968). The resolving power of gross Earth data. *Geophys. J., Roy. Astr. Soc.*, 16, 169–205.

Benioff, H. (1935). A linear strain seismograph. *Bull. Seismol. Soc. Amer.*, 25, 283–309.

Benioff, H., Press, F. and Smith, S. (1961). Excitation of the free oscillations of the Earth by earthquakes. *J, Geophys. Res.*, 66, 605–619.

Blackman, R. and Tukey, J. (1958). *The Measurement of Power Spectra*, Dover, New York.

Bogert, B.P. (1961). An observation of free oscillations of the Earth. *J. Geophys. Res.*, 66, 643–646.

Bolt, B.A. (1963). Recent information on the Earth's interior from studies of mantle waves and eigenvibrations. *Phys. and Chem. of Earth.*, 5, 55–119.

Bullen, K.E. (1972). Some problems connected with constructing Earth models. *Trans. Conference on Solid Earth Problems, Buenos Aires, 1970*, II, 63–77.

Bullen, K.E. and Haddon, R.A. (1967a). Earth oscillations and the Earth's interior. *Nature* (Lond.), 213, 574–576.

Bullen, K.E. and Haddon, R.A. (1967b). Derivation of an Earth model from free oscillation data. *Proc. Nat. Acad. Sci. USA.*, 58, 846–852.

Bullen, K.E. and Haddon, R.A. (1967c). Earth models based on compressibility theory. *Phys. Earth Planet. Interiors*, 1, 1–13.

Bullen, K.E. and Haddon, R.A. (1973). Some recent work on Earth models, with special reference to core structure. *Geophys. J., Roy. Astr. Soc.*, 35, 31–38.

Caputo, M. (1963). Free modes of layered oblate planets. *J. Geophys. Res.*, 68, 497–503.

Carr, R.E. (1961, 1963). Free oscillations of a gravitating solid sphere. *Jet Propulsion Laboratory, Calif. Inst. Tech.*, Technical Repts. 32–164 and 32–372.

Coulomb, J. and Jobert, G. (1973). *Traité de Géophysique Interne*, Tome I, Sismologie et Pesanteur, Masson, Paris.

Dahlen, F.A. (1968). The normal modes of a rotating elliptical Earth. *Geophys. J., Roy. Astr. Soc.*, **16**, 329–367.

Dahlen, F.A. (1969). The normal modes of a rotating elliptical Earth. *Geophys. J., Roy. Astr. Soc.*, **18**, 397–436.

Derr, J.S. (1969a). Free oscillation observations through 1968. *Bull. Seismol. Soc. Amer.*, **59**, 2079–2099.

Derr, J.S. (1969b). Internal structure of the Earth inferred from free oscillations. *J. Geophys. Res.*, **74**, 5202–5220.

Dziewonski, A.M. and Gilbert, F. (1971). Solidity of the inner core of the Earth inferred from normal mode observations. *Nature* (Lond.), **234**, 465–466.

Dziewonski, A.M. and Gilbert, F. (1972, 1973). Observations of normal modes from 84 recordings of the Alaskan earthquake of 28 March 1964. *Geophys. J., Roy. Astr. Soc.*, **27**, 393–446; **35**, 401–437.

Dziewonski, A. and Landisman, M. (1970). Great circle Rayleigh and Love wave dispersion from 100 to 900 seconds. *Geophys. J., Roy. Astr. Soc.*, **19**, 37–91.

Gilbert, F. and Backus, G. (1968). Approximate solutions to the inverse normal mode problem. *Bull. Seismol. Soc. Amer.*, **58**, 103–131.

Haddon, R.A. and Bullen, K.E. (1969). An Earth model incorporating free Earth oscillation data. *Phys. Earth Planet. Interiors*, **2**, 35–49.

Hoskins, L.M. (1920). The strain of a gravitating sphere of variable density and elasticity. *Trans. Amer. Math. Soc.*, **21**, 1–43.

Jeans, J.H. (1923). The propagation of earthquake waves. *Proc. Roy. Soc.*, A, **102**, 554–574.

Jeffreys, Sir H. (1967). Spheroidal oscillations of the Earth. *Geophys. J., Roy. Astr. Soc.*, **14**, 177.

Jeffreys, Sir H. (1970). *The Earth* (5th Ed.), Cambridge University Press.

Jeffreys, Sir H. and Vicente, R.O. (1966). Comparison of forms of the elastic equations for the Earth. *Mem. Acad. Roy. de Belgique*, **37**, 3–31.

Jobert, N. (1956). Évaluation de la période d'oscillation d'une sphere élastique héterogène par l'application du principe de Rayleigh. *C. R. Acad. Sci.* (Paris), **243**, 1230–1232.

Jordan, T.H. and Anderson, D.L. (1974). Earth structure from free oscillations and travel times. *Geophys. J. Roy. Astr. Soc.*, **36**, 411–459.

Lamb, H. (1882). On the vibrations of an elastic sphere. *Proc. Lond. Math. Soc.*, **13**, 189.

Landisman, M., Dziewonski, A. and Satô, Y. (1969). Recent improvements in the analysis of surface wave observations. *Geophys, J., Roy. Astr. Soc.*, **17**, 369–403.

Landisman, M., Satô, Y. and Nafe, J. (1965). Free vibrations of the Earth and the properties of its deep interior regions. Part 1: Density. *Geophys. J., Roy. Astr. Soc.*, **9**, 439–502.

Love, A.E.H. (1911). *Some Problems of Geodynamics*, Cambridge University Press.

REFERENCES

MacDonald, G.J.F. and Ness, N.F. (1961). A study of free oscillations of the Earth. *J. Geophys. Res.*, 66, 1865–1911.

Matumoto, T. and Satô, Y. (1954). On the vibration of an elastic globe with one layer. The vibration of the first class. *Bull. Earthq. Res. Inst.* (Tokyo), 32, 247–258.

Ness, N.F., Harrison, J.C. and Slichter, L.B. (1961). Observations of the free oscillations of the Earth. *J. Geophys. Res.*, 66, 621–629.

Pekeris, C.L. (1966). The internal constitution of the Earth. *Geophys. J., Roy. Astr. Soc.*, 11, 85–132.

Pekeris, C.L., Alterman, Z. and Jarosch, H. (1961). Terrestrial spectroscopy. *Nature* (Lond.), 190, 498–500.

Pekeris, C.L., Alterman, Z. and Jarosch, H. (1963). Studies in terrestrial spectroscopy. *J. Geophys. Res.*, 68, 2887–2908.

Pekeris, C.L. and Jarosch, H. (1958). The free oscillations of the Earth. *Contributions in Geophysics*, 171–192, Pergamon, London.

Slichter, L.B. (1966). A glimpse at the geophysical scene. *Trans. Amer. Geophys. Un.*, 47, 346–354.

Slichter, L.B. (1967). Free oscillations of the Earth. *International Dictionary of Geophysics*, 331–343, Pergamon, London.

Slichter, L.B., MacDonald, G.J.F., Caputo, M. and Hager, C.L. (1966). Comparison of spectra for spheroidal modes excited by the Chilean and Alaskan quakes. *Geophys. J., Roy. Astr. Soc.*, 11, 256.

Smith, S.W. (1966). Free oscillations excited by the Alaskan earthquake. *J. Geophys. Res.*, 71, 1183–1193.

Stoneley, R. (1961). The oscillations of the Earth. *Phys. and Chem. of Earth*, 4, 239–250, Pergamon, London.

Takeuchi, H. (1959). Torsional oscillations of the Earth and some related problems. *Geophys. J., Roy. Astr. Soc.*, 2, 89–100.

Usami, T. and Satô, Y. (1962). Torsional oscillations of a homogeneous elastic spheroid. *Bull. Seismol. Soc. Amer.*, 52, 469–484.

Wiggins, R.A. (1968). Terrestrial variational tables for the periods and attenuation of the free oscillations. *Phys. Earth Planet. Interiors*, 1, 201–266.

Zharkov, V.N. and Lyubimov, V.M. (1970). The theory of spheroidal vibrations for a spherically symmetrical model of the Earth. *Izvestiya, Acad. Sci. USSR* (English Ed.), 10, 613–618.

CHAPTER 15

Miscellaneous developments

The present chapter is concerned with a variety of developments that complement or supplement discussions in earlier chapters, or involve alternative approaches. Some of the developments are important, or at least potentially so. The symbols $\rho, k, \mu, p, g, \alpha, \beta, r$, and z will have their usual meanings; ϕ will denote k/ρ or $\alpha^2 - 4\beta^2/3$ (§11.2.1), and η will be as defined in §11.5.2. The Earth will be treated as spherically symmetrical except where otherwise stated.

15.1 Equations of state, and related equations, for the Earth's interior

Numerical tables or equations connecting ρ or k with p in particular internal zones of the Earth, or relations from which such connections can be readily derived, have the nature of equations of state for the Earth. Examples in tabular form are the Tables 10.2, 12.4, and 14.6, which relate to particular Earth models. Examples in explicit equation form are the equations (11.53,54) which fit the lower mantle and core of a B-type Earth model (§11.8.3). Examples of relations from which the required connections can be readily derived are (7.49,52,53) of the Murnaghan-Birch finite-strain theory, the corresponding explicit (ρ, p) relation being (7.51). A related type of equation is Birch's empirical equation (10.14) connecting ρ with α and the mean atomic weight w.

The equations of state and related equations also involve dependence on temperature and composition, but variables representing those properties do not always need to appear explicitly – for example, when it is understood that temperature gradients are being assumed adiabatic, or when a material of fixed composition is being considered.

[Refs. on p. 345]

As the above examples show, equations of state for the Earth may be largely empirical, or partly empirical and partly theoretically derived; in some cases, they are largely derived from considerations outside the Earth's interior. As already illustrated in Chapter 11, comparisons of equations derived by different routes can contribute to improved knowledge both of the Earth's interior and of the behaviour of certain materials at pressures up to nearly 4×10^{11} N/m². The subject of equations of state is a vast one in its own right and a comprehensive account would be outside the scope of this book. In the following subsections, a few miscellaneous comments will be made on equations of the type mentioned above. In Chapter 17, it will be seen that equations of state for the Earth can be usefully applied to the interiors of other planets.

15.1.1 *Assessment of Birch's finite-strain theory*

The importance, as a pioneering work, of Birch's adaptation of Murnaghan's finite-strain theory (Chapters 7 and 11) makes it desirable to comment a little further on the degree of precision attaching to it.

An important test was provided by the comparison already given of results on k and dk/dp in the Earth derived through Birch's theory and through seismology and related sources of evidence. The degree of agreement made it evident (§11.2.6) that the theory can be relied on to a useful first approximation. At the least, it provides a quantity of well-determined information about the general variation of k in the Earth below 1000 km depth. The particularly close agreement with the seismically based evidence on dk/dp in the lower mantle and upper core is an important success. The agreement was seen in §11.4.1 to be a little less close in respect of the jump Δk in k at the mantle-core boundary N and also the variation of dk/dp at depth in the core (§11.4.5). Even allowing for all uncertainties in the seismically estimated values of α, β, and $d\phi/dz$, there are strong suggestion that Birch's theory here needs some amendments.

A note of caution was also sounded in work of Thomsen and O.L. Anderson (1969) – see also Knopoff (1963) and Thomsen (1970). These authors drew attention to inconsistencies in applications to the Earth of equations of state for solids derived on different approaches to finite-strain theory. The Murnaghan-Birch approach is Eulerian (in the hydrodynamical sense – see Lamb,

1932), while Thomsen and Anderson considered a Lagrangian approach. They found that when approximations based on Taylor expansions of the free energy function [see e.g. (7.47)] are used, the two approaches give significantly different results when $\rho/\rho_0 \geqslant 1\cdot 2$ (ρ_0 here denoting the density at zero pressure); questions of slowness of convergence of series are (inter alia) involved. They stated that these differences could be resolved only through the theory of lattice dynamics and related experimental evidence, and that part of the evidence needed is outside present experimental capacity.

The analysis of Thomsen and Anderson suggested that certain of Birch's equations need substantial corrections in applications to the upper mantle, but smaller corrections in the lower mantle. These corrections are quite apart from uncertainties which may arise from the assumptions not yet experimentally tested. Thomsen and Anderson did not apply their analysis directly to the core, but it would appear that their work so far as it has gone would not require much larger corrections to Birch's equations in the core than in the lower mantle. Thus while, in respect of the mantle, their review substantiates the general accord between Birch's theory and results inferred using seismological data, it has not resolved the discrepancies on Δk at N and dk/dp in the core. In view of possible errors in the untested assumptions, it thus appears that the finite-strain theory cannot be relied upon to give more than general — though unquestionably important — support to the results on k and dk/dp derived on the Model A and B procedures.

Thomsen and Anderson have incidentally questioned the reliability of Birch's theory in applications to intense shock waves, a topic of relevance to §§11.4.5, 12.5.3, and 17.2.3. In a discussion of Thomsen's approach Sammis (1971) pointed out that, in spite of its advantages, the usual difficulties over convergence of series remain. Other critical assessments of Birch's theory are included in work of Davies and D.L. Anderson (1971) and Cook (1972).

[It may be remarked incidentally that Birch's famed 1952 paper, after expressing (p. 234) celebrated and commendable criticisms of language distortions under the phenomenon of 'high-pressure', itself contains, here and there, edicts which do not appear to be wholly untouched by the phenomenon — more particularly in some rather emphatic affirmations and rejections. Nevertheless, and notwithstanding the foregoing cautionary comments, the fact remains that

Birch's approach has contributed more to an understanding of physical behaviour in the Earth's deep interior than have practically all other positive approaches (see below) that have been based principally on laboratory and related theoretical evidence.]

15.1.2 *Other approaches through finite-strain theory*

Birch's version of finite-strain theory has been discussed at some length because it has been the most widely and successfully used in geophysical applications. Following are some other examples of adaptations of finite-strain theory.

Murnaghan himself (1944) made the assumption

$$k = k_0 + y_0 p, \qquad (15.1)$$

where τ denotes temperature and $y = (\partial k/\partial p)_\tau$. Throughout the present section, k denotes the isothermal incompressibility and the subscripts 0 and ∞ relate to zero and infinite pressure. The relations (15.1) and

$$(\partial \rho/\partial p)_\tau = \rho/k \qquad (15.2)$$

yield an equation of state

$$p(\rho, \tau) = (k_0/y_0)\{(\rho/\rho_0)^{y_0} - 1\}. \qquad (15.3)$$

Thomsen and Anderson (1969) showed that Murnaghan's equation (15.3) agrees fairly closely with Birch's equation (7.51) over a considerable range of p. This can be simply verified by putting $y_0 = 4$ in (15.3), corresponding to Birch's expression (7.53) for $(\partial k/\partial p)_\tau$, and substituting $\rho/\rho_0 = 1 \cdot 0, 1 \cdot 2, 1 \cdot 4, 1 \cdot 6, 1 \cdot 8, 2 \cdot 0$ on the right sides of (7.51) and (15.3), respectively; the percentage differences in the values of p thus calculated from the two equations are 0, 2, 7, 15, 24, 34.

Keane (1954) sought to restrict the behaviour of the equation of state at $p = \infty$ by assuming

$$y = (y_0 - y_\infty) k_0/k + y_\infty, \qquad (15.4)$$

which gives $(\partial k/\partial p)_\tau$ decreasing monotonely from y_0 to y_∞. The equations (15.2) and (15.4) yield

$$p/k_0 \doteq (y_0/y_\infty^2)\{(\rho/\rho_0)^{y_\infty} - 1\} - \{(y_0 - y_\infty)/y_\infty\} \ln(\rho/\rho_0), \qquad (15.5)$$

$$k/k_0 = 1 + (y_0/y_\infty)\{(\rho/\rho_0)^{y_\infty} - 1\}. \tag{15.6}$$

Keane's equations (15.5) and (15.6) reduce to Murnaghan's equations (15.1) and (15.3) if $y_\infty = y_0$. For a discussion of (15.5) and (15.6), see O.L. Anderson (1968).

Knopoff (1963) drew attention to a lack of uniqueness in the derivation of (7.51) and derived

$$p = (3k_0/2a)\{(\rho/\rho_0)^{(3+4a)/3} - (\rho/\rho_0)^{(1+4a)/3}\} \tag{15.7}$$

as an example of a more general finite-strain equation of state. Birch's equation is the particular case $a = 1$.

For some further detail on finite-strain equations of state, see Gilvarry (1957), Knopoff (*loc. cit.*), Anderson (*loc. cit.*), Boschi and Caputo (1969), Thomsen and Anderson (1969), Thomsen (1970), Macdonald and Powell (1971), Sammis (1971), Ahrens and Thomsen (1972), Davies (1973).

15.1.3 *Microscopic approaches*

Equations of state of the type mentioned in §15.1.2 are arrived at largely through macroscopic considerations resting on finite-strain theory and comparison with laboratory investigations. Other equations of state have been developed through microscopic approaches involving considerations of inter-atomic forces.

A typical procedure, based on work of Born and his colleagues (see e.g. Born, 1939), is to start with an assumed expression for the mutual potential energy Ψ of a pair of atoms or ions; examples are

$$\Psi = -ar^{-m} + br^{-n}, \quad \Psi = -ar^{-1} + br^{-2} + cr^{-3}, \tag{15.8, 9}$$

where r is the distance between the pair; a, b, c, m, and n are constants; in (15.8), $m < n$. Differentiating Ψ leads to equations of state. Those derived from (15.8) and (15.9) are (see Boschi and Caputo, 1969, p. 478)

$$p = \{3k_0/(n-m)\}\{(\rho/\rho_0)^{\frac{1}{3}n+1} - A(\rho/\rho_0)^{\frac{1}{3}m+1}\}, \tag{15.10}$$

and Bardeen's equation (1938)

$$p = (\rho/\rho_0)^{4/3}\{(\rho/\rho_0)^{2/3} - 1\}\{3k_0/2 + B[(\rho/\rho_0)^{2/3} - 1]\}, \tag{15.11}$$

where m, n, A, and B are to be found from experiment.

[*Refs. on p. 345*]

The expressions involved in this procedure are generally closed, so that the convergence problems facing finite-strain approaches do not arise. On the other hand, the available experimental evidence does not enable the constants in equations such as (15.10) and (15.11) to be estimated sufficiently accurately to yield closely reliable geophysical inferences. Moreover, as in science in general, unknown influences may affect the step from the microscopic to the macroscopic, and so add further to the uncertainties in geophysical applications. At the same time, microscopically derived equations may give useful broad guidance on some geophysical questions: they may assist in restricting the otherwise acceptable possibilities.

As instanced in §§10.1.3 and 11.2.4, Grüneisen's ratio is much used in expressing thermal aspects of equations of state. For a recent discussion of Grüneisen's and several other equations of state involving τ, see again Boschi and Caputo (1969).

For some further detail, see Shimazu (1954), O.L. Anderson (1965), Knopoff and Shapiro (1969), Macdonald (1969), Sammis (1970).

15.1.4 *Density-velocity relations*

Reference has already been made to Birch's empirical relation (10.14), namely,

$$\alpha \approx 3\cdot 31\rho - f(w). \qquad (15.12)$$

As pointed out in §10.4.2, the formulation would have been less limited if expressed in terms of ρ, ϕ, and w, where $\phi = k/\rho$. In high-pressure investigations, $\sqrt{\phi}$ or $\sqrt{(\partial p/\partial \rho)}$ is sometimes referred to as the hydrodynamic or bulk sound velocity. Examples of (ρ, ϕ, w) relations are given below.

D.L. Anderson (1967) inferred that (15.12) would yield for the density at the base of the Earth's mantle a value about 0·4 g/cm³ less than that for the original Model A, and sought an improved relation which would apply to close accuracy throughout the mantle. By a theoretical-empirical procedure, he arrived at the form

$$\rho = aw\phi^n, \qquad (15.13)$$

where the constant n, of order $\tfrac{1}{4}$ to $\tfrac{1}{3}$, is related to Grüneisen's ratio. Later (1970), he showed that a form analogous to (15.12),

[*Refs.* on p. 345]

namely,
$$\rho = a + b\sqrt{\phi}, \tag{15.14}$$

proposed by Wang (1968), fits the numerical evidence equally well.

The relations (15.12), (15.13), and (15.14) are of some special interest in that they have been applied in the course of constructing Earth models; in particular, they have contributed to the assessment of ρ in regions of the upper mantle where significant departures from the Williamson-Adams density gradient are suspected.

For further references to density-velocity relations and their applications, see Knopoff (1967), Birch (1969), Chung (1972).

15.1.5 *Questions of scientific inference*

Because of the many sources of uncertainty, the application of laboratory and theoretically based equations of state to the Earth requires more attention to questions of scientific inference than has usually been given. Following a statistical examination of methods used, Macdonald (1969) concluded that previous workers in this field "have not usually appreciated the need to find an equation of state that reduces least-squares residuals to approximate randomicity". He added: "From 50 to over 90 per cent of the data are unworthy of critical examination. In many cases, the lack of worth arises from loss of information through oversimplification. An example is the frequent failure to distinguish between random and systematic estimates of uncertainty." Macdonald and Powell (1971) have discussed criteria which they consider should be followed in assessing model equations of state.

The comments of the foregoing paragraph are not intended to deprecate the high quality of the progress that has been made in a difficult field. But they do draw attention to the need for caution in the finer geophysical inferences made. It is partly for this reason that, in the discussions in Chapter 11 and elsewhere, the present writer has preferred, for the Earth context, to attach primary weight to (ρ, p) and (k, p) relations considerably based on evidence provided by the deep Earth itself.

15.2   Some miscellaneous Earth models

I think it was Birch who remarked that the number of Earth models is now fast approaching the number of the world's seismologists. (One might add that the current over-production of Earth models is

15.2.1 Models using empirical density-velocity relations

Birch (1964) constructed a pair of models, I and II, in the following way. Distributions of α and β, fairly close to those of Jeffreys (§9.6.2) except for modifications above $z = 400$ km and inside the core, were assumed; sudden changes connected with the inner core were ignored. A uniform crust of thickness 33 km and density 2·79 g/cm^3 was assumed. Postulates on composition resulted in the density ρ' at the top of the mantle being taken as 3·425 and 3·320 g/cm^3 in the Models I and II, respectively. A value close to the revised value 0·3308 (§5.7) was used for the Earth's moment of inertia coefficient y. In the upper mantle, densities were derived using Birch's formula (15.12); in the lower mantle and core, the principles of construction of the original A-type models were followed.

Results for the Models I and II, respectively, include: density ranges 3·42–5·51 and 3·32–5·58 in the mantle, and 10·05–12·62 and 9·96–12·48 g/cm^3 in the core; pressures at N (the mantle-core boundary) 1·354 and 1·363; central pressures 3·64 and 3·61 $\times 10^{11}$ N/m^2; values of g at N, 10·81 and 10·70 m/s^2. (The central densities are minimum values.)

It is notable that the differences in ρ for either of Birch's models and Model A'' (§10.7.2), which also has $y = 0.3308$, are less than 0·1 g/cm^3 at nearly all depths and reach a maximum of only 0·12 g/cm^3. At most depths, the A'' values lie between those of I and II. This comparison makes it evident that the real importance of Birch's models is that they were the first published models to incorporate the revised y. The models I, II, and A'' are now of course superseded because they disregard free Earth oscillation data. [Model A'' was intended only as a step on the way towards incorporating oscillation data (§10.7.2). See also Bullen (1972).]

Although the degree of agreement between the models I, II, and A'' indicates that Birch's formula (15.12) does not add significantly to the precision of well-constructed Earth models, the agreement does provide an interesting measure of support to the utility of the formula in other geophysical problems, especially where phase changes are involved (see §10.4.2).

[*Refs. on p. 345*]

The model II has been more widely used than I, for example, by Wiggins (1968) in computing variational effects on free Earth oscillation periods.

Wang (1970) produced several variants of Birch's models, using (15.14) in place of (15.12); (15.14) has the advantage that it involves ϕ instead of α. The essential differences between Wang's and Birch's densities in the upper mantle appear, however, to arise from differences in the assumed velocity distributions rather than in the assumed density-velocity relations.

15.2.2 Model based on meteorite evidence

On the assumption that the distributions of composition found in meteorites yield a fair sample of the materials in the Earth, and further geochemical assumptions, it is possible to arrive at estimates of the Earth's density distribution. The estimates are guided by the seismic distributions of α and β and must of course fit the observational evidence on y, etc. An interesting early model derived in this way is that of Daly (1943). This model has a density range 3·0—6·3 in the mantle and 9·7—12·8 g/cm³ in the core.

15.2.3 Models based on petrological evidence

Numerous models have taken account of petrological and related geochemical evidence, especially in estimating density distributions in regions of the Earth where the Williamson-Adams equation does not provide a reliable first approximation to $d\rho/dz$. Examples are two models, I and II say, of Clark and Ringwood (1964). Model I assumes an overall 'pyrolite' (pyroxene-olivine rock) composition for the upper mantle, and model II an eclogite composition. For the two models, respectively: the density ranges are 3·28—5·71 and 3·54—5·49 in the mantle and 9·85—12·86 and 10·11—13·26 g/cm³ in the core; the pressures at N are 1·39 and 1·36; the central pressures are 3·57 and 3·68 $\times 10^{11}$ N/m²; the values of g at N are 10·56 and 10·87 m/s². (In these models, the depth of N was taken as 2920 km, one of Gutenberg's old estimates.)

15.3 Monte Carlo techniques

15.3.1 Background

The capacity of electronic computers to make trial and error methods feasible on a large scale has had a major impact on the evolution of Earth models. This is simply illustrated in the procedure of §13.2 where properties of large numbers of trial Earth model distributions of (α, β, ρ) are worked out and compared with a body of seismic surface-wave data with a view to selecting models which fit the data best. Another illustration is provided in the evolution, by trial and error, of Earth models fitting free Earth oscillation data (see §14.4).

In the case of the above illustrations, on the early procedures, the computer printed out the properties of all the trial models and the properties were subsequently compared with the observational data. Later, computer programmes were arranged so that models could be generated and tested inside the computer and the print-outs confined to those models which fitted the data within limits also prescribed in the programmes. In this way, models could be tested on a scale that could not otherwise be contrived. An early illustration is given in work of Press and Biehler (1964) where a correlation between delays in P wave arrivals at stations and gravity anomalies was sought.

A further development was to apply a Monte Carlo technique to generate trial models inside the computer. This technique uses random selection in generating the models: an auxiliary procedure for the purpose is part of the programme. The number of tested models can in this way be greatly extended. The first applications of a Monte Carlo technique in a seismic context were carried out by Keilis-Borok and colleagues (1966, 1967).

15.3.2 Application to derivation of Earth models

Press (1968) applied Monte Carlo techniques to find spherically symmetrical Earth models which would fit within prescribed limits a wider range of observational seismic and other data than Keilis-Borok had used. Press's data included bodily-wave travel-time data, surface-wave dispersion data, and free Earth oscillation data, in addition to data on the Earth's mass and moment of inertia. The prescribed limits were based on Press's assessment of the observational

[Refs. on p. 345]

evidence on hand at the time. Included in the prescribed limits were such features as restricting the number of discontinuities in $d\alpha/dz$ and $d\beta/dz$, and requiring the radius R_c of the core to lie in the range 3473 ± 25 km. Certain simplifying assumptions were made; e.g. the core was assumed fluid throughout.

In a first series of computations, Press (1968) was able to generate and test some 200 000 models per hour. Altogether, some five million models were tested and only six passed all the tests and were printed out. The principal results emerging were confirmatory evidence that: β diminishes with z near the top of the mantle; R_c exceeds 3473 km – the six successful models showed increases ΔR_c of 14 to 22 km (cf. §14.6.2) and Press estimated that ΔR_c must lie between 5 and 20 km; the density at the top of the core is close to 10 g/cm^3; the core is not homogeneous.

A surprising feature was that Press's procedure did not find a model anything like as simple as Model HB_1 (§14.7) or any of several other models later derived by Haddon and Bullen (1969). Inside the upper mantle, Press's successful models were much more complex than the HB models. Press ascribed the failure to find a model like HB_1 to differences in the P velocity distributions assumed. Haddon and Bullen (*loc. cit.*) showed, however, that this could not be the main cause and attributed the failure in part to a weakness of the Monte Carlo technique as first applied by Press: the procedure tended to favour over-parametrized models since a parametrically simple random walk would automatically be relegated to low probability. (Press had allowed random variation, within the prescribed limits, of parameters representing α, β, and ρ at 23 different values of z.) Notwithstanding the likelihood, thus illustrated, that some important classes of models are missed on the original procedure, Press's work demonstrated the considerable power of the Monte Carlo technique, a power that would be enhanced if the technique could be suitably modified to avoid undue weight being given to complex models.

In a later series of computations, Press (1970) applied a revised Monte Carlo technique and brought new observational data to bear. The surface-wave data were now restricted to oceanic paths. The revised technique speeded up the computations by a factor of more than ten and thereby enabled a larger number of successful models (twenty-seven) to be found; the successful models also fitted to better precision a more extensive set of data than the

[*Refs. on p. 345*]

data used previously. Important results were that all the successful models had low β zones centred at depths between 150 and 250 km and had densities in the range 9·9–10·2 g/cm³ at the top of the core; see also §15.3.3.

Wiggins (1969) applied a Monte Carlo technique in investigating limits to the permissible α and β distributions set by bodily-wave travel-time data.

15.3.3 *The question of a density inversion in the upper mantle*

Reference has been made in §13.7 to evidence favouring a density inversion near the top of the mantle. The evidence includes the results for Press's successful sub-oceanic models, all of which had values of ρ clustering between 3·5 and 3·6 g/cm³ near $z = 100$ km. Below this level, the clustering was less close, but there was a strong suggestion that $d\rho/dz$ turns negative with ρ falling to a minimum of about 3·3–3·4 g/cm³ where $z = 250$–350 km. Taking note of the models of Clark and Ringwood (§15.2.3), Press suggested an interpretation of his density results in terms of a predominantly eclogite composition for $z \approx 100$–200 km, with pyrolite below.

The density inversion, if substantiated, would carry some far-reaching implications. For example, others have drawn support from Press's results for theories of plate tectonics and continental drift. It is therefore desirable to examine how much weight can be attached to the evidence for inversion.

The investigations of Haddon and Bullen (1969) favoured an average value of η appreciably less than unity, but greater than zero, in the outermost 350 km of the mantle (§14.7.2). They would in fact permit a density inversion similar to that inferred by Press, but did not discriminate in favour of this.

The following further comments (see Bullen, 1972) may be made. Press's programme allowed for an uncertainty of only 0·4 per cent in the observational surface-wave phase velocities; this assessment of the uncertainty seems, however, rather too low. Unpublished calculations of Haddon showed that if the uncertainty is raised to 0·7 per cent, there need be no density inversion. Press also allowed little flexibility in his assumed α distribution: in a footnote, Press (1969) stated that varying the α distribution would not affect the density inversion. Haddon has questioned this, finding that a suitably chosen redistribution of α inside the outermost 500 km of the mantle involving changes nowhere exceeding 0·2 km/s would entirely remove the need for a density inversion. The question then arises as to how well the α distribution is determined below the oceans; the currently available evidence would appear to leave room for uncertainties of at least 0·2 km/s. A minor question is how far the suggested density inversion is reconcilable with, say, Birch's (ρ, α) formula (15.12) [though this is not intended to suggest that (15.12) is necessarily to be preferred to Press's findings]. For further critical comments, see Jordon and Anderson (1974).

The conclusion is that a density inversion can reasonably be regarded as quite possible on currently available data, but by no means yet established.

The pros and cons of the Monte Carlo technique will be referred to again in Chapter 16.

15.4 The general problem of 'inverting' observational data

15.4.1 *Preliminary sketch*

Procedures to derive distributions in the Earth of physical properties such as α, β, ρ, k, and μ from seismic and other observational data are nowadays commonly referred to as inversion procedures. An example is the early procedure (§9.7.1) involving the solution of an integral equation in deriving distributions of α and β from seismic bodily-wave travel-time (T, Δ) data.

Attempts to carry out inversions by direct mathematical processes, as in the successful case of §9.6.1, came to be largely superseded by trial-and-error methods making use of modern computing facilities (§15.3.1), with auxiliary techniques such as the Monte Carlo technique sometimes assisting. Recently, however, the problem of inversion has been approached in a more general way. Efforts involving considerable use of abstract algebra have been made to assess the maximum information on the Earth's internal physical properties that can, through the available computing facilities, be extracted from assigned widely ranging bodies of observational data. General theory is brought to bear to examine the fruits of large-scale trial-and-error methods with a view to encompassing all acceptable Earth models.

For several reasons, the employment of sophisticated mathematics in this way does not lead to exact results. The first obvious reason is that the observational data are necessarily uncertain to greater and less degrees. A second reason is that any body of observational data, however large, necessarily consists of a finite number of items (see §6.1.1; see also Backus and Gilbert, 1967). Thus in practice, a unique Earth model could not be derived from any set of actual observational data.

Putting aside for the moment these very vital questions of observational finiteness and uncertainty, there remains the question as to whether theoretically specified representations of data can be inverted to yield unique results. It transpires that the results may be unique, may be non-unique, or may fall into a third category sometimes referred to as quasi-unique; this will be illustrated in §15.4.2. When, as is usually the case, the results are not unique, there is a further question as to the extent to which bounds can be determined for various items in the results (for example, values of ρ at specified r or z). Part of the theory of inversion procedures is directed towards investigating these bounds, or investigating probability distributions over ranges of values of the items. Information thus derived is potentially important, notwithstanding the important questions left aside.

15.4.2 *Illustration in an elementary case*

Uniqueness, non-uniqueness, and quasi-uniqueness are illustrated in the comparatively simple inversion procedure of §9.6.1. Let a given (T, Δ) relation for (say) P waves be $T = F(\Delta)$. (For present purposes, it has to be stipulated that F conforms to certain mathematical requirements.) The analysis described in §9.6.1 shows that if a solution $\alpha = f(r)$ is yielded such that the condition (9.17), namely, $d\alpha/dr < \alpha/r$, is not violated, then that solution is unique. If (9.17) is violated, there can be no unique solution. If there is no violation for $r_1 < r < r_0$, where r_0 is the radius of the Earth model, but violation for $r_2 < r < r_1$, then α is yielded uniquely for $r > r_1$ but not for $r < r_1$. The term quasi-unique would apply in the last case, meaning that some but not all values of the sought physical parameter α are uniquely determined.

In the present illustration, the fact that the formal solution (9.18) of the integral equation (9.12) is available makes the uniqueness problem simply decidable. Suppose, however, that the solution (9.18) were not available and that by some other means (e.g. trial

[*Refs. on p. 345*]

and error) it had been found that $\alpha = f(r)$ is one relation compatible with the given relation $T = F(\Delta)$. Let $(\delta\alpha)'$ be an arbitrary variation of α from the value given by $\alpha = f(r)$ at $r = r'$; in general, this variation (if uncompensated) will remove compatibility with $T = F(\Delta)$. Then the uniqueness of $\alpha = f(r)$ might be tested by ascertaining whether a compensating single variation or set of variations, $(\delta\alpha)_i$ ($i = 1, 2, \ldots$) at $r = r_i$, can be found which will restore compatibility with $T = F(\Delta)$. If compatibility can be restored in this way, $\alpha = f(r)$ is not unique. If, for all r', it can be established that there exists no set $(\delta\alpha)_i$ compensating $(\delta\alpha)'$, then $\alpha = f(r)$ is unique. And similarly with quasi-uniqueness. [With the present illustration, such a procedure would of course lead indirectly to the criterion (9.17).]

15.4.3 *General procedures*

The principles illustrated in §15.4.2 are applicable, more generally, to much wider bodies Θ of observational data, including surface-wave data, free Earth oscillation data, data on the moment of inertia coefficient y, etc. Suppose a model representation Γ of Θ has yielded a set Ω of model distributions of (α, β, ρ) for the Earth. For any r, let any one of α, β, and ρ be arbitrarily varied from the value in Ω. Then the uniqueness of Ω as derived from Γ may, after the manner in §15.4.2, be tested by ascertaining whether or not a compensating set of variations, which may involve any or all of α, β, and ρ, exists.

As stated in §15.4.1, Ω cannot be uniquely derived from the observational data Θ. But the testing for uniqueness of Ω as derived from Γ can assist valuably in discriminating between contributions to non-uniqueness (if any) from Γ and the additional contributions arising from the finiteness and uncertainties of Θ.

The whole approach can be widened by reclassifying representations Γ according as the results Ω for α, β, and ρ lie or do not lie inside certain specified ranges for each r. In place of a set of ordinary (α, r) etc. curves, the graphs of Ω would now occupy bands or 'strips' of limited but (in general) variable width running across the paper. In this widened approach, the strict uniqueness requirement is relaxed. Non-uniqueness arising from the observational limitations of finiteness and uncertainty cannot, however, be dealt with in this way; the best that can be done in the presence of these limitations is to attempt to assess confidence levels attaching to whatever band widths are taken.

[*Refs. on p. 345*]

Although considerable progress has been made with the auxiliary mathematical theory, the above procedures have not yet added greatly to the information on Earth models already obtained by the methods of successive approximation discussed in preceding chapters. A source of some difficulty, common to all approaches, is the interplay between the distributions of α, β, and ρ (sometimes referred to as the 'trade-off' problem) when changes are made to meet given data requirements: on given Θ, modifying the band widths for ρ, for example, often has marked effects on the widths for β and may have significant effects on α (see e.g. §15.3.3).

One achievement of the general theory has been confirmation that the outermost 300 km of the mantle must be regarded as appreciably more uncertainly determined than has been claimed by many investigators. In this connection, it may be remarked that some confusion has been brought about through the publication of numerous distributions (especially of β) which are neither parametrically simple nor near-unique. [In a number of cases, a semblance of near-uniqueness has resulted from an unwarrantably narrow selection of additional assumptions (sometimes concealed).] The general theory has the potentiality of putting such distributions in proper perspective.

For information on various analytical aspects of general inversion procedures, see Slichter (1954), Backus and Gilbert (1967, 1968, 1970). Takeuchi and Sudo (1968), Smith and Franklin (1969), Dziewonski (1970), Jordon and Franklin (1971), Keilis-Borok (1971, 1972), Johnson and Gilbert (1972), Knopoff and Jackson (1973), Gilbert, Dziewonski, and Brune (1973), Wiggins, McMechan, and Toksöz (1973). (References to other important Soviet papers on the subject are included by Keilis-Borok.)

General inversion theory may also serve as a background from which, through the introduction of additional devices, a single preferred Earth model may be derived from an assigned set of data. (See §16.3.3.)

15.5 Density and seismic wave amplitudes

Båth (1954) drew attention to the possibility of estimating density jumps at boundaries inside the Earth from observations of the ratios of the amplitudes of incident and reflected seismic waves. He carried out a model calculation for *PcP* waves involving the ratio σ of the density jump at the mantle-core boundary N. Let

[*Refs. on p. 345*]

γ be the ratio of the amplitudes of the upward reflected P wave at N and the downward incident P wave, corresponding to an angle of incidence i. Let α' and β' be the P and S velocities just above N, and α'' the P velocity just below. Båth assumed a fully symmetrical Earth model and took $\alpha', \beta', \alpha'' = 13\cdot 7, 7\cdot 25, 8\cdot 0$ km/s. For $\sigma = 1$, he calculated that γ becomes zero and changes sign at $i \approx 35°$ and at $i \approx 80°$; the corresponding Δ for PcP are about $30°$ and $82°$. For $\sigma = 1\cdot 7$, γ changes sign only at $i \approx 84°$ ($\Delta \approx 90°$).

Båth remarked that observed PcP amplitudes are often, though by no means always, very strong near $\Delta = 30°$ and interpreted this as providing a measure of independent evidence that $\sigma > 1$, i.e. that ρ jumps discontinuously at N.

Reference has been made in §12.2.2 to Buchbinder's different observational interpretation (1968) in which he estimated that the PcP amplitudes pass through a minimum at $\Delta \approx 32°$ and inferred that σ is actually close to unity. As stated in §12.2.2, quite apart from the conflict with independent evidence on σ, Buchbinder's interpretation would entail an implausible change in k across N. For this reason, as well as the discord between Båth's and Buchbinder's observations, it seems probable that PcP amplitudes are influenced by complicating factors not considered in Båth's calculations. Possible complications include small irregularities in the shape of N or in the region immediately above N (see §§15.6, 15.7). In another study, Buchbinder (1968b) suggested that the amplitude spectrum of PcP may be strongly affected by conditions near the source point of an earthquake or explosion. Until finer knowledge of such factors becomes available, it seems that close estimates of σ cannot be derived from amplitude data. See also Ibrahim (1971a, 1971b) and Wright (1973).

Attempts to estimate bounds to density jumps inside the core by analogous procedures have been referred to in §12.5.4. For similar reasons to the above, these estimates also cannot yet be relied on.

In an analogous type of procedure using data on ScS and $ScSScS$ amplitudes, Press (1956) estimated 10^9 N/m² as an upper bound to the mean rigidity of the outer core (cf. §10.2). See also Honda, Sima, and Nakamura (1956).

15.6 Implications of wave-scattering investigations

15.6.1 *Introductory discussion*

Starting from an early investigation of Jeffreys (1931), the scattering of seismic waves due to small-scale irregularities has been a topic of intermittent interest — see e.g. Knopoff (1959), Gilbert and Knopoff (1959).

The following gives a general idea of the scattering processes to be considered. Let M be a medium which includes in its interior a region R which has small irregularities in its physical properties, causing small scattering of waves which pass through R. Let A and B be points of M outside R, and consider waves travelling from A to B, passing through R en route. Let A and B be such that, when scattering is neglected, a ray, σ_{AB} say, goes from A to B; then the wave energy arriving at B may be regarded as travelling along σ_{AB}. This still applies to most of the energy when the scattering is taken into account; the waves thus carrying most of the energy will be referred to as the primary waves. But small fractions of the energy will now travel inside R otherwise than along σ_{AB}; the corresponding waves are referred to as scattered waves. Since scattered waves once generated are subject to further scattering during the passage through R, there are first-order scattered waves, second-order scattered waves, and so on; the energy associated with the second-order waves is a small fraction of that associated with the first order, etc. When A and B cannot be joined by a ray (e.g. when B is in a shadow zone for the primary waves), the first-order scattered waves will be the strongest arrivals at B.

An important development (see Chernov, 1960) was an acoustical theory of wave scattering in which the relevant physical properties, ρ_j say, of M (here a fluid, so that $\rho_j = \rho, k$) are specified so that

$$\rho_j = (\rho_j)_0 + \Delta\rho_j, \qquad (15.15)$$

where each $(\rho_j)_0$ varies at most slowly with distance, and $(\Delta\rho_j)$ is a small random fluctuation varying rapidly inside R; outside R, $\Delta\rho_j = 0$.

This theory was extended by various authors (see e.g. Knopoff and Hudson, 1964–8) to the case of elastic media and applied to seismological and other problems (see §15.6.4). Independently, Haddon (see Haddon and Cleary, 1974; Haddon, 1974) extended the theory to the case of elastic waves in a spherically symmetrical

[*Refs. on p. 345*]

Earth. Haddon's work bears on certain large-scale features of the Earth's interior and so will be prominent in the following discussion.

15.6.2 Theory for a symmetrical Earth model

In the elastic medium to be considered, the ρ_j are ρ, k, μ. In general, the scattering of P waves inside the region R generates both P and S scattered waves, but the S waves are not relevant in the applications here made and will be ignored. Only primary and first-order scattered waves will be considered. Effects of higher-order scattered waves though possibly detectable are minor. The subscript 0 will indicate values (e.g. of α, θ – see below) corresponding to ρ_0, k_0, μ_0.

Haddon assumed

$$\Delta\rho/\rho_0 = -\Delta k/k_0 = -\Delta\mu/\mu_0 = cH, \qquad (15.16)$$

where c is a small constant and H is a stationary isotropic random function of position with an auto-correlation function K given by

$$K = \exp(-s^2/\sigma^2). \qquad (15.17)$$

In (15.17), s is the distance between a pair of points in R; σ, the 'correlation distance', characterizes the scale of the random fluctuations. The assumption (15.16) corresponds to a root-mean-square change proportional to c in each of ρ, k, and μ.

Consider a packet of P waves issuing at time $t = 0$ spherically symmetrically from a point source A in the medium M outside R and propagating through M. For the simplified case in which α_0 is constant, the dilatation θ_0 associated with the packet was expressed by Haddon in the form

$$\theta_0 = q^{-1}E(q - \alpha_0 t)G(q - \alpha_0 t), \qquad (15.18)$$

where q denotes the distance from A, E the modulation or envelope shape of the packet, and G a random function with auto-correlation function K' given by

$$K' = \exp(-z^2/\lambda^2). \qquad (15.19)$$

In (15.19), z is the distance between two points belonging to the wave packet and λ characterizes the wave lengths.

In Haddon's main applications, the point A was taken as an earthquake focus, and R as the Earth region D''. The theory was applied to scattered waves generated by a P wave packet W which issues from A and passes down through the mantle into the core and up again. Scattered waves would be generated during the two passages through D''. The constant c was put equal to 0·01. Having regard to observational data on short-period seismic P waves recorded at the Earth's surface, λ was taken as 4 km (corresponding to predominant periods of 1 s) and E was taken to satisfy (where A_0 is a constant)

$$E(\alpha_0 t) = 0 \, (t < 0, t > 4 \text{ s}); \; = A_0 \, (0 \leqslant t \leqslant 4 \text{ s}). \quad (15.20)$$

Let B be a point of the Earth's surface at which scattered waves arising from W arrive (B may or may not be inside a shadow zone for the primary waves). These scattered waves would be generated at various points Q inside D'' and would arrive superposed at B with a range of slownesses $(dT/d\Delta)$ and within a range of azimuths from A. The energy in the scattered wave packet generated at a particular scattering point Q would travel along P rays from A to Q and from Q to B. Corresponding to (15.20), the ground motion at B for this packet would be spread over a time interval 4 s. For the whole set of scattered wave packets the ground motion at B is of course spread over a greater interval. Inside this interval, there arrive two classes of scattered waves: those with Q (inside D'') between (a) A and the core, (b) the core and B.

Theory based on equations (15.16)–(15.20) was applied by Haddon to estimate the amplitude distributions, slownesses, and azimuthal variations of the scattered waves arriving at B. He allowed for energy losses due to reflexion and partial conversion into S waves at the mantle-core boundary N, but ignored various other effects such as attenuation and scattering outside D'' as probably minor. He also developed theory to assist in narrowing the location of the effective scattering region inside D''. An important finding was that the classes (a) and (b) have markedly different slowness and azimuthal characteristics.

15.6.3 Application to core structure

Prior to 1972, precursors to the seismic phase *PKIKP* (§ § 12.3.1, 12.3.2) had usually been interpreted in terms of purely ray theory. Such interpretations require the core to contain one or more discontinuities additional to the discontinuity at the inner core boundary. Haddon (1972) put forward an alternative interpretation

[*Refs. on p. 345*]

of the precursors in terms of the foregoing scattering theory. Cleary and Haddon (1972) showed that the theoretical least-time curve for first-order scattered waves was consistent with the available travel-time and slowness data (see also Bullen and Haddon, 1973). Doornbos and Husebye (1972) independently questioned the purely ray-theory interpretations, which could not be reconciled with travel-time and slowness data gathered at the NORSAR array station, and suggested diffraction from the *PKP* caustic as another alternative.

Subsequently, Doornbos and Vlaar (1973) gathered a large body of pertinent data using the NORSAR array, and King, Haddon, and Cleary (1973a, 1973b) published further data obtained at the Warramunga array station (Australia). These data all agreed well (see Haddon and Cleary, 1974) with the results of the theory of §15.6.2, including the variations with Δ of the times and slownesses of the precursor onsets, and the variations of amplitude, azimuth, and slowness in the precursor wave train. None of the purely ray-theory interpretations gives the same degree of overall agreement. Haddon concluded (among other things) that the markedly different characteristics of the two classes (a) and (b) of the scattered waves (§15.6.2) gave, on comparison with observation, a crucial test in favour of the scattering interpretation.

The results strongly confirmed that it is no longer necessary to postulate the presence of transition layers between the outer and inner core proper (cf. §12.3.2). The evidence from the scattering theory does not preclude such layers, but it annuls the evidence for them. (See also King, Haddon, and Cleary, 1973a).

Doornbos and Vlaar (1973) inferred significant scattering at points up to 900 km above N, but Haddon and Cleary (1974) found this to be unnecessary and in fact suggested that the main scattering may occur inside a layer above N of thickness considerably less than 200 km (the postulated thickness of D'').

Further support to the scattering interpretation was given by Müller (1973) from amplitude studies of long-period core phases from five earthquakes recorded on long-period instruments in the Worldwide Network of Standardized Seismographs. He inferred smooth variation of α from N to the inner core boundary L and a jump of 0·6-0·7 km/s in α at L. He also inferred a pronounced P velocity gradient and an S velocity of 3−4 km/s in the outermost part of the inner core, and suggested that $d\beta/dz$ may be negative in the inner core (cf. §§12.6.2, 12.7.2). See also Ansell (1973).

[Refs. on p. 345]

15.6.4 Other applications

Nguyen Hai (1963) examined observations of precursors to the seismic phase *PP* and suggested that certain of the precursors arriving at $\Delta = 105-110°$ are associated with reflexions on the undersides of boundaries inside the upper mantle. The reflexion points being below the Earth's crust, the waves would arrive before the regular *PP* phase. Bolt, O'Neill, and Qamar (1968) supported the interpretation and referred to the *PP* precursors as phases *PdP*. Adams (1968, 1971) suggested a similar interpretation of certain observed precursors to the phase *PKPPKP* in terms of underside reflexions at depths near 80, 400, and 650 km. In respect of the 650 km reflexion, which would correspond to the $C'-C''$ boundary (§14.7.2), this interpretation was strongly supported by slowness data of Engdahl and Flinn (1969) and nuclear-explosion data of Bolt and Qamar (1972). King, Cleary, and Haddon (1973, 1974) examined these precursors in the light of Haddon's scattering theory and concluded that scattering in the crust and its vicinity provides an adequate mechanism for the precursors that had been assumed reflected at levels higher than 650 km. For another alternative to the *PdP* interpretation, see Whitcomb (1973). See also Wright (1972).

Further applications of scattering theory have been made to problems of primarily seismological interest. For example, Hudson and Knopoff (1966, 1967) applied scattering theory to problems of seismic noise. Aki (1969) in seeking to extract information from the seismic coda of local earthquakes analysed the coda in terms of scattered waves. Cleary, King, and Haddon (1974) also ascribed the *P* wave coda to scattering.

15.7 Deviations from spherical symmetry

The smallness of the deviatoric stress components P_{ij} compared with the mean $(\Sigma p_{kk})/3$ of the principal stresses everywhere in the Earth except near the surface (§7.8.1) justifies the procedure generally followed in the book of treating ρ, k, and μ as functions predominantly of r. The largest deviations from spherical symmetry, the ellipticity effects arising from the Earth's rotation, can be reliably allowed for according to the theory given in §§5.9, 9.2, and 14.2.2.

[*Refs. on p. 345*]

The remaining deviations, generally called lateral differences, have been referred to intermittently in earlier chapters. The lateral differences are greatest in the crust and are appreciable for some distance below the crust, as indicated in Brune's summary of P_n and S_n velocities in Table 12.1 and discussed in §12.1.1. For additional data on S_n, see Huestis, Molnar, and Oliver (1973). The existence of significant lateral differences at greater depths is evidenced, among other things, by the unsymmetrical distributions of earthquake epicentres and of deep-focus earthquakes originating at depths down to 700 km (see B, §§15.6.1, 15.6.2).

Work on surface waves dating as far back as 1920 (§13.5) had revealed lateral differences in the outer part of the Earth, and lateral differences have been a major complicating factor in deriving fine information about the Earth's structure from surface-wave data and free Earth oscillations with periods less than about 3 minutes (§§13.6.5, 14.2.3, 14.5.2). The overall evidence indicates that lateral differences persist significantly to a depth of at least 400 km and probably 1000 km. Some investigators have suggested that lateral differences occur down to the bottom of the mantle. Buchbinder (1971) inferred from data on the seismic phases $PmKP$ that there is negligible lateral variation of α in the outer core. [The notation $PmKP$ is shorthand for $PKK..KP$ where K occurs m times – see Engdahl (1968).] For a useful list of references to seismic evidence on lateral differences, see Brown (1972).

Except near the Earth's crust, the quantitative study of lateral deviations is as yet only in its infancy. Brune's table indicates lateral differences in the P_n velocity of order 10 per cent. Differences as large as this are not likely to persist very far down into the mantle, though some investigators have tentatively suggested lateral differences of a few per cent in α at depths of several hundred kilometres (see e.g. Wiechert, 1972). Little reliable evidence is available on lateral differences in β except near the crust.

A direct study of lateral variations in ρ was made by Arkani-Hamed (1970), using evidence from artificial satellites on variations in the Earth's external gravitational field. He inferred that the lateral variations of ρ are of order 0·3 g/cm^3 in the crust, 0·2 g/cm^3 in the upper mantle, and 0·04 g/cm^3 in the lower mantle.

Implications of lateral differences in the construction of Earth models will be discussed in §16.1.2.

15.8 Changes in gravitational constant

Some attention has lately been devoted to an idea put forward by Dirac in 1938 that the constant of gravitation G has been decreasing over geological time. The idea is mentioned here because it carries implications very pertinent to the history of the Earth's density. In accordance with

$$dp/dz = Gm\rho/r^2 \qquad (15.21)$$

(§10.1.2), a decrease in G during the Earth's lifetime would entail that the pressures reached in the interior were formerly greater than they are now, and hence that the internal densities were greater and the Earth smaller.

In accordance with the simplicity postulate of scientific inference, it is appropriate to treat entities like G as constant so long as the available evidence does not suggest otherwise. But that is not to assert that G is necessarily independent of time. Dirac's idea, based on a consideration of large numbers in cosmology and atomic physics (see Dirac, 1972, 1973) led him to propose that G is proportional to t^{-1}, where t is the age of the Universe. He also suggested that matter is being continually created inside the Earth.

Egyed (see Egyed, 1969) put forward in 1957 a theory of an expanding Earth which assumed Dirac's diminution in G and also assumed a phase transition at the mantle-core boundary N (see §17.2.3). Egyed's theory was one of several expanding Earth theories put forward as alternatives to the usual theories of continental drift. Egyed inferred an expansion in the Earth's radius of 0.65 ± 0.15 mm/year.

The hurdles confronting expanding Earth theories (the requirements of both a diminishing G and a phase-change at N) are considerable, and some writers (see e.g. Birch, 1968) have asserted that the hurdles are insuperable. Jeffreys (1962) commented that Egyed's theory fits the present disposition of land masses better than he would have expected. A recent discussion of expanding Earth theories, including references to alternative approaches to Dirac's on changes in G, has been published by Wesson (1973).

REFERENCES

Adams, R.D. (1968). Early reflections of $P'P'$ as an indication of upper mantle structure. *Bull. Seismol. Soc. Amer.*, **58**, 1933–1947.

MISCELLANEOUS DEVELOPMENTS

Adams, R.D. (1971). Reflections from discontinuities beneath Antarctica. *Bull. Seismol. Soc. Amer.*, 61, 1441–1451.

Ahrens, T.J. and Thomsen, L. (1972). Application of the fourth-order anharmonic theory to the prediction of equations of state at high compressions and temperatures. *Phys. Earth Planet. Interiors*, 5, 282–294.

Aki, K. (1969). Analysis of seismic coda of local earthquakes as scattered waves. *J. Geophys. Res.*, 74, 615–631.

Anderson, O.L. (1965). Lattice dynamics in geophysics. *Trans. N.Y. Acad. Sci.* II, 27, 298–308.

Anderson, O.L. (1968). On the use of ultrasonic and shock-wave data to estimate compressions at extremely high pressures. *Phys. Earth Planet. Interiors*, 1, 169–176.

Ansell, J.H. (1973). Precursors to *PKP*: seismic wave scattering and spectral data. *Geophys. J., Roy. Astr. Soc.*, 35, 487–489.

Arkani-Hamed, J. (1970). Lateral variations of density in the mantle. *Geophys. J., Roy. Astr. Soc.*, 20, 431–455.

Asbel, I. Ja., Keilis-Borok, V.I. and Yanovskaya, T.B. (1966). A technique of a joint interpretation of travel-time and amplitude-distance curves in the upper mantle studies. *Geophys. J., Roy. Astr. Soc.*, 11, 25–55.

Backus, G. and Gilbert, F. (1967). Numerical applications of a formalism for a geophysical inversion problem. *Geophys. J., Roy. Astr. Soc.*, 13, 247–276.

Backus, G. and Gilbert, F. (1968). The resolving power of gross Earth data. *Geophys. J., Roy. Astr. Soc.*, 16, 169–205.

Backus, G. and Gilbert, F. (1970). Uniqueness in the inversion of inaccurate gross Earth data. *Phil. Trans. Roy. Soc.* A, 266, 123–192.

Bardeen, J. (1938). Compressibilities of the alkali metals. *J. Chem. Phys.*, 6, 408–414.

Båth, M. (1954). The density ratio at the boundary of the Earth's core. *Tellus*, 6, 408–414.

Birch, F. (1952). Elasticity and constitution of the Earth's interior. *J. Geophys. Res.*, 57, 227–286.

Birch, F. (1968). On the possibility of large scale changes in the Earth's volume. *Phys. Earth Planet. Interiors*, 1, 141–147.

Birch, F. (1969). Density and composition of the upper mantle: first approximation as an olivine layer. In *The Earth's Crust and Upper Mantle*, Geophysical Monograph, Amer. Geophys. Un., 13, 18–36.

Bolt, B.A. (1970). *PdP* and *PKiKP* and diffracted *PcP* waves. *Geophys. J., Roy. Astr. Soc.*, 20, 367–382.

Bolt, B.A., O'Neill, M. and Qamar, A. (1968). Seismic waves near 110°: is structure in core or upper mantle responsible? *Geophys. J., Roy. Astr. Soc.*, 16, 475–487.

Bolt, B.A. and Qamar, A. (1972). Observations of pseudo-aftershocks from underground nuclear explosions. *Phys. Earth Planet. Interiors*, 5, 400–402.

REFERENCES

Born, M. (1939). Thermodynamics of crystals and melting. *J. Chem. Phys.*, **7**, 591–603.

Boschi, E. and Caputo, M. (1969). Equations of state at high pressure and the Earth's interior. *Rev. Nuovo Cimento* I, **1**, 441–513.

Brown, R.J. (1972). Lateral inhomogeneity in the crust and upper mantle from P-wave anomalies. *Pure Appl. Geophys.*, **101**, 102–154.

Buchbinder, G.G.R. (1968a). Properties of the core-mantle boundary and observations of PcP. *J. Geophys. Res.*, **73**, 5901–5923.

Buchbinder, G.G.R. (1968b). Amplitude spectra of PcP and P phases. *Bull. Seismol. Soc. Amer.*, **58**, 1797–1819.

Buchbinder, G.G.R. (1972). Travel times and velocities in the outer core from PmKP. *Earth Planet. Sci. Letters*, **14**, 161–168.

Bullen, K.E. (1972). Some problems connected with constructing Earth models. *Proc. Conference on Solid Earth Problems, Buenos Aires*, II, 63–77.

Bullen, K.E. and Haddon, R.A.H. (1973). Some recent work on Earth models, with special reference to core structure. *Geophys. J., Roy. Astr. Soc.*, **35**, 31–38.

Chernov, L.A. (1960). *Wave Propagation in a Random Medium*, 168 pp., McGraw-Hill, New York.

Chung, D.H. (1972). Birch's Law. *Science*, **177**, 261–263.

Clark, S.P., Jr., and Ringwood, A.E. (1964). Density distribution and constitution of the mantle. *Rev. Geophys.*, **2**, 35–88.

Cleary, J.R. and Haddon, R.A.W. (1972). Seismic wave scattering near the core-mantle boundary: a new interpretation of precursors to PKP. *Nature* (Lond.), **240**, 549–551.

Cleary, J.R., King, D.W. and Haddon, R.A.W. (1974). Seismic wave scattering in the Earth's crust and mantle. In course of publication.

Cook, A.H. (1972). The dynamical properties and internal structure of the Earth, the Moon and the planets. *Proc. R. Soc.* (Lond.), A, **328**, 301–336.

Daly, R.A. (1943). Meteorites and an Earth-model. *Bull. Geol. Soc. Amer.*, **54**, 401–456.

Davies, G.F. (1973). Quasi-harmonic finite strain equations of state of solids. *J. Phys. Chem. Solids*, **34**, 1417–1429.

Davies, G. and Anderson, D.L. (1971). Revised shock-wave equations of state for high-pressure phases of rocks and minerals. *J. Geophys. Res.*, **76**, 2617–2627.

Dirac, P.A.M. (1972). Evolutionary cosmology. *Commentarii Pontif. Acad. Sci.*, **2**, No. 46, 16 pp.

Dirac, P.A.M. (1973). Long range forces and broken symmetries. *Proc. Roy. Soc.*, (Lond.), A, **333**, 403–418.

Doornbos, D.J. and Vlaar, N.J. (1972). Regions of seismic wave scattering in the their precursors. *Phys. Earth Planet. Interiors*, **5**, 387–399.

Doornbos, D.G. and Vlaar, N.J. (1972). Regions of seismic wave scattering in the Earth's mantle and precursors to PKP. *Nature* (Lond.), **243**, 58.

Dziewonski, A.M. (1970). Correlation properties of free period partial derivatives and their relation to the resolution of gross Earth data. *Bull. Seismol. Soc. Amer.*, **60**, 741–768.

Egyed, L. (1969). The slow expansion hypothesis. In *The Application of Modern Physics to the Earth and Planetary Interiors* (ed. S.K. Runcorn), 65–75, Wiley-Interscience, London.

Engdahl. E.R. (1968). Seismic waves within the Earth's outer core: multiple reflection. *Science*, **161**, 263.

Engdahl, E.R. and Flinn, E.A. (1969). Seismic waves reflected from discontinuities within the Earth's upper mantle. *Science*, **163**, 177–179.

Gilbert, F., Dziewonski, A. and Brune, J. (1973). An informative solution to the inverse problem. *Proc. Nat. Acad. Sci. USA*, **70**, 1410–1473.

Gilbert, F. and Knopoff, L. (1959). Scattering of impulsive elastic waves by a rigid cylinder. *J. Acoust. Soc. Amer.*, **31**, 1169–1175.

Gilvarry, J.J. (1957). Temperature dependent equations of state of solids. *J. Appl. Phys.*, **28**, 1253–1261.

Haddon, R.A.W. (1972). Corrugations on the mantle-core boundary or transition layers between inner and outer cores? *Trans. Amer. Geophys. Un.*, **53**, 600.

Haddon, R.A.W. (1974). Scattering of seismic body waves by small random inhomogeneities in the Earth. In course of publication.

Haddon, R.A. and Bullen, K.E. (1969). An Earth model incorporating free Earth oscillation data. *Phys. Earth Planet. Interiors*, **2**, 35–49.

Haddon, R.A.W. and Cleary, J.R. (1974). Evidence for scattering of seismic *PKP* waves near the mantle-core boundary. *Phys. Earth Planet. Interiors*, **8**, 211–234.

Honda, H., Sima, H. and Nakamura, K. (1956). The *ScS* wave, the mechanism of deep earthquake and the rigidity of the inner core. *Sci. Repts., Tohoku Univ.*, Ser. 3, **7**, 169–179.

Hudson, J.A. (1968). The scattering of elastic waves by granular media. *Q. J. Mechs. Appl. Math.*, **21**, 487–502.

Hudson, J.A. and Knopoff, L. (1966). Signal generated seismic noise. *Geophys. J., Roy. Astr. Soc.*, **11**, 19–24.

Hudson, J.A. and Knopoff, L. (1967). Stationary properties of Rayleigh waves due to scattering by topography. *Bull. Seismol. Soc. Amer.*, **57**, 83–90.

Huestis, S., Molnar, P. and Oliver, J. (1973). Regional *Sn* velocities and shear velocity in the upper mantle. *Bull. Seismol. Soc. Amer.*, **63**, 469–475.

Ibrahim, A.K. (1971a). Effects of a rigid core on the reflection and transmission coefficients from a multi-layered core-mantle boundary. *Pure Appl. Geophys.*, **91**, 95–113.

Ibrahim, A.K. (1971b). The amplitude ratio *PcP/P* and the core-mantle boundary. *Pure Appl. Geophys.*, **91**, 114–133.

Jeffreys, H. (1931). On the cause of oscillatory movement in seismograms. *Mon. Not. Roy. Astr. Soc., Geophys. Suppl.*, **2**, 318–323.

REFERENCES

Jeffreys, H. (1948). *Theory of Probability* (2nd Ed.), 411 pp., Cambridge University Press.

Jeffreys, Sir H. (1962). *Nature* (Lond.), **195**, 448.

Johnson, L.E. and Gilbert, F. (1972). Inversion and inference from teleseismic ray data. In *Methods in Computational Physics* (ed. B.A. Bolt and others), **12**, 231–266.

Jordan, T.H. and Anderson, D.L. (1974). Earth structure from free oscillations and travel times. *Geophys. J., Roy. Astr. Soc.*, **36**, 411–459.

Jordan, T.H. and Franklin, J.N. (1971). Optimal solutions to a linear inverse problem in geophysics. *Proc. Nat. Acad. Sci. USA*, **68**, 291–293.

Keane, A. (1954). An investigation into finite strain in an isotropic material subjected to hydrostatic pressure and its seismological applications. *Austral. J. Phys.*, **7**, 322–333.

Keilis-Borok, V.I. (1971). The inverse problem of seismology. In *Mantle and Core in Planetary Physics*, 242–274, Academic Press, New York.

Keilis-Borok, V.I. (ed.) (1972). *Computational Methods in Seismology* (English translation by E.A. Flinn), 227 pp., Consultants Bureau, New York.

Keilis-Borok, V.I. and Yanovskaya, T.B. (1967). Inverse problems of seismology (structural review). *Geophys. J., Roy. Astr. Soc.*, **13**, 223–234.

King, D.W. and Cleary, J.R. (1974). A note on the interpretation of precursors to PKPPKP. In course of publication.

King, D.W., Haddon, R.A.W. and Cleary, J.R. (1973). Evidence for seismic wave scattering in the D'' layer. *Earth Planet. Sci. Letters*, **20**, 353–356.

King, D.W., Haddon, R.A.W. and Cleary, J.R. (1974). *Geophys. J., Roy. Astr. Soc.*, in course of publication.

Knopoff, L. (1959). Scattering of waves by spherical obstacles. *Geophysics*, **24**, 30–39, 209–219.

Knopoff, L. (1963). Solids: Equations of state of solids at moderately high pressures. In *High Pressure Physics and Chemistry*, Vol. 1 (ed. R.S. Bradley), 227–247, Academic Press, New York.

Knopoff, L. (1967). Density-velocity relations for rocks. *Geophys. J., Roy. Astr. Soc.*, **13**, 1–8.

Knopoff, L. and Hudson, J.A. (1964). Scattering of elastic waves by small inhomogeneities. *J. Acoust. Soc. Amer.*, **36**, 338–343.

Knopoff, L. and Hudson, J.A. (1967). Frequency dependence of amplitudes and scattered elastic waves. *J. Acoust. Soc. Amer.*, **42**, 18–20.

Knopoff, L. and Jackson, D.D. (1973). The analysis of undetermined and overdetermined systems. Preprint privately distributed.

Knopoff, L. and Shapiro, J.N. (1970). Pseudo-Grüneisen parameter for liquids. *Phys. Rev.*, B, **1**, 3893–3895.

Lamb, Sir H. (1932). *Hydrodynamics* (6th Ed.), Cambridge University Press.

Macdonald, J.R. (1969). Review of some experimental and analytical equations of state. *Rev. Mod. Phys.*, **41**, 316–349.

MISCELLANEOUS DEVELOPMENTS

Macdonald, J.R. and Powell, D.R. (1971). Discrimination between equations of state. *J. Res. Nat. Bur. Standards*, 75A, 441–453.

Müller, G. (1973). Amplitudes of core phases. *J. Geophys. Res.*, 78, 3469–3490.

Murnaghan, F. (1944). The compressibility of media under extreme pressures. *Proc. Nat. Acad. Sci. USA*, 30, 244–255.

Nguyen Hai (1963). Propagation des ondes dans le noyau terrestre. *Ann. Géophys.*, 19, 285–346.

Press, F. (1956). Rigidity of the Earth's core. *Science*, 124, 1204.

Press, F. (1968). Earth models obtained by Monte Carlo inversion. *J. Geophys. Res.*, 73, 5223–5234.

Press, F. (1970). Earth models consistent with geophysical data. *Phys. Earth Planet. Interiors*, 3, 3–22.

Press, F. and Biehler, S. (1964). Inferences on crustal velocities and densities from P wave delays and gravity anomalies. *J. Geophys. Res.*, 69, 2979–2995.

Sammis, C.G. (1971). Theoretical equations of state. *Trans. Amer. Geophys. Un.*, 52, 122–126.

Shimazu, Y. (1954). Equation of state of materials composing the Earth's interior. *J. Earth Sciences* (Nagoya), 2, 15–172.

Slichter, L.B. (1954). Seismic interpretation theory for an elastic Earth. *Proc. Roy. Soc.* (Lond.), A, 224, 43–63.

Smith, M.L. and Franklin, J.N. (1969). Geophysical application of a generalized inverse theory. *J. Geophys. Res.*, 74, 2783–2785.

Takeuchi, H. and Sudo, K. (1968). Partial derivatives of free oscillation period with respect to physical parameter changes within the Earth. *J. Geophys. Res.*, 73, 3801–3806.

Thomsen, L. (1970). On the fourth-order anharmonic equation of state of solids. *J. Phys. Chem. Solids*, 31, 2003–2016.

Thomsen, L. and Anderson, O.L. (1969). On the high-temperature equation of state of solids. *J. Geophys. Res.*, 74, 981–991.

Wang, C.-Y. (1970). Density and constitution of the mantle. *J. Geophys. Res.*, 75, 3264–3284.

Wesson, P.A. (1973). The implications for geophysics of modern cosmologies in which G is variable. *Q. J. Roy. Astr. Soc.*, 14, 9–64.

Whitcomb, J.H. (1973). Asymmetric $P'P'$: an alternative to $P'dP'$ reflections in the uppermost mantle (0 to 110 km). *Bull. Seismol. Soc. Amer.*, 63, 133–143.

Wiechert, D.H. (1972). Anomalous azimuths of P: evidence for lateral variations in the deep mantle. *Earth Planet. Sci. Letters*, 17, 181–188.

Wiggins, R.A. (1968). Terrestrial variational tables for the periods and attenuation of the free oscillations. *Phys. Earth Planet. Interiors*, 1, 201–266.

Wiggins, R.A. (1969). Monte Carlo inversion of body-wave observations. *J. Geophys. Res.*, 74, 3171–3181.

Wiggins, R.A., McMechan, G.A. and Toksöz, M.N. (1973). The range of Earth structure non-uniqueness implied by body-wave observations. *Rev. Geophys. & Space Phys.*, 11, 87–113.

REFERENCES

Wright, C. (1972). Array studies of seismic waves arriving between P and PP in the distance range $90°$ to $115°$. *Bull. Seismol. Soc. Amer.*, **62**, 385–400.

Wright, C. (1973). Array studies of P phases and the structure of the D'' region of the mantle. *J. Geophys. Res.*, **78**, 4965–4982.

CHAPTER 16

Optimum and standard Earth models

The preceding chapters give an indication of the prodigious, wide-ranging, and occasionally brilliant efforts of numerous investigators to add refinements to Earth models giving distributions of the density ρ and related physical properties $- k$, μ, p, g, etc. $-$ in the Earth's interior. The refinements made during the period since World War II, though sometimes carrying very important implications outside the main Earth model problem, have mostly been fairly small, the net changes in the numerical estimates of ρ etc. over this period having been less than 5 per cent at most depths. The stage has now been reached where many Earth models are being produced which are, at best, minor variants of others.

The time is therefore ripe to look at what may be said about models giving the best ('optimum') representations of the overall data. A related but by no means identical problem is that of formulating a standard Earth model to serve for general reference purposes both inside and outside geophysics. These two topics form the subject-matter of the present chapter. The chapter is to some extent a summarizing chapter, being the last chapter concerned solely with the planet Earth.

16.1 General requirements of Earth models

16.1.1 *The observational evidence*

For models intended to represent the distributions of ρ etc. in the Earth closely, data on the Earth's mean radius R, mass M, and moment of inertia coefficient y constitute the first set of data to be fitted. The uncertainties of these data are sufficiently small, compared with other uncertainties, to be neglected.

[Refs. on p. 365]

The second set consists of data derived from records of seismic bodily and surface waves and of free oscillation data. These data occupy a dominant, though not exclusive, place in constructing the models.

Thirdly, account needs to be taken of a large body of evidence from other sources, including data on Earth tides, thermal data, investigations on the variation of k with p, finite-strain and solid-state theory, laboratory experiments on rocks, including shock-wave experiments at pressures up to 4×10^{11} N/m², and evidence from geodesy, planetary physics, geology, and geochemistry. This third body of evidence, though not in general as precisely determined as the seismic data, assists in assessing the plausibilities of models which fit the seismic data and usefully supplements the seismic data where the latter are more than usually uncertain.

The relative significances of the seismic and non-seismic data have been illustrated in preceding chapters.

16.1.2 *Dependence on space variables*

The ultimate aspiration is of course an Earth model which, within certain specified limits of precision, would show values of ρ etc. in terms of three space variables, e.g. (r, ϕ, λ) of §3.1. The results could be expressed in terms of a spherically symmetrical (SS) Earth model, which would be a first approximation giving ρ etc. in terms of r alone, together with supplementary tables showing variations with the latitude ϕ and longitude λ for each r. In the following discussion, it will be assumed that SS models have had due allowance made for ellipticity effects (§5.9) after the manner described in preceding chapters. (The ellipticity effects would also be involved in the supplementary tables.)

Any SS model gives, in some sense, laterally averaged values of ρ etc. The question immediately arises as to how the averaging should be carried out. The main lateral differences occur in the outer part of the Earth and especially in and near the crust. It would be out of perspective to try to take account of all the minutiae of geological crustal information in Earth models intended to describe the whole interior. On this count alone, some artificiality is inevitable in forming an SS model: simplifying conventions (see e.g. §§10.4.1, 14.7.2) have to be made. Complications arise from the fact that, while sub-oceanic structure (since it involves the larger fraction of the Earth's

[*Refs. on p.* 365]

surface area) is more characteristic of the Earth as a whole than subcontinental structure, the suboceanic data are appreciably the more uncertain. A further point is that global averages implicit in the use of seismic bodily-wave data are influenced by the locations of earthquake foci and observing stations, which are by no means randomly distributed. Such averages are expected to differ from averages implicit in the use of free Earth oscillation data [see Bullen (1972), Dziewonski and Gilbert (1973)].

Some investigators concerned with purely seismological aspects of Earth models have questioned the feasibility of seeking an optimum SS model because of the various complications on lateral differences and the extent of uncertainties about them. But this attitude appears to overlook several things. First, it overlooks the fact that statistical theory has the capacity (see Jeffreys, 1948) to determine weighted means in contexts where the relevant evidence has gaps or is substantially uncertain — that indeed a central purpose of statistical theory is to provide procedures for arriving at most suitable results in circumstances where there are degrees of ignorance. Secondly, the demand for an SS model involves wider contexts than seismology (see § 16.5). Thirdly, with non-SS models (ignoring ellipticity effects), there are such complications as that stresses would not be adequately represented in terms of the single parameter p; current evidence does not permit a useful assessment of the distribution of deviatoric stresses P_{ij} to be made.

The view taken in this book is therefore that SS models are necessary and that an appropriate goal for the present is an SS model which is near an optimum. That is not to say that the evidence on lateral differences already mentioned should not be fully heeded when inferences are being drawn about the Earth from SS models. The existence of the lateral differences and the uncertainties about their details of course add to the uncertainties in determining Earth model distributions, whether SS or not. Attention may be appropriately given to the supplementary tables expressing variations with ϕ and λ (or equivalent space variables) as finer observational details become available.

16.2 Consequence of non-uniqueness

As made clear in § 15.4.1, there is a variety of reasons why the overall data cannot yield a unique Earth model. Attempts have been made to determine 'strips' (§15.4.3) which ascribe (at specified confidence levels) bounds within which the properties of an acceptable model should lie. Because of numerous complications, including the difficulties of the 'trade-off' problem (§15.4.3), these attempts have not as yet been notably successful, although they have brought to light useful information relating to bounds to ρ, k and μ in some parts of the Earth. For example, an inversion of seismic data relevant to core structure by Keilis-Borok and Yanovskaya (see Yanovskaya, 1972) successfully found two classes of model fitting all the travel-time data (as then interpreted), yet did not find models of the type now preferred (see §§12.3.2, 15.6.3).

Even if it is assumed that a set of strips has been reliably and comprehensively determined, there remains the problem of formulating criteria whereby models whose properties lie inside the strips can be sorted out in some order of preference. This and some further aspects of non-uniqueness are touched on in the following section.

16.3 Approaches to the optimum model problem

An Earth (or any other) model could be an optimum only in relation to a given body of evidence. On improved evidence, what was an optimum model would probably no longer be so. Thus the optimum model problem is the problem of finding the best model, or models, on present evidence. Various approaches to the problem are discussed below.

16.3.1 *Successive approximation*

Over a long period of years and until quite recently, most improvements in Earth models were made through successive approximation. This is well illustrated in the sequence starting from the density distributions in models such as the original models A and A', and taking account successively of evidence on the variation of k with p to form the first B-type model with solid inner core, evidence from shock-wave data to improve the lower-core density distribution, revised data on y to form the improved model A'', data on free Earth oscillations to form the further improved model HB_1, and so on. The application of successive approximation in estimating the P and S velocity distributions α and β goes back to the first decade of the century: the evolution of the Zöppritz-Turner tables, correction of the Z.T. tables by successive approximation to form the J.B. tables of 1940 along with the Jeffreys 1939 distributions of α and β, and subsequent modifications of the latter.

On several occasions in the past, Earth models derived through successive approximation have been near the optimum on the then available data. The present position is that successive approximation has brought the reliabilities of models to the point where procedures towards further improvements are often reducible to linear theory. With this stage reached, the way is open for a variety of approaches towards the optimum model.

An advantage of well-conducted successive approximation over some other procedures is that, broadly speaking, new parameters are introduced into model representations only where statistically demanded by the observational evidence. Thus at each stage the models are approximately the simplest which fit the data on hand. A further advantage is that the user of successive approximation can always be in close touch with the fine details contributing to the improvements being made.

16.3.2 *Monte Carlo procedures*

One of the newer approaches is through Monte Carlo procedures (§15.3). These procedures have an advantage that huge numbers of Earth models can be tested in a short time. In consequence, important suggestions may come to light, that otherwise might not readily do so, on properties required of models to fit an assigned set of data.

A further advantage sometimes claimed for Monte Carlo procedures is that, by virtue of the random processes involved, the models are unbiased, in contrast to models derived through procedures which start from a model not randomly selected. However, the Monte Carlo procedures generally result in models which are biased in other ways, e.g. models which are biased against simplicity (see §15.3.2). This bias may obscure the process of drawing reliable inferences from a set of data. For example, it may prove difficult to discriminate between genuine rapid changes of property with r and spurious changes arising from this bias.

Experience to date shows further that the Monte Carlo procedures may fail to find significant classes of models (see §15.3.2). Another disadvantage is that, since the testing is mostly done inside a computer and details of the rejection of models are not revealed, the user is out of touch with important detail.

[*Refs. on p, 365*]

For further comments on Monte Carlo procedures, see Jordan and Anderson (1974).

16.3.3 *General inversion procedures*

A third approach is through general inversion procedures (§ 15.4). As pointed out, the bounds to values of ρ etc. so far derived through the general theory have not been narrow enough to be immediately useful. At the same time, the results have had some value in indicating ranges of depth in the Earth where the uncertainties of determinations are relatively high or low.

A recent innovation is to adapt the general theory to produce single models from assigned sets of data by introducing additional criteria. For example, Dziewonski and Gilbert (1973) arrived at their model B497 by including a 'credibility' criterion (see Backus, 1971) and a formalized procedure to meet simplicity requirements. Jordan and Anderson (1974) used general inversion theory and a variety of other considerations in producing their preferred model B1.

An advantage of proceeding in this way is that the full might of modern computing facilities can be brought to bear with the avoidance of much drudgery. A corresponding disadvantage is that only a limited quantity of numerical detail is revealed to users of this procedure.

16.3.4 *Comments*

All three approaches of §§ 16.3.1–16.3.3 have yielded Earth models which fit assigned sets of data. Until the disadvantages of the Monte Carlo approach are more fully overcome than is the case now, models derived through the other approaches are likely to be nearer the optimum. Successive approximation is likely to continue to be indispensable towards securing the most reliable models, but with general theory and its adaptations of the type mentioned in § 16.3.3 playing an increasing auxiliary role.

Among presently existing Earth models, those fitting the most comprehensive sets of data have been derived through general theory adapted as in § 16.3.3.

16.4 Progess towards an optimum Earth model

The models B497 of Dziewonski and Gilbert (1973) and B1 of Jordan

[*Refs. on p. 365*]

and Anderson (1974) mentioned in §16.3.3 are examples of SS models constructed using the most comprehensive data currently available. Both are preferred models according to criteria laid down by their authors and both reasonably satisfy simplicity requirements. Having regard to the comprehensiveness of the data used, the substantial application of general theory, and the powerful computing processes brought to bear, these models can be regarded as useful steps towards an optimum.

Table 16.1 shows values of ρ, α, and β for the two models at various depths. Values of p, k, μ, g, and Poisson's ratio σ for the model B497 have not so far been published, but are shown for the model B1 in Table 16.2. In order to assist in assessing the extent of recent progress, the tables also show an earlier set of values taken from the model HB_1 (§14.7.2) down to 2000 km depth and from the model B_2 (as modified in Table 12.4, §12.7.2) at depths greater than 2000 km; this set will be referred to as the model HB'_1. Being a mixture of two models, HB'_1 does not entirely meet the requirements of a single Earth model, but only minor adjustments would be needed to make it do so; thus HB'_1 serves as a useful base for comparison with advances since around 1969. The values for B497 and the Jordan-Anderson model B1 are shown in the tables as differences against HB'_1.

A comment is desirable on the selection of depths in the tables. In Table 16.1, the depths 2886 and 2889 km both appear because they straddle the range of depths (2886 for HB'_1 and B1; 2889 for B497) of the mantle-core boundary N in all three models; and similarly with 5120 and 5160 km for the depth of the inner-core boundary L (5160 for HB'_1; 5121 for B497 and 5156 km for B1). Closer values of z are included for the upper mantle than elsewhere because of the extent of fluctuations in ρ, α, and β there occurring. Similar remarks apply to Table 16.2.

The crustal thicknesses are 15 km (HB'_1), 10 km (B497), and 21 km (B1). The crustal values of ρ, α, and β in B497 are markedly less than those of the other models. This is principally a consequence of the relative weighting attached by Dziewonski and Gilbert to the free Earth oscillation data and the seismic bodily-wave data. They pointed out that the oscillation periods are more closely related to the properties of an SS 'average' Earth than 'the nearly exclusively land-based travel-time observations' (cf. §16.1.2) and took this strongly into account in arranging the weightings. The difference

TABLE 16.1

Estimates of ρ (in g/cm^3), and α and β (km/s) in some spherically symmetrical Earth models; z is the depth in km. For each of ρ, α, and β, column 1 shows values for the model HB'_1. Columns 2 and 3 show the increases necessary to give values for the model B497 of Dziewonski and Gilbert and the model B1 of Jordan and Anderson, respectively. The units for columns 2 and 3 are 0·01 of those for column 1.

z	ρ	α	β
0	2·84; −68, −05	6·30; −164, −10	3·55; −99, −15
30	3·32; +03, −02	7·74; −11, +21	4·62; −04, +20
100	3·35; 00, −03	7·95; −26, +13	4·50; 00, +03
200	3·39; −01, −02	8·26; −06, +05	4·50; −08, −16
300	3·42; +05, +05	8·59; +10, −07	4·50; +10, +06
400	3·77; −16, −21	8·92; +26, −21	4·72; +18, −07
650	4·17; +01, −14	10·48; −16, −46	5·80; −16, −59
1000	4·54; +06, +04	11·44; −05, +04	6·36; −02, +03
2000	5·09; 00, +02	12·79; −07, −02	6·92; −03, −04
2700	5·40; +08, +08	13·61; −01, −02	7·26; −12, −06
2886	5·69; −13, −11	13·64; +11, +03	7·30; −17, −03
2889	9·95; +02, −05	8·12; −07, −10	0; 00, 00
4000	11·39; −01, +02	9·53; −02, +11	0; 00, 00
5120	12·70; −36, −60	10·33; +03, −19	0; 00, 00
5160	12·74; +02, −46	11·25; −18, −03	3·86; −17, −40
6371	13·03; −13, −45	11·25; +10, −05	2·91; +62, +59

between their crust and others provides a measure of the contribution of lateral differences to the uncertainties of SS models.

In the upper mantle, HB'_1 has discontinuities at 350 and 650 km; B1 at 420 and 671 km; B497 has rapid but continuous changes instead of these discontinuities. The differences between the three models in the upper mantle arise principally from differences in treating the finer details of evidence on rapid changes of property.

Throughout the lower mantle and most of the outer core, the differences in the values of ρ etc. for the different models are fairly small. It needs to be pointed out that the uncertainty in α at the top of the core is somewhat greater than the table indicates. In and near the inner core the differences, especially in ρ, β, k, and μ, become greater.

The values of k just above and just below the mantle-core boundary N are 6·74 and 6·46 for the model B497; 6·49 and

[*Refs. on p. 365*]

TABLE 16.2

Estimates of p, k, and μ (in $10^{11}\,\text{N/m}^2$), σ, and g (in m/s^2) at depths z km. For each item, column 1 shows values for the model HB'_1. Column 2 shows the increases necessary to give values for the model B1 of Jordan and Anderson. The units for column 2 are 0·01 of those for p, k, μ, and g in column 1, and 0·001 for σ

z	p		k		μ		σ		g	
0	0;	0	0·65;	−01	0·36;	−4	0·267;	+18	9·82;	0
30	0·01;	0	1·04;	+03	0·71;	+6	0·222;	−10	9·84;	+1
100	0·03;	0	1·21;	+05	0·68;	0	0·264;	+08	9·86;	0
200	0·06;	0	1·39;	+09	0·69;	−5	0·289;	+23	9·90;	0
300	0·10;	0	1·60;	+05	0·69;	+3	0·311;	−12	9·93;	0
400	0·14;	0	1·89;	−18	0·84;	−6	0·306;	−04	9·96;	+1
650	0·23;	0	2·68;	−10	1·43;	−33	0·273;	+41	9·99;	+2
1000	0·39;	0	3·49;	+07	1·84;	+2	0·276;	+02	9·96;	0
2000	0·87;	0	5·07;	+04	2·44;	+2	0·293;	+02	10·02;	−3
2700	1·24;	+1	6·20;	+13	2·85;	−1	0·301;	+04	10·53;	−9
2886	1·35;	0	6·54;	−05	3·04;	−9	0·299;	+04	10·77;	−9
2886	1·35;	0	6·54;	−17	0;	0	0·5;	00	10·77;	−9
4000	2·48;	+7	10·34;	+27	0;	0	0·5;	00	7·93;	−42
5155	3·34;	−6	13·59;	−113	0;	0	0·5;	00	4·37;	−12
5160	3·34;	−6	13·60;	−11	1·90;	−43	0·43;	+17	4·35;	−12
6371	3·67;	−7	15·00;	−151	1·11;	+43	0·46;	−14	0;	0

$6.37 \times 10^{11}\,\text{N/m}^2$ for B1. Near-continuity of k at N is again confirmed, but both these models give a small drop across N and so deviate a little further from the indications of the Murnaghan-Birch theory than HB'_1, for which $k = 6.54 \times 10^{11}\,\text{N/m}^2$ on both sides of N. The models HB'_1 and B497 have k continuous at the inner-core boundary L, the values being 13·60 and $13.24 \times 10^{11}\,\text{N/m}^2$, respectively; B1 has $k = 12.46$ and $13.49 \times 10^{11}\,\text{N/m}^2$ just above and just below L. Since B497 shows that a model can be found which fits all the data and has k continuous at L, the model B1 appears to introduce an unnecessary parameter in having a jump in k at L. On the other hand, B1 appears to be slightly preferable to B497 in respect of k at N.

An incidental point is that the changes from HB'_1 to B497 and the Jordan-Anderson B1 are remarkably small at most depths: most differences from HB'_1 are of the same order as the differences between B497 and B1.

[Refs. on p. 365]

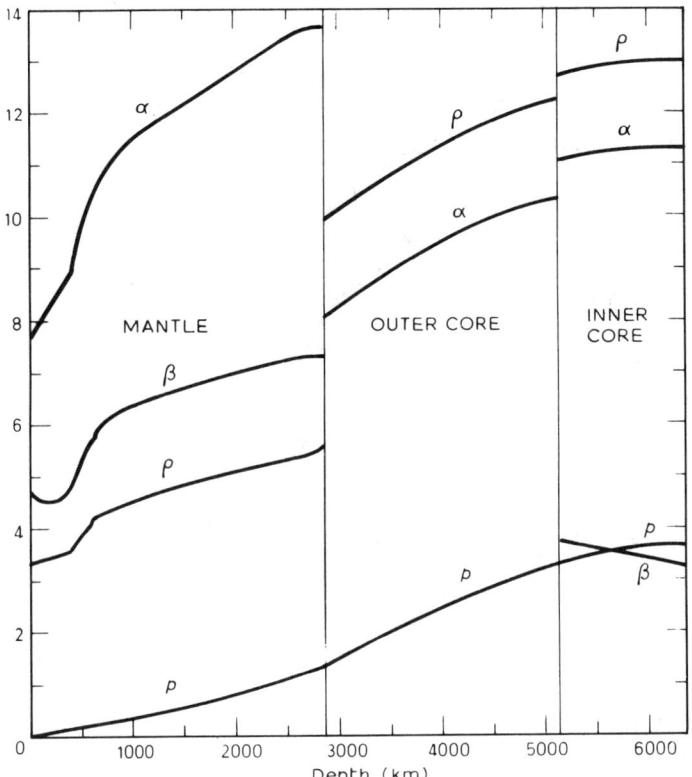

Fig. 16.1. Distributions in the Earth of the density ρ (g/cm^3), pressure p (10^{11} N/m^2), and P and S velocities α and β (km/s)

The above brief survey suggests that both the models B497 and B1 are capable of improvement. For example, the crustal thickness of B1 appears rather too great for a laterally averaged SS model, and a modification of the procedure which gave the unrealistically low values of ρ, α, and β in the B497 crust would seem to be desirable if it could be contrived without doing too much violence to the postulates underlying B497. The differences between B1 and B497 in the upper mantle, especially near $z = 400$ and 650 km, might be reducible by formal optimization procedures. In most of the lower mantle and outer core, both models seem well determined. Differences in the inner core are to be expected because of limitations on the resolving power of the available data. A further point is that factors not yet taken into account, e.g. the effect of imperfect elasticity

[Refs. on p. 365]

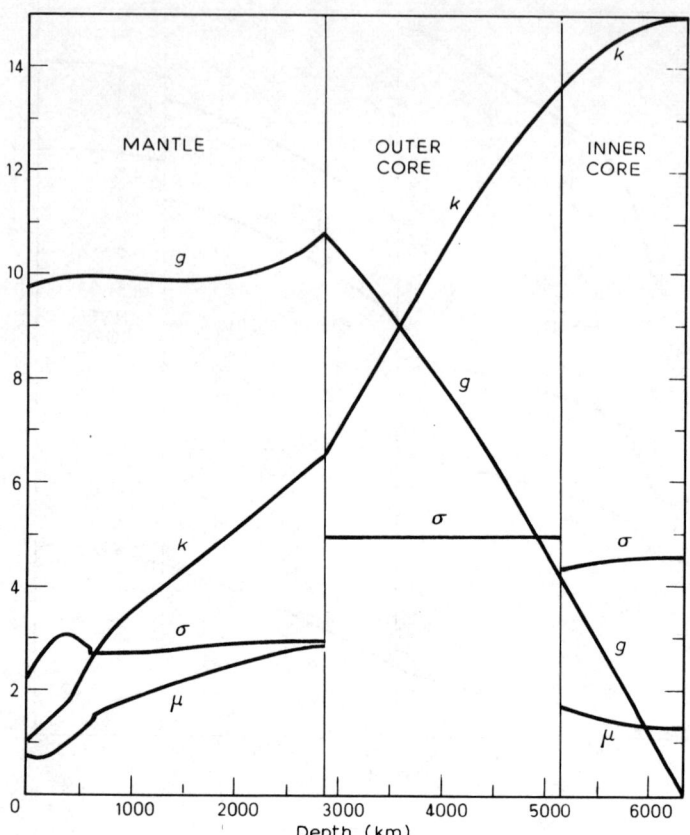

Fig. 16.2. *Distributions in the Earth of the incompressibility k and rigidity μ (10^{11} N/m²), gravitational intensity g (m/s²), and Poisson's ratio σ. (The scale unit for σ corresponds to 0·1.)*

on free Earth oscillation periods (see §12.4), may require both models to be modified.

Thus notwithstanding the quality of the data now brought to bear and the considerable attention given to general theory and computing adjuncts, it would appear that the goal of an optimum model is still some distance away. Tables 16.1 and 16.2 may be regarded as indicating the present state of progress towards the optimum goal, and the order of magnitude of remaining uncertainties. Figures 16.1 and 16.2 exhibit what appear at the moment to be approximately the most likely distributions of ρ, p, α, β, k, μ, σ, and g.

[*Refs. on p.* 365]

16.5 The problem of a standard Earth model

There has long been a demand for a well-chosen internationally agreed standard or reference Earth model, giving distributions of ρ etc., which would serve two main purposes. The first purpose is to provide a reference base against which new data and results could be tidily presented (cf. the presentation of seismic travel times in the International Seismological Summary as residuals against the J.B. tables). Lack of an agreed reference base has been responsible for much recent untidiness in the presentation of numerical seismological and other results — an untidiness which, inter alia, greatly complicates the task of putting strands of evidence together as in reviews and books like this one. Added confusion has been caused through the intermittent use of hybrid models for temporary reference purposes, e.g. combinations of the original Model A or B density distributions with one of Gutenberg's seismic velocity distributions — combinations which have non-matching discontinuities and other discordancies. The second purpose is to satisfy demands for a standard Earth model which can serve in contexts outside seismology, e.g. in geodesy and planetary physics.

The demand resulted in an international standard Earth model committee being set up in 1971 at the instigation of the International Association of Geodesy. The task of the committee has proved to be more complex than had been anticipated, partly for the reason that prima donnas performing simultaneously on different keys are slow to produce harmony, partly because of widespread failure to appreciate the difference between a standard and an optimum model.

The aspiration of an optimum model is to fit all the available data. Thus an optimum model would involve as many parameters as the data demand, with due allowance for uncertainties. A cardinal requirement of a standard model is simplicity, and a central problem is deciding on the degree of simplicity to be permitted. The requirement of simplicity is so paramount that a standard model is likely to have fewer than the minimum parameters demanded: some of its properties are therefore likely to be incompatible (though hopefully only to a small extent) with parts of the data. (Ill fitting with parts of the data may be unavoidable even in attempts towards constructing optimum models, as with the crust of the model B497.) Subject to the simplicity demands, a standard model should of

[Refs. on p. 365]

course fit the data closely enough for residuals against values of the model properties to be amenable to linear theory, so far as possible. There is no question of uniqueness: a variety of significantly different models may qualify equally as standard models; it does not much matter which model is selected as the standard, provided that the broad criteria are met. For reasons given in §16.1.2, it is appropriate at present to seek a spherically symmetrical standard model.

In practice, a number of questions, some of them subtle, have to be answered in the course of applying the criteria. Following are examples:

(i) What is the appropriate representation in a range of depth where there is evidence of a rapid or sudden change of property? Should the representation be in terms of a mathematical discontinuity or a rapid continuous change? If the representation is continuous, how rapidly should the change be taken? It is usual to treat the Mohorovičić discontinuity and the mantle-core and inner-core boundaries as mathematical discontinuities. Elsewhere, e.g. at depths near 400 and 650 km, the decision may be difficult (cf. the models HB'_1, B497, and B1 of §16.4). "When in doubt, smooth" has often been stated as a guiding principle, but the difficult question is really "how smooth?".

(ii) How should relative degrees of smoothing be determined among parameters between which there is appreciable interplay (trade-off) when models are perturbed? For example, should priority be given to simplicity in the variation of ρ or of β over a range of depth where appreciable interplay occurs? Should the parameter η (§11.5.2) be kept constant inside a region such as the region D', or should η be let fluctuate as a consequence of smoothing procedures applied in the representations of α and β?

(iii) There is good evidence that the region D'' is somewhat abnormal (see §11.2.2). Should a significant rise in $d\rho/dz$, or a fall in μ, or both, be allowed for in order to incorporate the abnormality, or should this abnormality be ignored in a standard model?

(iv) What should be done about the crust (cf. the models B497 and B1 of §16.4)?

(v) Should any provision be made for possible transition layers between the outer and inner core proper? (In the light of §15.6.3 and the simplicity requirement, it now seems that a simple two-layer core should be adopted.)

(vi) Should α be taken constant in the inner core?

(vii) Should β be taken constant in the inner core? It is to be noted that if β is taken constant, μ cannot be constant since ρ is not constant.

(viii) Having regard to the observational uncertainties, what value should be taken for the central density?

Further details on the standard Earth model problem are given in papers published under the auspices of the Standard Earth Model committee by Adams and Engdahl (1974), Bullen (1974a, 1974b), Cleary (1974), Dziewonski and Haddon (1974), Hales (1974), Hales, Lapwood, and Dziewonski (1974), and Marussi, Moritz, Rapp, and Vicente (1974).

Should a later edition of this book eventuate, perhaps it will be possible to take the history of knowledge of the Earth's density distribution to the point where an optimum and a standard Earth model have both been determined.

REFERENCES

Adams, R.D. and Engdahl, E.R. (1974). *P*-wave velocities in the Earth's core: an interim report. *Phys. Earth Planet. Interiors*, 9, 36–40.

Backus, G. (1971). *Amer. Math. Soc., Lect. Appl. Math.*, 14, 1–7.

Bullen, K.E. (1972). Some problems connected with constructing Earth models. *Proc. Conference on Solid Earth Problems, Buenos Aires, 1970*, II, 63–77.

Bullen, K.E. (1974a). Introductory remarks on standard Earth model. *Phys. Earth Planet. Interiors*, 9, 1–3.

Bullen, K.E. (1974b). Standard Earth model requirements in respect of density and rigidity in the inner core. *Phys. Earth Planet. Interiors*, 9, 41–44.

Cleary, J. (1974). The D'' region. *Phys. Earth Planet. Interiors*, 9, 13–27.

Dziewonski, A.M. and Gilbert, F. (1973). observations of normal modes from 84 recordings of the Alaskan earthquakes of 1964 March 28 – II. Further remarks based on new spheroidal overtone data. *Geophys. J., Roy. Astr. Soc.*, 35, 401–437.

Dziewonski, A.M. and Haddon, R.A.W. (1974). The radius of the core-mantle boundary inferred from travel time and free oscillation data; a critical review. *Phys. Earth Planet. Interiors*, 9, 28–35.

Hales, A.L. (1974). Crustal structure modelling in a spherically symmetrical Earth. *Phys. Earth Planet. Interiors*, 9, 7–8.

Hales, A.L., Lapwood, A.R. and Dziewonski, A.M. (1974). Parameterization of a spherically symmetrical Earth model with special references to the upper mantle. *Phys. Earth Planet. Interiors*, 9, 9–12.

Jeffreys, H. (1948). *Theory of Probability* (2nd Ed.), 411 pp., Cambridge University Press.

Jordan, T.H. and Anderson, D.L. (1974). Earth structure from free oscillations and travel times. *Geophys. J., Roy. Astr. Soc.*, 36, 411–459.

Marussi, A., Moritz, H., Rapp, R.H. and Vicente, R.O. (1974). Ellipsoidal density models and hydrostatic equilibrium. *Phys. Earth Planet. Interiors*, 9, 4–6.

Yanovskaya, T.B. (1972). Determination of a set of velocity-depth curves for the transition zone in the Earth's core from travel times of successive arrivals of *PKP* waves. *Geophys. J., Roy. Astr. Soc.*, 29, 227–235.

CHAPTER 17

Application to other planets and the Moon

The distributions of the density ρ, pressure p, and incompressibility k, as derived for the Earth in the preceding chapters, bear strongly on the estimation of the distributions in other planets. The present chapter is principally concerned with applying to other planets equations of state for particular internal zones of the Earth (see §15.1), along with pertinent planetary observational data and some limited supplementary evidence on composition. The results indicate broadly the inferences that can be made about internal planetary structures with reliabilities not too low. Occasional reference will be made to other theories of planetary structure, e.g. geochemical theories, but deep incursions into such theories lie outside the scope of this book. (The highly complex assumptions of some planetary theories, incidentally, tend to fluctuate rather rapidly with time.)

The bodies to be considered are the Moon, Mercury, Venus, Mars, Earth, Jupiter, Saturn, Uranus, Neptune, Pluto. The individual bodies will sometimes be indicated by their initial letters (Mo, Me, V, Ma, E, J, S, U, N, P) and their properties by corresponding subscripts.

In general, equations of state involve a variable representing temperature. For planets other than the Earth, reliable evidence on internal temperature distributions is, however, very meagre. In astrophysical theory, planets are classed as 'cold bodies' in which temperature influences in equations of state are treated as small. Since, further, only fairly rough approximations to the internal density distributions of other planets are in any case as yet derivable, differences in internal temperatures are usually ignored when applying (ρ, p), etc. relations for the Earth to another planet. It may sometimes be desirable, however, to take account of super-adiabatic gradients (see §17.2.1). (When, in the future, improved approximations to the density distributions in other planets are sought, it

[Refs. on p. 398]

may be feasible to take some further account of temperature.)

The equations of state also depend on the materials involved; so consideration has to be given to questions of chemical composition and phase in internal zones of both the Earth and other planets. The uncertainties are such that, even for the Earth, the implications of differing assumptions about composition or phase in some regions, especially the outer core, have to be examined, leading, as will be seen, to series of alternative models for other planets. With the other planets, inferences on composition are mainly drawn from observational data on the mean densities $\bar{\rho}$, in conjunction with comparisons with the Earth.

The terrestrial planets (Me, V, E, Ma) and the Moon have mean densities of the same order (see §17.1.1). Having regard to theories of the origin of the solar system, it is therefore strongly probable that the compositions of this group of bodies have much in common. Hence for some members of the group, useful progress can be made by applying equations of state for the Earth fairly directly. The possibility has to be allowed for, however, that a member may have a less differentiated interior than the Earth so that, even should the overall compositions be about the same, the internal variations of composition with p may be noticeably different. As will be seen, not only do details for the Earth provide evidence on other planetary interiors, but also findings on other planets have the possibility of feeding back evidence on some aspects of the Earth's composition.

The internal compositions of the major planets (J, S, U, N) differ too much from that of the Earth (see §17.1.1) to fit the specific equations of state for the main materials of the Earth's interior. Nevertheless, certain of the principles involved in investigating the terrestrial planets are applicable to the major planets. Since, further, it is desirable to make certain comparisons of the structures of various groups of planets, the chapter will include a short discussion of the major planets. Brief reference will also be made to Pluto.

Direct observational evidence on the planets includes estimates of the masses M, mean radii R, and hence also $\bar{\rho}$; with several planets, this is the sole direct observational evidence. With some others, estimates of the mean moment of inertia I and hence the coefficient y $(= I/MR^2)$ are also available; a prerequisite for this is a sufficient axial angular speed.

In general, hydrostatic stress will be assumed throughout the interiors, so that the representation of stress will be solely in terms of the pressure p. Spherical symmetry will be assumed except so far as ellipticities need to be taken into account in estimating I and y. Thus the main internal properties will be treated as functions of the distance r from the centre, or depth z below the surface. In general, the larger the planet, the more accurate is the hydrostatic assumption, but the less accurate is Radau's approximation (§5.4.4).

17.1 Planetary observational data

17.1.1 *Masses, radii, and mean densities*

Table 17.1 shows observational values of M, M/M_E, R, and $\bar{\rho}$ for the planets and Moon. The values are preferred values derived from data assembled by Pecker (1966), Bullen (1969), Cook (1972), and Duncombe, Seidelmann, and Klepczynski (1973). Except for the Earth and Moon, the last digit in each table entry is uncertain by at least one unit.

Table 17.1
Masses (M), mean radii (R), and mean densities ($\bar{\rho}$) of the planets and Moon

Planet	M	M/M_E	R	$\bar{\rho}$
Moon	7.35×10^{22} kg	0.01230	1.738×10^3 km	3.34 g/cm^3
Mercury	3.3×10^{23}	0.055	2.43×10^3	5.5
Venus	4.87×10^{24}	0.815	6.05×10^3	5.25
Earth	5.976×10^{24}	1	6.371×10^3	5.517
Mars	6.42×10^{23}	0.107	3.39×10^3	3.93
Jupiter	1.90×10^{27}	318	7.1×10^4	1.27
Saturn	5.69×10^{26}	95	6.0×10^4	0.63
Uranus	8.74×10^{25}	14.6	2.5×10^4	1.33
Neptune	1.029×10^{26}	17.2	2.5×10^4	1.57
Pluto	6×10^{23}	0.1	3×10^3	5

Table 17.1 shows immediately that there must be significant differences in the present internal compositions among various groups of planets. The pressures reached are much higher inside the major than inside the terrestrial planets; yet the mean densities are appreciably less. Hence the major planets must be composed principally of materials of lower atomic numbers than the terrestrial planets. Further, for Jupiter and Saturn, $M/M_E = O(100)$, while for Uranus

[Refs. on p. 398]

and Neptune, $M/M_E = O(10)$; yet Jupiter and Saturn have smaller $\bar{\rho}$. So Uranus and Neptune must be intermediate in composition between the terrestrial planets and the pair Jupiter and Saturn.

For reasons to be given, each of the Moon, Mercury, and Pluto will be considered separately from the other planets. So it is convenient to examine the various planetary interiors in six groups: E, V, Ma; Me; Mo; J, S; U, N; P.

17.1.2 *Flattenings and moments of inertia*

Theory given in Chapter 5 is relevant to estimating the flattenings f (where f corresponds to ϵ_a for the Earth, §5.2), and y and I for other planets.

A first step is to estimate for each planet the constant h, defined by equation (5.8), using data on M, R, and the axial angular speed Ω. Data on M and R are shown in Table 17.1 and Ω is derived from observations of the period of rotation, D sidereal days (say): $\Omega D = 2\pi/(24 \times 60^2) = 7 \cdot 2722 \times 10^{-5}$ radians. Table 17.2 shows the observational D and the corresponding computed Ω and h.

Table 17.2
Rotational periods and the constants h of planets

Planet	D	Ω	h
Mercury	58·6 sid. days	$1 \cdot 24 \times 10^{-6}$ rad/s	$1 \cdot 00 \times 10^{-6}$
Venus	243	$2 \cdot 99 \times 10^{-7}$	$6 \cdot 10 \times 10^{-8}$
Mars	1·0259	$7 \cdot 089 \times 10^{-5}$	$4 \cdot 570 \times 10^{-3}$
Jupiter	0·410	$1 \cdot 77 \times 10^{-4}$	0·089
Saturn	0·426	$1 \cdot 71 \times 10^{-4}$	0·166
Uranus	0·448	$1 \cdot 62 \times 10^{-4}$	0·071
Neptune	0·658	$1 \cdot 10 \times 10^{-4}$	0·028
Pluto	6·4	$1 \cdot 15 \times 10^{-5}$	8×10^{-6}

For several of the planets, estimates of the coefficient J_2 (§5.3.3) are available from observations of the orbits of their natural satellites. The flattenings f can then be estimated using the values of h in Table 17.2 and equation (5.15), which gives

$$2f = 3J_2 + h, \qquad (17.1)$$

correct to the first order. Assuming Radau's approximation, y can

[*Refs. on p. 398*]

then be estimated using

$$2\eta = 5h/f - 4, \qquad (17.2)$$

$$2 - 3y = 0 \cdot 8\sqrt{(1 + \eta)}; \qquad (17.3)$$

(17.2) corresponds to (5.29), with η written for η_a; (17.3) is approximately equivalent to (5.49).

For Uranus, J_2 is not directly derivable from observations. [For J_2 to be derivable from observations of a satellite orbit, the term containing J_2 in (5.14) must vary significantly during the orbit. Uranus has five satellites but their orbits lie so close to the equatorial plane of Uranus that the factor $P_n(\mu)$ and hence the term containing J_2 is nearly constant throughout each orbit.] But a visual observation of Luoff (1932) which gave $f^{-1} = 18 \cdot 0 \pm 0 \cdot 6$ may be used to provide rough estimates of J_2, η, and y. The estimates are rough because, for any planet, visual observations of f are usually fairly imprecise and, further, the surface configuration may be markedly affected by deviatoric stresses near the surface.

The assumed values of J_2 [taken, except for Uranus, from data assembled by Cook (1972)] are shown in Table 17.3. It may be remarked that as early as 1944, Woolard, studying motions of the satellites Phobos and Deimos, had obtained a result equivalent to $J_2 = 0 \cdot 00195$ for Mars. Table 17.3 gives the values of f, η, and y derived from the stated data and procedures.

Table 17.3
Values of J_2, f, η, and y for various planets

Planet	J_2	f	η	y
Mars	0·001966	0·00523	0·183	0·376
Jupiter	0·01471	0·0665	1·34	0·26
Saturn	0·01667	0·1079	1·84	0·22
Uranus	0·014	0·056	1·18	0·27
Neptune	0·0050	0·0215	1·25	0·27

The results for y in Table 17.3 agree fairly well with results of Cook (1972); the agreement is very close for Mars and within about 0·01 for the major planets. But, like Cook's, they can be regarded only as first approximations requiring correction for errors in the assumptions made.

[*Refs.* on p. 398]

With Mars, which is much smaller than the major planets, the principal error is likely to arise from departures from the assumed hydrostatic stress. The most recent visual estimate of f for Mars is 1/188, derived from observations of the occultation of an artificial satellite; this gives $f = 0.00532$, which differs by only 2 per cent from the table estimate of 0.00523. Cook thence inferred that Mars does not depart very far from hydrostatic stress. But it needs to be noted that near agreement of the two estimates, while compatible with nearly hydrostatic stress, by no means establishes that this is the case. Calculations by Bullen (1957) and later by MacDonald (1962) indicate that deviatoric stresses inside Mars can well result in an error exceeding 0.01 in the computed y. According to Binder and Davis (1973), an analysis of recent evidence makes it now likely that the dynamical and optical flattenings of Mars are significantly different.

With the major planets, significant errors can arise from several sources, especially inadequacies in the Radau approximation, neglect of second-order terms in the ellipticity theory, and the simplifying of equation (5.49). Miles and Ramsey (1952) estimated that the Radau approximation yields values of y too small by 8 and 18 per cent for Jupiter and Saturn. Application of these corrections would raise the estimates of y for these two planets to 0.28 and 0.26, respectively. The corrections depend, however, on model density distributions assumed for the purpose and are therefore not finely determined. Thus while 0.28 and 0.26 may be regarded as perhaps the best values for Jupiter and Saturn on present knowledge, they remain uncertain by at least 0.02. The table values of y for Uranus and Neptune are likewise probably a little too low. Ramsey (1967) gave $y = 0.305$ for both these planets, but the observational data he used have since been substantially revised.

In respect of the five planets in Table 17.3, the essential summary is that the inferred values of y provide useful restrictions on possible internal density distributions, but that caution must be taken to allow for the sizeable uncertainties.

The axial rotation periods of Venus and Mercury are 243.1 and 58.6 sidereal days. These periods are too long for f or J_2 to be large enough to be measurable at present. Observations of a space vehicle round Venus have indicated that $J_2 < 10^{-5}$; no observational information of J_2 is available for Mercury. Thus y cannot be estimated in the above way for Venus or Mercury. For Pluto, there are no data on J_2, so that again f and y cannot be estimated.

[Refs. on p. 398]

For the Moon, Michael, Blackshear, and Gapcynski (1969), using data on perturbations of artificial satellite orbits, gave $(A, B, C)/MR^2 =$ (0·4002, 0·4003, 0·4005), where A, B, C are principal moments of inertia corresponding to the axes Ox_1, Ox_2, Ox_3 of Chapter 5. (The Moon is not quite axially symmetrical, so that $A - B$ differs detectably from zero.) In a re-estimation, using calculations of Jeffreys (1971) on the Moon's librations, Cook (1970) gave (0·4029, 0·4030, 0·4032), with formal uncertainties of about 0·0026; the real uncertainties are likely to be greater. (Librations are irregularities in the Moon's motion which manifest themselves in 'rocking motions' whereby features near the edge of the Moon's disc become alternately visible and invisible.) Hence the observational data on the Moon are compatible with a nearly constant internal density.

17.2 Assumptions on the Earth's internal composition

Some details on the Earth's internal composition and phase changes have been already given in subsections of §11.6. The present section carries the discussion further on aspects which bear especially on the interiors of terrestrial-type planets.

In applying equations of state derived for the Earth to regions of these planets, the following simplifications will be made. The crust (region A) of the Earth being comparatively small, planetary crusts will be ignored except where specially mentioned. Abnormalities connected with the subregion D'' will be ignored, partly because D'' is relatively small and partly because details of the abnormalities remain somewhat uncertain (§ § 11.2.2, 11.6.4); thus D (including D' and D'') will be treated as a single region having the properties of D'. The Earth's core will be treated as consisting of only the outer core (region E) and inner core (region G), evidence on the properties (and even on the existence) of the transition zone F being now somewhat uncertain (§ § 12.3.2, 15.6.3).

17.2.1 *Some general considerations*

When the (ρ, p) relation for a particular region of the Earth is applied directly to a region of another planet, the chemical compositions are assumed to be the same for each value of p inside the two regions. Each of the regions B and D of the Earth may be treated for present purposes as having uniform composition and phase. Apart from the effect of possible super-adiabatic gradients ϑ (see below), the coefficient η (§11.5.2) is then equal to unity. The region C has η noticeably greater than unity, but this is attributed principally to phase changes though there are probably small composition changes as well (§11.6.3). The presence of phase changes inside a region does not affect the direct application of the (ρ, p) relation to another planet, and the small composition changes inside C are likely to be ignorable in planetary applications. Since, further, any changes in composition at the B–C and C–D boundaries are likely to be small (because of the continuity or near-continuity of the P and S seismic velocities), it is appropriate for present purposes to treat the entire mantle of the Earth as a single region conforming to a single (ρ, p) equation of state.

It is sufficient for present purposes to treat each of the regions E and G of the Earth as having uniform composition and phase and negligible ϑ (see §§11.6.5, 12.2.1, 12.6.1). But the possible differences in their composition (see §§17.2.2–17.2.4) make it desirable to treat the (ρ, p) relations for the two regions as distinct.

In general, to begin with, regions of planets will be assumed to have uniform compositions, so that (ρ, p) relations for the Earth may be applied directly to appropriate planets. But when a planet is assumed to be less differentiated than the Earth, thus having continuous variation of composition inside one or more regions, the applicability of the (ρ, p) relations may be affected.

The possibility that a boundary may be predominantly associated with a phase change rather than a change of chemical composition can have an important effect in applying the (ρ, p) relations. For example, if the change at the Earth's mantle-core boundary should be primarily a phase change, a single (ρ, p) relation would be relevant for the whole Earth down to the outer-inner-core boundary; otherwise, two separate relations are necessary (see §§17.2.2, 17.2.3).

The dependence of the (ρ, p) equation on temperature would need to be considered in practice only when there is evidence that the values of ϑ in a region of the Earth and the corresponding region of a planet differ significantly. For example, ϑ appears to be marked in at least part of the region B of the Earth, and it may therefore sometimes be desirable to separate out the contribution due to ϑ from other contributions to $d\rho/dz$ inside B before applying the (ρ, p) relation; in practice, this is sometimes done by using a model (ρ, p)

[Refs. on p. 398]

relation for B which neglects the contribution from ϑ (as, for example, in the original A-type models).

Some writers have used Emden's equation (6.19) in applying data on the Earth to other planets. It is to be noted that this procedure, in addition to assuming uniform composition and neglecting ϑ, assumes (k, p) relations of the form $k = a + bp$. This last assumption is fairly reliable for the regions D, E, G (see §11.8.3), but can be a source of confusion when applied without suitable precautions above D (see Bullen, 1966a). It has frequently been overlooked that the k-p hypothesis (§11.4), involving fairly simple dependence of k on p, does not become relevant until pressures of order $0 \cdot 4 \times 10^{11}\,\text{N/m}^2$ are reached. The use of (k, p) equations of state can, however, be quite useful when needed precautions are taken.

17.2.2 *The iron-core theory*

At the time of its discovery, the Earth's core was considered from meteorite evidence to be predominantly composed of iron or iron-nickel.

The finding in 1936 that the core has outer and inner parts did not immediately affect this conclusion. But evidence over the period 1952–61 (see §11.6.5) drawn from the (ρ, p) core distributions and shock-wave experiments on iron indicated that the outer core is about 15 per cent less dense than iron at the pressures involved. Subsequently, the generally favoured theory has been that the outer core consists of iron alloyed with one or more elements such as silicon, sulphur, and carbon. This theory is incompatible with the terrestrial planets having closely similar overall compositions. For example, assuming the Earth, Mars, and Venus to have similar mantle and similar core compositions, the entailed mantle-core mass ratios would be about 3·6 for Venus and 5·5 for Mars, as against 2·1 for the Earth (Jeffreys, 1937; Bullen, 1937; see also below).

In the inner core, the (ρ, p) distributions are quite compatible with an iron-nickel composition. The distributions preclude a representative atomic number Z noticeably less than that of iron (26), and those elements for which Z exceeds the value for nickel (28) are much less abundant. Hence it remains strongly probable that the inner core is in fact composed predominantly of iron-nickel.

[*Refs. on. p. 398*]

17.2.3 *The mantle-core phase-transition theory*

The predominantly iron core theory had some early doubters. L.H. Adams (1937) referred to Ono and others who preferred the idea of a core identical in composition with that of the mantle, the higher density of the core being brought about by pressure.

A later sequence of events led Ramsey (1948) and Bullen (1949a) independently to a similar idea. Kuhn and Rittmann (1941) had put forward the theory that the Earth's core consists predominantly of hydrogen. That theory had been quickly shown to be untenable on several grounds, but had led Kronig, de Boer, and Korringa (1946) to calculate that at a pressure of about 0.7×10^{11} N/m² (reached in the Earth at $z \approx 1600$ km), hydrogen would become metallic (with high electrical conductivity), the density jumping from about 0.4 to 0.8 g/cm³. Ramsey and Bullen then proposed as a compromise between the iron and hydrogen core theories what is now referred to as the phase-transition theory: at the mantle-core boundary (N say), where the pressure has a critical value, p_c say, the lower mantle material undergoes a phase transformation to a metallic high-density modification with essentially unchanged composition. Both authors perceived that this proposal would act in the direction of bringing the Earth, Venus, and Mars into line with a common overall composition.

The approaches of Ramsey and the writer, while similar on major aspects of the theory, were not identical. Ramsey made an argument from solid-state physics central in his proposal, whereas I preferred to restrict the evidence to the planetary fit entailed. Ramsey used the original Model A (Bullen, 1942) to represent the Earth, whereas I used the later Model B (1950a); it transpired that Model B gave a noticeably better fit. Ramsey assumed a second phase transition at the boundary (L say) of the inner core, whereas I assumed an iron-nickel inner core. The numerical detail given below on the phase-transition theory is drawn mainly from the writer's calculations (Bullen, 1949a, 1949b, 1950b), but does not differ greatly from Ramsey's results.

On the phase-transition theory, the composition of the Earth is approximately uniform from the top of the Earth's mantle to the bottom of the outer core. Having regard to details in §17.2.1, the theory therefore permits the numerical (ρ, p) relations for this whole region to be applied in effect as a single equation of state to

[*Refs. on p. 398*]

terrestrial-type planets. Ramsey's version would include the inner core in the single equation as well. In the writer's version, the (ρ, p) details for the Earth's inner core are treated as a separate equation of state, which is extrapolated where necessary in the planetary applications. See also Levin (1971).

Given the mass of a terrestrial-type planet, and the equations of state as described, the phase-transition theory yields closely determined values of the radius R, the density distribution, and the moment of inertia coefficient y. For a model planet with the mass of Venus, the theory gave $R_V = 6270$ km (Bullen, 1950b). The observational estimate of R_V at the time was 6200 ± 50 km. As stated in §17.1.2, there is no observational estimate of y_V. Thus the theory appeared to fit the observational data on Venus fairly well.

For a planet with the mass of Mars, then taken as 6.44×10^{23} kg, the results were $R_{Ma} = 3390$ km, $y_{Ma} = 0.382$ (Bullen, 1949b). The result for R_{Ma} gave excellent agreement with the then observational estimate of 3396 ± 8 km. Assuming hydrostatic stress throughout Mars, the result for y_{Ma} gave for the flattening $f_{Ma} = 1/188$, in close agreement with the then observational estimate of $1/192$.

Separate discussion on Mercury and the Moon will be given later.

Notwithstanding the seemingly good agreement with Venus and Mars (which was in strong contrast to results on the iron-core theory, §17.2.2), the phase-transition theory was soon attacked on several grounds. Kuiper and Urey held the theory to be untenable (see Bullen, 1957) on the ground that their preferred observational estimates, 3330 and 3310 km, of R_{Ma} were significantly less than the predicted value of 3390 km. As it happened, this objection has since been totally removed. Data of Kliore, Cain, and Levy (1967) on the occultation of the Mariner IV satellite by Mars yielded an estimate (Bullen, 1966b) of 3390 ± 6 km for R_{Ma}, and later observations have not significantly changed this estimate (see e.g. Dollfus, 1970).

Changes in estimates of R_V have, however, proved to be somewhat more serious, the current estimate being 6052 ± 10 km (Dollfus, 1970).

Two other arguments against the theory are also rather serious: (i) the difficulty of reconciling the jump in ρ (in the ratio 0·7 or more) at N with geochemical theory, a principal factor being the space occupied by oxygen atoms in the assumed lower mantle composition; (ii) the failure to find positive evidence of the requisite

[Refs. on p. 398]

transition in shock-wave experiments. These arguments are not, as many writers have claimed, conclusive: definitive calculations on (i) have not yet been made, and inferences from (ii) are somewhat uncertain because the high pressures realized are extremely short-lived and also because there are certain difficulties of interpretation (see §15.1.1). But the arguments at least throw the theory into considerable doubt.

The apparently excellent fit obtained around 1950 was also slightly worsened by revisions of the original Earth models used. The more important effects arose from reductions in the estimated y_E from 0·3335 to 0·3308 and in the estimated central density of the Earth from about 18 to 13 g/cm^3 (see §§11.8.2, 12.7).

An extended version of the phase-transition theory, adding temperature considerations and involving the notion of an expanding Earth's core and a theory of mountain building, was put forward by Lyttleton (1965). But Lyttleton's version faces the same difficulties as the original theory.

17.2.4 The Fe_2O theory

The treatment of the pressure p_c at N as a critical pressure was an important feature in securing the planetary fit of §17.2.3. The Fe_2O theory to be now considered also treats p_c as a critical pressure, though in a different way, and achieves an improved planetary fit while at the same time avoiding the main difficulties of the phase-transition theory.

Calculations of Soroktin (1971) had indicated that the iron oxide Fe_2O, unstable at Earth mantle pressures, becomes stable at core pressures, and that its density ranges from 9·6-10·1 g/cm^3 at the pressure at N to 10·9-12·0 g/cm^3 at L. These values fit recent Earth models fairly well and Soroktin had therefore suggested that the Earth's outer core consists of Fe_2O.

The theory to be referred to as the Fe_2O theory (Bullen, 1973a, 1973b, 1973c, 1974) follows Soroktin to this point but then deviates. Soroktin had associated the occurrence of Fe_2O in the Earth's outer core with a relation essentially equivalent to $2FeO \rightarrow Fe_2O + O$, and worked out an Earth model in which the oxygen thus released is a major determining factor. This type of Earth model does not closely fit a uniform overall composition for Earth, Venus, and Mars. The Fe_2O

[Refs. on p. 398]

theory achieves a fairly close fit by replacing Soroktin's assumed relation by $Fe_2O \rightleftharpoons FeO + Fe$.

Stripped to essentials, the Fe_2O theory postulates a set of planets, all with the same overall composition, consisting of: (i) a basic mantle material, X say; (ii) Fe_2O. The equivalent masses of (i) and (ii) are in the same ratio for all the planets, But the Fe_2O exists as such only in those planets where the internal pressure exceeds p_c. In planets where the pressure is not high enough, some or all of the Fe_2O that would otherwise exist has broken down into FeO and Fe, the FeO going into the mantle to be mixed or combined with X, and the Fe sinking to form an inner core. [For present purposes, the composition of X need not be specified; X may include some FeO (e.g. as part of olivine), and the FeO arising from the breakdown of Fe_2O would be additional to any FeO already part of X.]

The circumstances entail three planetary subsets, H, J, K say, in which, respectively, all, some, and none of the Fe_2O has broken down into FeO and Fe. The subset H consists of the smallest planets in which the mantle pressures do not reach p_c; the mantles are composed of a mixture of X and additional FeO derived from the breakdown of Fe_2O, and the cores (which correspond to the Earth's inner core) purely of Fe also derived from the breakdown. The subset J consists of intermediate planets which have mantles of X plus some additional FeO, outer cores of Fe_2O, and inner cores of Fe. The subset K consists of the largest planets which have mantles composed purely of X, and cores (which correspond to the Earth's outer core) purely of Fe_2O. For convenience, all Fe_2O zones will be referred to as outer cores, whether or not an inner core exists; likewise, all Fe cores will be referred to as inner cores. Members of H thus have no outer cores, and K no inner cores. (See Fig. 17.1.) The mantle-core mass ratios are the same for all members of H and of K (but the ratio is different for H and K).

Mars would correspond to a member of H; Earth and Venus, of J. No known planet belongs to K. From evidence for the Earth, all Fe_2O zones (outer cores) would be expected to be fluid, and all Fe zones (inner cores) solid.

The (ρ, p) relations applying inside the Earth to the materials X, Fe_2O, and Fe are postulated to apply also inside the other planets. The masses of Fe and additional FeO inside any planet are taken to be in the ratio $56/(56 + 16)$, or $7/9$. Suitable further postulates are made on such questions as the density of FeO at the requisite

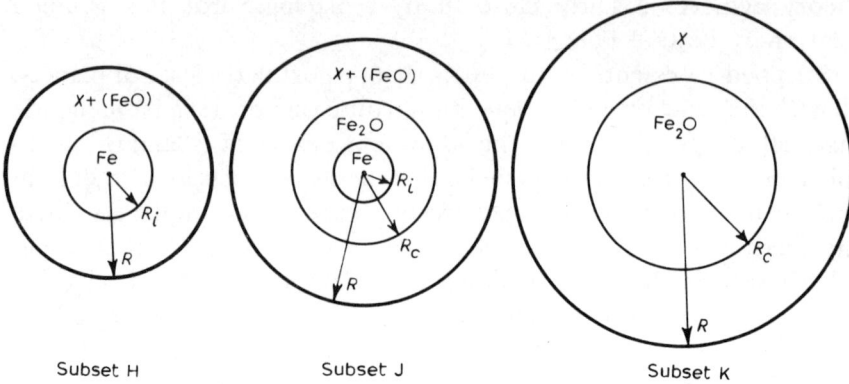

Fig. 17.1. *Types of planetary interiors on Fe_2O theory*

pressures. Starting from an Earth model treated as a member of *J*, the theory and postulates then determine *R*, the density distribution, and *y* for any planet of assigned mass.

A first approximation in which density variations inside particular zones of planets are ignored (the mean density being used for each zone) has been derived for planets with the masses of Venus and Mars. The results are indicated in §17.3, together with tentative corrections taking account of density variations inside zones. At a later stage, allowances for such complications as the presence of nickel in cores and possible volume changes and/or chemical interactions in the mixing of X and additional FeO in mantles can be considered; provisional calculations indicate that the effects on the essential results are fairly slight.

17.3 Earth, Venus, and Mars

As already stated, the observational values of $\bar{\rho}$ in Table 17.1 suggest that the internal compositions of the terrestrial planets have much in common. Jeffreys (1934) drew attention to the likelihood that, having regard to ideas on the origin of the solar system, all the planets originally had closely similar compositions, present differences arising principally through the escape of lighter ingredients. He also noted that, on the observations available by 1934, $\bar{\rho}$ decreases with *M* for the terrestrial planets and drew attention to possible implications on their internal structures. Mercury has probably had greater losses through escape than Earth, Venus, and Mars (see §17.4). Furthermore, the more recent data of Table 17.1 indicate that Mercury is apparently out of line with the general trend of $\bar{\rho}$ and M. It is therefore appropriate to treat Earth, Venus, and Mars as a single group, separately from Mercury. In §§17.3.1 and 17.3.2, possible internal structures of Venus and Mars are examined in the light of results for the Earth derived in preceding chapters.

17.3.1 Venus

Jeffreys (1937) constructed a model for the interior of Venus by applying the (ρ, p) distributions derived for the Earth by Bullen (1936) and fitting observational values taken as $M_V = 4 \cdot 91 \times 10^{24}$ kg and $R_V = 6150$ km. The Earth was assumed to have a core of distinct composition from the mantle, and Venus to have mantle and core compositions as in the Earth. Inner cores were ignored. The Venus model had the following properties: ρ ranged from 3·29 to 5·44 in the mantle and from 9·6 to 11·1 g/cm^3 in the core; p was 1·24 at the mantle-core boundary and $2 \cdot 4 \times 10^{11}$ N/m^2 at the centre; the mass and radius of the core were $1 \cdot 06 \times 10^{24}$ kg and 2910 km. Allowance for an inner core could be made in a variety of ways; plausible allowance for an inner core would result in an outer-core thickness not much less than 2000 km. Although the numerical data assumed in deriving the model are now superseded, the model nevertheless gives a useful broad indication of the likely internal structure of Venus when the iron-core theory of §17.2.2 is assumed.

With the phase-transition theory, the outer cores of Earth and Venus are treated as having the same ingredients as the mantles, the density jumping suddenly at the critical pressure p_c existing at the mantle-core boundaries of both planets. Two Venus models (Bullen 1950b), (a) with an inner core assumed composed of iron-nickel, (b) with no inner-core allowance, indicate the broad implications of the phase-transition theory on the structure of Venus. In constructing these models, M_V was taken as $4 \cdot 88 \times 10^{24}$ kg, and numerical (ρ, p) data for the original Earth model B (§11.8.1) applied. The yielded R_V were 6270 and 6310 km, respectively; the model (a) thus gave the better observational fit (see §17.2.3).

The model (a) had the following further properties: the mantle and outer-core thicknesses and inner-core radius were 4020, 920, and 1330 km; ρ ranged from 3·32 to 5·57 inside the mantle and from 9·7 to 10·5 g/cm^3 inside the outer core; p had the critical value, then

[*Refs. on p. 398*]

taken as 1·33, at the mantle-core boundary, and $1·91 \times 10^{11}$ N/m² at the inner-core boundary. Values of ρ and p for the inner core are no longer of interest, needing substantial correction because of the subsequent reduction from around 18 to 13 g/cm³ in preferred estimates of the Earth's central density. The density distribution for the model (a) gave $y_V = 0·362$. Assuming hydrostatic stress and taking h_V as in Table 17.2, the relations (17.1), (17.2), and (17.3) would then give $\eta_V = 0·305$, $(h/f)_V = 0·92$, $f_V = 7 \times 10^{-8}$, and $(J_2)_V = 2 \times 10^{-8}$. These estimates of f_V and $(J_2)_V$ are far below observability.

With the model (b), the main differences from (a) were: a mantle thickness and core radius of 4240 and 2070 km; central density and pressure of 10·6 g/cm³ and $1·97 \times 10^{11}$ N/m²; $y_V = 0·368$. It will be noticed that the phase-transition models for Venus have much smaller cores (less than half by volume and mass) than models based on the iron-core theory. This is an immediate and important consequence of treating the pressure at the mantle-core boundary as a critical pressure.

With the Fe_2O theory, first-approximation calculations (Bullen, 1973b), taking $M_V = 4·86 \times 10^{24}$ kg, yielded $R_V = 6010$ km, with mantle and outer-core thicknesses and inner-core radius of 3290, 600, and 2120 km. Tentative second-approximation corrections would increase R_V to nearly the current observational value of 6050 km, and slightly decrease the outer-core thickness. Thus the fit with R_V is decidedly superior to that on the phase-transition theory.

The Fe_2O theory entails for Venus the largest inner core of all theories, but the thinnest outer core. Assuming that the Earth's main magnetic field arises from convection currents in the outer core, and that the outer core is the sole internal fluid zone in Venus as well as in the Earth, the Fe_2O theory thus goes the closest of all theories to fitting the observational result that the magnetic field around Venus is extremely small.

In due course it may be possible to discriminate between the three theories discussed above by locating the upper and lower boundaries of the Venus outer core by seismic means. But this must await the solution of the problem of gathering seismic data on a planet with surface temperature as high as on Venus. Some other investigations will be referred to in §17.9.

[Refs. on p. 398]

17.3.2 Mars

Two Mars models, (a) and (b), were constructed by Jeffreys (1937), taking $M_{Ma} = 6.43 \times 10^{23}$ kg and $R_{Ma} = 3385$ km as observational data and applying the 1936 Earth (ρ, p) distributions. The model (a) followed the procedures Jeffreys had used for Venus (§17.3.1), so that its properties indicate broadly the likely internal structure of Mars when the iron-core theory is assumed and the mantle and core of Mars are assumed to have the same compositions as in the Earth. The model (a) had the following properties: ρ ranged from 3·29 to 4·37 in the mantle and from 8·28 to 8·60 g/cm^3 in the core; p was 0·248 at the mantle-core boundary and 0.444×10^{11} N/m^2 at the centre; the mass and radius of the core were 1.02×10^{23} kg and 1420 km; $y_{Ma} = 0.356$; $f_{Ma} = 0.00487$ (assuming hydrostatic stress). These results showed that the iron-core theory entails with Mars a mantle-core mass ratio appreciably larger than with Venus, and larger still than with the Earth (§17.2.2). Thus the iron-core theory does not permit an identical overall composition for any pair of the three planets.

The procedures for the model (b) differed from those for (a) by assuming a change of composition (instead of phase) in the Earth near 400 km depth, and in providing for no core in Mars. In the model (b): ρ ranged from 3·29 to 4·41 g/cm^3; p was 0.274×10^{11} N/m^2 at the centre; $y_{Ma} = 0.386$; $f_{Ma} = 0.00566$.

Kovach and Anderson (1965) investigated whether a common overall composition for Earth and Mars might be secured on the iron-core theory by a suitable mixing inside Mars of the Earth's mantle and core materials. They concluded that this could be contrived, within the uncertainty of the observational M_{Ma}, only if $R_{Ma} \leqslant 3310$ km.

A first Mars model (Bullen, 1949b) based on the phase-transition theory, taking $M_{Ma} = 6.44 \times 10^{23}$ kg, assuming the Earth's inner core to be composed of iron-nickel, and applying (ρ, p) distributions of the original Earth model B, had the following properties: $R_{Ma} = 3390$ km; mantle thickness 2260 and (inner) core radius 730 km; ρ ranging from 3·3 to 4·25 in the mantle and from 10·1 to 10·3 g/cm^3 in the core; pressure equal to 0·28 at the mantle-core boundary and 0.35×10^{11} N/m^2 at the centre; $y_{Ma} = 0.382$; $f_{Ma} = 0.00534$ (assuming hydrostatic stress). The values of f_{Ma} and y_{Ma} in Table 17.3 also assume hydrostatic stress; permissible deviatoric stresses could allow the table value of y_{Ma} to lie anywhere between about 0·363 and

0·389, the corresponding range for f_{Ma} as derived using (17.1)–(17.3) being about 0·0050 to 0·0055 (Bullen, 1957). (Assuming $R_{Ma} = 3390$ km, the excesses of the equatorial over the polar radius of Mars are 16·9, 17·7, 18·6 km when $f_{Ma} = 0·00500$, 0·00523, 0·00550, respectively.) The above model thus fits the observational R_{Ma} excellently, and f_{Ma} and y_{Ma} well within the uncertainties.

A second phase-transition model, differing from the first only in following Ramsey's assumption that the inner core of the Earth is associated with a second phase-transition gave: $R_{Ma} = 3410$ km; central density and pressure 4·23 g/cm^3 and 0·26 × 10^{11} N/m^2; $y_{Ma} = 0·391$; $f_{Ma} = 0·00552$. This Mars model has no core. A series of curves (Bullen and Low, 1952) shows values of $\bar{\rho}$ against M for terrestrial-type planets on various versions of the phase-transition theory. [Revised calculations (Bullen, 1957) designed to fit Kuiper's estimate of 3330 km for R_{Ma} are now obsolete because of the observational reversion to 3390 km.]

With the Fe_2O theory, a first approximation taking $M_{Ma} = 6·42 \times 10^{23}$ kg, ignoring density variations inside mantle and core, and applying (ρ, p) relations for a recent Earth model, gave R_{Ma} as lying between 3350 and 3370 km (the particular value depending on certain secondary assumptions) and gave $y_{Ma} = 0·376$ and (assuming hydrostatic stress) $f_{Ma} = 0·00523$. A tentative second approximation allowing for likely density variations inside layers lowered the estimates of y_{Ma} and f_{Ma} to about 0·364 and 0·00500. Thus the Fe_2O theory entails a value of R_{Ma} somewhat less than the currently preferred observational value, and a value of f_{Ma} involving deviations from hydrostatic stress close to the estimated limits; in the absence of allowance for deviatoric stresses, the excess of the equatorial over the polar radius of Mars would thus be about 17 km on the Fe_2O theory. The Fe_2O theory tentatively gives 1300–1400 km for the radius of an iron core in Mars.

Binder and Davis (1973) constructed models of Mars in which special consideration is given to possible types of Martian crust in interpreting data on the coefficient J_2 for Mars. Their findings are compatible with Mars having an iron core as large as that indicated on the Fe_2O theory and y_{Ma} as small as 0·364.

Practically all theories of planetary composition, including the above theories, are compatible with the whole interior of Mars being essentially solid; if Mars has a core, it is likely to be of similar composition to the Earth's inner core and no hotter, and therefore solid.

[Refs. on p. 398]

Absence of a fluid zone inside Mars, is, moreover, compatible with the failure to detect a significant magnetic field in artificial satellite observations of Mars.

Other theories have treated Mars as less differentiated than the invoked ad hoc chemical arguments to fit the observational data. Brief reference to these theories will be made in §17.9.

Observational evidence on the seismic velocity distributions in Mars may be forthcoming before too long, and, as with Venus, would assist very much in discriminating between various ideas on the internal structure. It seems desirable to await well-based evidence of this type rather than clutter up the literature with excessive complex speculation. The legions of speculative papers on properties of the Moon, later shown to be quite futile by evidence from artificial satellite observations, provide a salutary illustration. (Also pertinent is a recent geochemical scramble to produce papers about the Moon's deep interior using data from early samples of materials gathered at the surface. Perhaps it is not too great an exaggeration to liken the scramble to aspiring to infer the Earth's internal structure by digging up one's backyard and performing chemical experiments on the diggings.)

The possibility of gathering seismic data on Mars and Venus incidentally illustrates how planetary data may feed back information on the Earth's internal properties.

17.4 Mercury

A rough estimate of the central pressure p_0 of a planet, ignoring internal density variation and deviatoric stress, is obtainable using

$$3p_0 = \int_0^R 4\pi G \bar{\rho}^2 r \, dr = 2\pi G R^2 \bar{\rho}^2, \qquad (17.4)$$

where G is the gravitation constant. On the data of Table 17.1, (17.4) formally gives $p_0 = 0.25$ and 1.7×10^{11} N/m^2, respectively, for Mercury and the Earth (see Bullen, 1971, p. 338). (Finer calculations indicate that these estimates need increasing by factors less than 2 for Mercury and a little greater than 2 for the Earth.) Thus the pressure reached inside Mercury is less than one-sixth of that inside the Earth. Also the table gives $\bar{\rho}$ nearly the same for both planets. On this evidence, the ingredients of Mercury if reduced to zero pressure are on the average appreciably denser than those of the Earth; presumably, Mercury has an appreciably greater iron content. The inference that Mercury in its present state is thus distinctly different from the Earth in overall composition is largely independent of particular theories of the Earth's composition, and could be shaken only if the observational estimate of $\bar{\rho}_{Me}$ in Table 17.1 is considerably too high.

A model of Mercury constructed by Jeffreys (1937), using procedures as for Venus and Mars, now has little interest because the preferred observational value of M_{Me} has since been much increased. (The model assumed $M_{Me} = 2\cdot65 \times 10^{23}$ kg, $R_{Me} = 2450$ km, $\bar{\rho}_{Me} = 4\cdot30$ g/cm^3, the stated uncertainties of M_{Me} and $\bar{\rho}_{Me}$ being of order 20 per cent.) A model also applying (ρ, p) relations for the Earth, but using more recent observational data, was constructed by Koszlovskaya (1969). This model assumed $M_{Me} = 3\cdot30 \times 10^{23}$ kg and $R_{Me} = 2440$ km, and assumed a mantle composed of the Earth's outer-mantle material, and a core composed of iron. Its calculated properties include: mantle thickness, 710 km; core radius, 1730 km; central density, $9\cdot8$ g/cm^3; mantle-core mass ratio, 0·67; pressure at mantle-core boundary, $0\cdot10 \times 10^{11}$ N/m^2; $p_0 = 0\cdot46 \times 10^{11}$ N/m^2; $y_{Me} = 0\cdot324$. A model of Lyttleton (1969), based on the same assumptions but using an Emden equation procedure, gave fairly similar results. The chief interest of these models lies in the smallness of the mantle-core ratio (0·67) compared with the ratio 2·1 for the Earth, indicating again that the proportion of iron in Mercury in its present state is considerably higher than in the Earth. Another model of Koszlovskaya, assuming a uniform composition throughout Mercury, has: ρ ranging from 5·0 at the surface to 6·3 g/cm^3 at the centre; $p_0 = 0\cdot26 \times 10^{11}$ N/m^2; $y_{Me} = 0\cdot388$.

The phase-transition theory would allow Mercury to have at most only a very small iron core and no other free iron. The maximum possible mass of this core would be about $0\cdot05 \times 10^{23}$ kg (which bears to M_{Me} the same ratio as does the mass of the Earth's inner core to M_E) and is much too low to fit the observational $\bar{\rho}_{Me}$. Thus, in contrast to the cases of Venus and Mars, the phase-transition theory is far from being compatible with the notion of a common overall composition for Mercury and the Earth.

The Fe_2O theory would, for a planet with the mass given for Mercury in Table 17.1, entail a core with mass about $0\cdot50 \times 10^{23}$ kg (see Bullen, 1973b), i.e. ten times the maximum on the phase-transition theory. Even so, this still falls short of fitting requirements for a common overall composition with the Earth.

Independently of the above calculations, there is strong likelihood

[Refs. on p. 398]

that Mercury is exceptional among the terrestrial planets in the extent to which it has suffered internal changes since it was originally formed. Mercury is not only the nearest planet to the Sun, but rotates so slowly that the temperatures over a sizeable area surrounding the sub-solar point are likely to be much higher than on other planets. Furthermore, it is likely that Mercury would originally have had a much more eccentric orbit and so have been much closer to the Sun at perihelion than now. A consequence is that losses by volatilization during the lifetime of Mercury are likely to have included, not merely a normal atmosphere, but also an appreciable proportion of some mantle-type ingredients (Bullen, 1952). On this ground, it is fully to be expected that Mercury now differs markedly from other planets in containing relatively more iron. At the same time, the evidence is quite compatible with the primitive Mercury having had the same overall composition as the other terrestrial planets.

A calculation assuming the Fe_2O theory is of some interest in this connection. Assuming the present Mercury to be composed of a mantle and core of mean densities 3·5 and 8·5 g/cm³, the data of Table 17.1 yield $1·85 \times 10^{23}$ kg for the core mass. On the Fe_2O theory, the mantle-core mass ratio is about 5·5 (Bullen, 1973b) for planets of the subset H (§17.2.4). If the primitive Mercury (with a much larger mantle than now) belonged to this subset and had the same core mass as now, its implied mass would therefore have been more than 10^{24} kg and nearly double the mass of Mars. This extremely simplified calculation ignores many possible complications, including the likely production of additional free iron during volatization. But it suggests the possibility that the primitive Mercury may have been somewhat larger than Mars. An incidental important advantage of the Fe_2O over the phase-transition theory is that it provides far more readily for a sizeable proportion of free iron in Mercury.

Because the pertinent observational data on Mercury are so meagre, the main interest in the above discussion attaches to the broad qualitative indications, rather than the numerical detail. Further theorizing has been directed to questions such as the extent of differentiation and chemical changes that may have occurred in Mercury during its lifetime. Data gathered through artificial satellites may soon lead to a sharpening of the precision of estimates of M_{Me} and R_{Me}, enabling some inferences on the internal structure to be made more definitely than is possible now.

[Refs. on p. 398]

17.5 Moon

Although the evidence is not clear-cut as to whether the moon should be regarded as having the structure of a normal planet, there is sufficient in common with other planets to warrant some discussion here.

Since $y_{Mo} \approx 0.4$ (§17.1.2), a good approximation to $(p_0)_{Mo}$ is yielded by treating the internal density of the Moon as uniform and applying (17.4). The result is 0.05×10^{11} N/m² (approx.), which is equal to the pressure near 150 km depth in the Earth. In the region B of the Earth, an increase in pressure of this amount would, due to compression alone [data for Model A'' or B_2 (Tables 10.2 and 12.4) are here appropriate], cause a density increase of about 0.12 g/cm³. Assume that ρ increases by this amount from the surface of the centre of the Moon and fits the form $(a - br^2)$. Then the data in Table 17.1 formally yield

$$\rho = 3.412 - 0.043 r^2 \qquad (17.5)$$

for the Moon's density distribution, the unit for r being here 10^3 km. The surface and central densities are then 3.29 and 3.41 g/cm³ (cf. Jeffreys, 1970, p. 197).

The distribution (17.5) entails $y_{Mo} = 0.3983$. The differences between this and the observational estimates in §17.1.2 could be due either to observational uncertainties or to deviations from the assumptions underlying (17.5), for example, deviations from hydrostatic stress or significant temperature effects. Artificial satellite observations have indicated the presence of mascons (local mass concentrations of relatively high density) in the outer part of the Moon. The mascons are not large enough to affect the above calculations significantly, but their presence confirms the existence (see Jeffreys, 1970, §4.08) of sizeable deviatoric stresses inside the Moon. Thus the average value of $d\rho/dz$ in the outer part of the Moon could be noticeably less than that given by (17.5) and even be slightly negative, so that y_{Mo} could well be nearer 0.400 than 0.398. Nevertheless, in the absence of detailed evidence, (17.5) serves as a useful reference model.

The values of $\bar{\rho}_{Mo}$ and y_{Mo} are fully compatible with the Moon having predominantly the same composition as the region B of the Earth. The closeness of y_{Mo} to 0.400 also severely limits the size of any iron core in the Moon.

The hypothesis Y that the Moon has the same overall composition

[Refs. on p. 398]

as the Earth is untenable on most theories. Consider, for example, the iron-core theory of the Earth and the Fe_2O theory. For the purpose of rough calculations, the density of any iron core in the Moon will be taken as 8 g/cm³. Then, assuming Y, the two theories would require the Moon to contain about 2·4, 1·1 × 10²² kg of free iron, respectively. This free iron could not be concentrated in a Moon core since the radius of the core would then be about 900, 700 km, giving, on the data in Table 17.1, $y_{Mo} \approx 0.334, 0.367$. By mixing the iron with the mantle material, y_{Mo} could be raised to 0·400, but $\bar{\rho}_{Mo}$ would then be about 4·1, 3·8 g/cm³, which significantly exceed the observational value of 3·34 g/cm³.

The phase-transition theory has two versions (see §17.2.3) in which the Earth's inner core is presumed (a) to be composed of iron (or iron-nickel), (b) to be associated with a second phase transition. On the hypothesis Y, the version (a) would require the Moon to contain about 0·10 × 10²² kg of free iron. The inclusion of this amount of free iron would increase $\bar{\rho}_{Mo}$ by only about 0·03 g/cm³. If the iron were concentrated in a core, the core radius would be about 300 km and y_{Mo} would be reduced by only about 0·003. Thus the version (a) goes fairly close to fitting a common overall composition. With (b), the fit is practically complete since (b) would require the Moon to consist predominantly of the material of the region B of the Earth.

Since only the phase-transition theory, which is now very doubtful, agrees with the hypothesis Y, it is strongly probable that the Moon has an overall composition distinctly different from that of the Earth (and other planets) and therefore has had an exceptional origin. The observational data on M_{Mo}, R_{Mo}, and y_{Mo} are well fitted by the theory that the Moon was ejected from a primitive Earth-Moon planet soon after formation, in which case the Moon would be expected to be composed predominantly of Earth uppermantle material. But this theory is not currently favoured on other grounds. Incidentally, one item of evidence often quoted as a ground for rejecting the theory is the calculation by Jeffreys (1930) that tidal resonance would have been inadequate to cause the Moon's ejection; the possibility that resonance could have contributed to the Moon's ejection without being the sole cause (see §17.9.2) has not received sufficient consideration (see Bullen, 1967a). To go much further into the problem of the Moon's origin is, however, beyond the scope of this book.

A little evidence on the Moon's interior has come from lunar seismic data. An analysis by Toksöz, Latham, and others (1972) gave a tentative distribution of the P velocity α in the outermost 100 km below a limited part of the Moon's surface (the Fra Mauro region of Oceanus Procellarum). The analysis suggested a two-layered crust, with α ranging inside the uppermost layer ($0 < z < 25$ km) from less than 1 km/s to about 6 km/s, then jumping fairly suddenly at $z = 25$ km and remaining constant at 6·8 km/s throughout the second layer ($25 < z < 65$ km). Evidence on correlations between α and ρ (see e.g. §10.4.2) would suggest that the densities inside this crust, as in the Earth's crust, are less than 3·0 g/cm^3. If sufficient further seismic observations sustain these tentative findings on α and show the crust to be Moon-wide, then some allowance will need to be made for the crust in estimating the density distribution below. The analysis also suggested that α jumps fairly suddenly at $z = 65$ km from about 7 to 9 km/s; but the data are as yet very meagre and it is more likely that the value of α below the crust does not much exceed the value 8 km/s at the top of the Earth's mantle. On the formal assumption of hydrostatic stress, $p \approx 3\cdot5 \times 10^8$ N/m^2 at $z = 65$ km. Since the Moon may have a much colder interior than the Earth, its internal strength may appreciably exceed $3\cdot5 \times 10^8$ N/m^2; hence deviatoric stresses may have significant effects down to depths well below 65 km. This may be a serious complicating factor in the seismic interpretations. Hence the seismic evidence cannot be applied very far as yet.

17.6 Jupiter and Saturn

The four major planets were thought to have hot gaseous interiors until Jeffreys (1923) showed that their surface temperatures are probably of order 100°K, as would be maintained by the Sun's radiation. This result was verified soon afterwards by direct observation. It thus became evident that the major planets are essentially cold bodies to which equations of state for solids and liquids are applicable. The comparatively low mean densities (Table 17.1) would therefore require the major planets to be composed predominantly of materials whose mean atomic numbers are much less than for the terrestrial planets.

Jeffreys (1924) arrived at upper bounds of 0·4 and 0·8 g/cm^3 for the densities at the effective surfaces of Jupiter and Saturn if assumed solid, and Wildt (1934) interpreted these results as implying predominantly hydrogen compositions in the outer layers.

Russell (1935) classed the major planets and the white dwarfs as a single family of cold bodies for which R and $\bar{\rho}$ are principally determined by M alone: for the planets R increases with M; for the white dwarfs it decreases. Wigner and Huntington (1935) estimated that, at a critical pressure p_c, hydrogen changes from its molecular phase to a metallic phase with an approximate doubling of density; Wildt (1938) and Kothari (1938) independently suggested that metallic hydrogen might be present in the major planets. Later calculations confirmed that this phase transition does occur with hydrogen; estimates of p_c have ranged from 0·7 to 3·5 × 10^{11} N/m² [see Kronig, de Boer, and Korringa (1946), Wildt (1963)]. Assuming a (ρ, p) equation of state in which $p_c = 0·8 \times 10^{11}$ N/m², Ramsey (1950a) calculated for cold, purely hydrogen bodies an M-R curve which showed a maximum radius $R_{max} = 85\,000$ km (at $M \approx 1000 M_E$). [Using a modified (ρ, p) relation, DeMarcus (1958) gave $R_{max} = 79\,000$ km; see also Kothari (1938).] For the planets as well as the white dwarfs, the observed R are all less than this calculated R_{max}. For Jupiter and Saturn, $R \approx 71\,000$ and 60 000 km. The calculations also showed that the atmospheric masses of hydrogen planets are negligible compared with the total masses.

On comparing the observational data with his M-R curve, Ramsey inferred that Jupiter and Saturn consist predominantly but not exclusively of hydrogen (see also Wildt, 1947). Proceeding to quantitative detail, and assuming hydrogen and helium to be the principal constituents, Ramsey (1951) and Miles and Ramsey (1952) gave the proportion of hydrogen by mass as 80 ± 4 per cent in Jupiter and 65 ± 4 per cent in Saturn. DeMarcus (1958) arrived at nearly the same proportions but suggested somewhat greater uncertainties.

Ramsey estimated the masses of the metallic cores (assumed to include helium) in Jupiter and Saturn as 92 and 67 per cent of the whole masses. These estimates are especially sensitive to the value adopted for p_c. Assuming $p_c = 1·93 \times 10^{11}$ N/m², DeMarcus gave 84 and 55 per cent. The possibility has to be allowed for, moreover, that the transition to metallic hydrogen may be more complicated than a single sudden change.

Making plausible assumptions on helium content and bringing moment-of-inertia criteria to bear, Miles and Ramsey also showed

[Refs. on p. 398]

that the observational data on Jupiter and Saturn permit the presence of dense innermost cores of terrestrial composition. Formal calculations gave $(11 \pm 10)M_E$ and $(5 \pm 4)M_E$ for the masses of these cores; the uncertainties are of course too great for a definite conclusion to be drawn.

Ramsey (1950a) showed that there is a range, $\gamma_1 < M/M_E < \gamma_2$ say, for which a hydrogen planet can exist in two configurations — with and without a metallic core. He gave $\gamma_1, \gamma_2 = 88, 95$; DeMarcus gave 70, 77. On these results alone, Jupiter ($M/M_E = 318$), which has a high hydrogen content, must have a (predominantly hydrogen) metallic core. The argument for a metallic core in Saturn ($M/M_E = 95$) takes the observational value of R as well as of M into account.

Some Jovian and Saturnian models fitting the available evidence on M, R, and I were produced by Miles and Ramsey (1952). Tables 17.4 and 17.5 show various values of ρ and p for their most plausible models J_3 and S_2, and corresponding values for a pair of models later produced by DeMarcus (1958). For the ranges of values of r/R shown in these tables, the differences for the two sets of models arise principally from different assumptions on p_c and on the helium distributions. For values of r/R outside those shown, ρ is substantially affected by assumptions on conditions near the surfaces and centres, and the assessments are too uncertain to serve any useful purpose. A further survey, including some thermodynamical discussion, was made by DeMarcus (1967); see also Peebles (1964).

Table 17.4

Densities (ρ) and pressures (p) in Jovian models of Miles and Ramsey (MR, Model J_3) and DeMarcus (DM)

r/R	ρ (g/cm^3)		p (10^{11} N/m^2)	
	MR	DM	MR	DM
0·87	$\begin{cases} 0·40 \\ 0·85 \end{cases}$	0·56	0·80	0·95
0·80	1·11	$\begin{cases} 0·78 \\ 1·08 \end{cases}$	2·24	1·93
0·60	2·0	2·1	9·5	9·5
0·40	2·9	3·1	21	22
0·20	3·9	4·1	40	38

[*Refs. on p. 398*]

Tables 17.5
Densities and pressures in Saturnian models of Miles and Ramsey (MR, Model S_2) and DeMarcus (DM)

r/R	ρ (g/cm^3)		p (10^{11} N/m^2)	
	MR	DM	MR	DM
0·80	0·34	0·40	0·52	0·40
0·674	{0·40, 0·86}	0·53	0·80	0·90
0·523	1·17	{0·72, 1·00}	2·45	1·93
0·40	1·5	1·3	4·9	4·0
0·30	1·8	4·2	7·3	8·7

17.7 Uranus and Neptune

For Uranus and Neptune, Table 17.1 gives $M/M_E \approx 14.6$ and 17.2, and $\bar{\rho} \approx 1\cdot 33$ and $1\cdot 57$ g/cm^3, respectively. On these figures, Uranus and Neptune are both too dense to consist predominantly of hydrogen.

Ramsey (1963, 1967) noted that a group of materials, consisting of the hydrogen-saturated compounds of carbon, nitrogen, and oxygen – methane (CH_4), solid ammonium (NH_4), and water (H_2O) – and neon (Ne), have cosmic abundances about one order of magnitude greater and less than those of solid terrestrial materials and of hydrogen and helium, respectively. He referred to this group as 'chonne material' and postulated that this is the predominating material in Uranus and Neptune. The molecular weights of the chonne ingredients range from 16 to 20·4, and in calculations on the internal structures of these two planets Ramsey treated the group as a single constituent of molecular weight 18. He allowed for the presence of some hydrogen, taken as uniformly mixed with chonne, and of cores of terrestrial composition.

As observational data he took: $M_U, M_N = (14\cdot 50, 17\cdot 36) M_E$; $R_U, R_N = (2\cdot 37, 2\cdot 15) \times 10^4$ km; $\bar{\rho}_U, \bar{\rho}_N = 1\cdot 55, 2\cdot 49$ g/cm^3; $y_U = y_N = 0\cdot 305$. Using theoretically derived (ρ, p) relations for chonne-hydrogen mixtures with different assigned proportions of free hydrogen, he thence inferred that the proportion of hydrogen by mass is at least 10 per cent in Uranus, is smaller in Neptune than in

Uranus, and does not exceed 15 per cent in either planet. He derived model (ρ, p) relations for both planets, and also estimated that the mass of a terrestrial core would probably not exceed $2M_E$ in Uranus and would lie between $5\cdot4M_E$ and $6\cdot9M_E$ in Neptune.

The above results give a general indication of the likely internal constitutions of Uranus and Neptune, though not much weight can be attached to the numerical detail. The observed R, even though substantially revised since 1967, remain considerably uncertain; the observationally based y likewise remain considerably uncertain (see § 17.1.2); there are sizeable uncertainties as to phase changes inside the mantles; etc. (See also § 17.9.3.)

17.8 Pluto

Observational estimates of the mean density of Pluto have fluctuated wildly, but the most recent estimate of 5 g/cm^3 (Table 17.1) suggests that Pluto may be a terrestrial-type planet.

If M_P and $\bar\rho_P$ could be determined to an accuracy of order 5–10 per cent, the result might bear usefully on some of the alternative theories put forward on the internal structures of Venus, Mars, and Mercury. For example, should it be sustained that Pluto has nearly the same mass as Mars but a mean density near that of Mercury, that could be evidence against a common overall composition for the terrestrial planets in their primitive states. Cook (1972) has offered the guess that Pluto is made of rather heavy material of terrestrial type, but even this goes rather further than the present uncertainties permit. Lyttleton (1936) and Kuiper (1956) suggested (on different theories) that Pluto is a former satellite of Neptune; in that event, Pluto would not be likely to provide a useful test on the compositions of the terrestrial planets.

17.9 Further remarks

17.9.1 *Geochemical theories*

Numerous attempts have been made to infer the internal constitutions of the planets with the help of geochemical theories. In the main, these theories involve such questions as the possibly varying extents of differentiation in different planets, details of the chemical processes contributing to the differing mean densities, and losses by volatilization, etc. For a variety of reasons, it is inappropriate to incorporate these theories into the approach of the present chapter: many of the underlying assumptions are complex, whereas in the present stage of limited reliable evidence it is desirable for the purpose of this chapter to present simple models so far as possible; important sections of the geochemical assumptions differ considerably from author to author, and in many cases the results of the varying assumptions as yet appear to offer little prospect of discriminatory tests. Ultimately, of course, all planetary models must be compatible with pertinent chemical evidence: when the observational data used in deriving models of the type discussed in this chapter become a little more precisely determined, and when reliable seismic data become available on (for example) the Moon, Mars, and Venus, a role of geochemistry may be to add important refinements to the models. Examples of differing geochemical assumptions and their implications may be read in work of Urey (1960, 1966), Ringwood (1966), and Anderson (1973).

17.9.2 *Further implications of phase-transition theory*

On the phase-transition theory, a planet can in certain circumstances exist in more than one stable configuration, the different configurations having different self-energies E; E includes the gravitational energy lost in assembling the planet from complete dispersal at zero pressure at infinity and the compressional energy acquired in forming the high-density phase.

Consider a simplified model set of terrestrial-type planets each of which has a mantle and, when the internal pressure is sufficient, a core whose material is a phase transition of the mantle material (inner cores are here being ignored); and let the same (ρ, p) relation apply to each member. Ramsey (1950b) showed that members whose masses lie inside a certain range $M' < M < M''$ can have three possible equilibrium configurations C_1, C_2, C_3 with (in general) different self-energies E_1, E_2, E_3. The configurations C_1 and C_3 are stable, and C_2 unstable; one of C_1 and C_3 has and the other lacks a core. In certain circumstances a planet could pass from C_1 to C_3 (or vice versa) with a virtually explosive release of energy. For this to happen, some additional source of energy is required to carry the planet over the potential barrier which corresponds to C_2. If M is close to M' or M'', the barrier is slight and the circumstances accordingly the most favourable; also the energy difference $|E_1 - E_3|, = \Delta E$ say, is greater than when M is not close to M' or M''.

Taking the (ρ, p) relation as that for the Earth model A (§10.7.1), Ramsey calculated that M', $M'' \approx (0.78, 0.80)M_E$ and that ΔE reaches 10^{29}-10^{30} J. He pointed out that $M_V \approx 0.80 M_E$ and suggested that, since allowance for temperature (not made in the calculations) might affect M' and M'' by up to 10 per cent, Venus might in the past have undergone an explosive transition of the above type, possibly giving rise to the asteroids.

The calculations were extended by Bullen (1951, 1967a) and Datta (1954) to the case of a model set of planets which, when the internal pressure is sufficient, have three zones involving two successive phase transitions of the mantle material. In this case, it is possible on appropriate assumptions to contrive an explosive energy release ΔE of order 10^{31} J from a planet of mass near M_E. The assumptions involve a density jump as in the Earth at the pressure at the Earth's mantle-core boundary and a second density jump in about the same ratio at a second critical pressure p'_c. An energy release of 10^{31} J would be sufficient to eject the Moon from a primitive Earth-Moon body and the mechanism was suggested (Bullen, 1951) as a possible one for the origin of the Moon. Resonance was envisaged as filling the auxiliary role of carrying the primitive body across the potential barrier separating an original three-zone state from a two-zone state in which the pressure p'_c is not reached.

These suggestions on Venus and the Moon are not currently favoured, a principal objection being (as already stated in §17.2.3) the sizes of the density jumps required at the transitions. The calculations nevertheless retain interest in indicating how ΔE can be estimated for cases of smaller density jumps at phase transitions inside planets. Smaller explosive releases of energy seem quite feasible in this way and may have contributed to some planetary changes during the lifetime of the solar system.

17.9.3 *Remark on primitive compositions of the planets*

Not all theories assume that the original overall compositions of the planets were approximately identical, but on the limited evidence available this may well have been the case. It is then tempting to assume that the present differences in the overall compositions of the planets are, to a first approximation, due to differing losses by past escape of lighter ingredients. The losses would depend predominantly on the surface temperatures and gravitational intensities, g say.

Thus the four large outer planets, which have very cold surfaces, would be expected to have lost much smaller proportions of lighter ingredients than the terrestrial planets and so be now appreciably less dense. On the data in Table 17.1, $g_J \approx 25, g_S \approx 10.5, g_U \approx 9.3$, and $g_N \approx 11$ m/s². The surface temperatures T are currently estimated as: $T_J \approx 135°K$, $T_S \approx 105°K$, $T_U \approx 65°K$, and $T_N \approx 50°K$. The large surface gravitational intensity of Jupiter would account well for Jupiter's retention of the greatest proportion of hydrogen and for the differences between the internal compositions of Jupiter and Saturn. Differences in g could account for internal differences between Saturn and Uranus.

On current observational data, Neptune does not fit well into the scheme: the data give $g_N > g_S$ and $T_N < T_S$; on these data, Neptune would be expected to have retained relatively more hydrogen than Saturn; since $\bar{\rho}_N$ is much greater than $\bar{\rho}_S$ (Table 17.1), this is almost certainly far from being the case. (The data also give $g_N > g_U$ and $T_N < T_U$, similarly suggesting that Neptune has retained relatively more hydrogen than Uranus. But in this case the difference between $\bar{\rho}_N$ and $\bar{\rho}_U$ is fairly small.) The apparently large discrepancy with Saturn could possibly arise from the considerable uncertainties of the observational data, especially for R_N. In this connection, it may be noted that the change in the observational R_N from the value 2·15 current at the time of Ramsey's (1967) calculation to the value 2.5×10^4 km in Table 17.1 entails a reduction in the corresponding g_N from about 15 to 11 m/s². This change is so large that one must allow for the possibility of further sizeable changes in g_N. For the present, however, Neptune appears to be the most difficult planet to reconcile with a common primitive composition. (The observational data on Pluto are as yet much too uncertain to provide any useful test.)

The notion of a common primitive composition would require each major planet to contain an innermost core composed principally of solid-Earth materials, the mass of the core being approximately proportional to the primitive mass of the planet. The meagre results so far available on such cores (§ § 17.6, 17.7) seem compatible with this when account is taken of the considerable uncertainties.

Because their surface temperatures are higher, the terrestrial

[Refs. on p. 398]

planets would have lost most of their atmospheric-type ingredients. As seen in §17.3, the present overall compositions of Earth, Venus, and Mars may be approximately identical. Mercury appears to have suffered greater losses than the Earth, Mars, and Venus because his surface temperature is higher still (§17.4). Moreover, as already stated, it is probable that the temperature at the sub-solar point of Mercury at perihelion was formerly much greater than now.

17.10 Further references

The following are a few additional references to investigations, not already mentioned, which either supplement the information given in this chapter, give detail on other approaches to the problems of planetary interiors, or contain useful selections of further references.

The solar system in general: Berlage (1951, 1953); Kaula (1968); Kovalevsky (1972); Kuiper and Middlehurst (1961); McCrea (1960); Runcorn (1967, 1969); Urey (1952); Wood (1971).

The terrestrial planets: Anderson (1972); Bullen (1967b, 1972); Cain and others (1973); Goody (1974); Lewis (1972); Lighthill (1950); MacDonald (1962); Öpik (1955); Reynolds and Summers (1969); Runcorn (1972); Schatzman (1951); Shimazu (1967).

The major planets: Hide (1974); Hubbard (1968); Freeman and Lyngå (1970); Kieffer (1967); Öpik (1962); Reynolds and Summers (1965); Smoluchowski (1967); Zharkov and Trubitsyn (1974).

The Moon: Alfvén (1967); Berlage (1967); Eckert (1967); Gavrilov and Yanovitskaya (1974); Gilvarry (1969, 1970); Jeffreys (1967); Kaula (1969); Kopal (1966); Lyttleton (1967); Mizutani, Matsui, and Takeuchi (1972); Nakamura and Latham (1969); Press (1971); Solomon and Toksöz (1968); Urey (1968, 1969).

REFERENCES

Adams, L.H. (1937). The Earth's interior, its nature and composition. *Sci. Monthly*, **44**, 206.

Alfvén, H. (1967). Origin of the Moon. In *Mantles of the Earth and Terrestrial Planets* (ed. S.K. Runcorn), 235–240, Interscience, London.

Anderson, D.L. (1972). Internal constitution of Mars. *J. Geophys. Res.*, **77**, 789–795.

Anderson, D.L. (1973). The composition and origin of the Moon. *Earth Planet. Sci. Letters*, **18**, 301–316.

REFERENCES

Anderson, D.L. and Kovach, R.L. (1972). The lunar interior. *Phys. Earth Planet. Interiors*, 6, 116–122.

Berlage, H.P. (1951). Some remarks on the internal constitution of the bodies of of the Solar System. *Proc. Kon. Ned. Akad. Wet.*, B, 54, 344–349.

Berlage, H.P. (1953). On the composition of the bodies of the Solar System. *Proc. Kon. Ned. Akad. Wet.*, B, 56, 45–55.

Berlage, H.P. (1967). Origin of the Moon. In *Mantles of the Earth and Terrestrial Planets* (ed. S.K. Runcorn), 241–250, Interscience, London.

Binder, A.B. and Davis, D.R. (1973). Internal structure of Mars. *Phys. Earth Planet. Interiors*, 7, 477–485.

Bullen, K.E. (1936). The variation of density and the ellipticities of strata of equal density within the Earth. *Mon. Not. R. Astr. Soc., Geophys. Suppl.*, 3, 395–401.

Bullen, K.E. (1937). The constitution of the Earth and certain of the planets. *Austral. N. Z. Assoc. Adv. Sci. Rept.*, 23, 25.

Bullen, K.E. (1942). The density variation of the Earth's central core. *Bull. Seismol. Soc. Amer.*, 32, 19–29.

Bullen, K.E. (1949a). On the constitution of Venus. *Mon. Not. R. Astr. Soc.*, 109, 457–461.

Bullen, K.E. (1949b). On the constitution of Mars. *Mon. Not. R. Astr. Soc.*, 109, 688–692.

Bullen, K.E. (1950a). An Earth model based on a compressibility-pressure hypothesis. *Mon. Not. R. Astr. Soc., Geophys. Suppl.*, 6, 50–59.

Bullen, K.E. (1950b). Venus and the Earth's inner core. *Mon. Not. R. Astr. Soc.*, 110, 256–259.

Bullen, K.E. (1951). Origin of the Moon. *Nature* (Lond.), 167, 29.

Bullen, K.E. (1952). Cores of terrestrial planets. *Nature* (Lond.), 170, 363.

Bullen, K.E. (1957). On the constitution of Mars (second paper). *Mon. Not. R. Astr. Soc., Geophys. Suppl.*, 7, 271–278.

Bullen, K.E. (1966a). On the constitution of Mars (third paper). *Mon. Not. R. Astr. Soc.*, 133, 229–238.

Bullen, K.E. (1966b). Implications of the revised Mars radius. *Nature* (Lond.), 211, 396.

Bullen, K.E. (1967a). Origin of the Moon. In *Mantles of the Earth and Terrestrial Planets* (ed. S.K.Runcorn), 261–264, Interscience, London.

Bullen, K.E. (1967b). Models of the internal density in the Earth, Mars and Venus. *Ibid.*, 127–138.

Bullen, K.E. (1969). The interiors of the planets. *Ann. Rev. Astron. Astrophys.*, 7, 177–200.

Bullen, K.E. (1971). *Theory of mechanics* (8th Ed.), Science Press, Sydney.

Bullen, K.E. (1972). Compressibility and planetary interiors. *Phys. Earth Planet. Interiors*, 6, 131–135.

Bullen, K.E. (1973a). Cores of the terrestrial planets. *Nature* (Lond.), 243, 68–70.

Bullen, K.E. (1973b). On planetary cores. *The Moon*, **7**, 384–395.
Bullen, K.E. (1973c). Structure of the terrestrial planets: reply to R.A. Lyttleton. *Nature* (Lond.), **246**, 85–86.
Bullen, K.E. (1974). On the structures of the terrestrial planets. In course of publication, *CERESIS*, Lima.
Bullen, K.E. and Low, A.H. (1952). Planetary models of terrestrial type. *Mon. Not. R. Astr. Soc.*, **112**, 637–640.
Cain. D.L. and others (1973). Approximations to the mean surface of Mars and Mars atmosphere during Mariner 9 occultations. *J. Geophys. Res.*, **78**, 4352–4354.
Cook, A.H. (1970). The moments of inertia and the density distribution of the Moon. *Mon. Not. R. Astr. Soc.*, **150**, 187–194.
Cook, A.H. (1972). The dynamical properties and internal structures of the Earth, the Moon and the planets. *Proc. Roy. Soc.* (Lond.), A, **328**, 301–336.
Datta, A.N. (1954). On the energy required to form the Moon. *Mon. Not. R. Astr. Soc., Geophys. Suppl.*, **6**, 535–539.
DeMarcus, W.C. (1958). The constitution of Jupiter and Saturn. *Astron. J.*, **63**, 1–28.
DeMarcus, W.C. (1967). Models of Jupiter and Saturn. In *Magnetism and the Cosmos* (ed. S.K. Runcorn), 352–364, Elsevier, Amsterdam.
Dollfus, A. (ed., 1970). *Surfaces and Interiors of Planets and Satellites*, Academic Press, London.
Duncombe, R.L., Seidelmann, P.K. and Klepczynski, W.J. (1973). Dynamical astronomy of the Solar System. *Ann. Rev. Astron. Astrophys.*, **11**, 135–154.
Eckert, W.J. (1967). The moment of inertia of the Moon determined from its orbital motion. In *Mantles of the Earth and Terrestrial Planets* (ed. S.K. Runcorn), 97–106, Interscience, London.
Freeman, K.C. and Lyngå, G. (1970). Data for Neptune from occultation observations. *Astrophys. J.*, **160**, 767–780.
Gavrilov, I.V. and Yanovitskaya, G.T. (1974). Comparison of dynamical and geometrical shape of the Moon. *Phys. Earth Planet. Interiors*, **8**, 102–104.
Gilvarry, J.J. (1969). Nature of the lunar mascons. *Nature* (Lond.), **221**, 732.
Gilvarry, J.J. (1970). Thermal history of the Moon. *Nature* (Lond.), **225**, 623.
Goody, R.M. (1974). Mars and Venus. *Proc. Roy. Soc.* (Lond.), A, **336**, 35–61.
Hide, R. (1974). Jupiter and Saturn. *Proc. Roy. Soc.* (Lond.), A, **336**, 63–84.
Hubbard, W.B. (1968). Thermal structure of Jupiter. *Astrophys. J.*, **152**, 745–754.
Jeffreys, H. (1923). The constitution of the four outer planets. *Mon. Not. R. Astr. Soc.*, **83**, 350–354.
Jeffreys, H. (1924). On the internal constitution of Jupiter and Saturn. *Mon. Not. R. Astr. Soc.*, **84**, 534–538.
Jeffreys, H. (1930). The resonance theory of the origin of the Moon (second paper). *Mon. Not. R. Astr. Soc.*, **91**, 169–173.

REFERENCES

Jeffreys, H. (1934). The constitutions of the inner planets. *Mon. Not. R. Astr. Soc.*, **94**, 823–824.

Jeffreys, H. (1937). The density distributions of the inner planets. *Mon. Not. R. Astr. Soc., Geophys. Suppl.*, **4**, 62–71.

Jeffreys, Sir H. (1967). Figure and desnity of the Moon. In *Mantles of the Earth and Terrestrial Planets* (ed S.K. Runcorn), 93–96, Interscience, London.

Jeffreys, Sir H. (1970). *The Earth* (5th Ed.), Cambridge University Press.

Jeffreys, Sir H. (1971). Dynamics of the Moon. *Phys. Earth Planet. Interiors*, **4**, 153–155.

Kaula, W.M. (1968). *An Introduction to Planetary Physics*, Wiley, New York.

Kaula, W.M. (1969). Interpretation of lunar mass concentrations. *Phys. Earth Planet. Interiors*, **2**, 123–137.

Kieffer, H.H. (1967). Calculated properties of planets in relation to composition and gravitational layering. *J. Geophys. Res.*, **72**, 3179–3197.

Kliore, A., Cain, D.L. and Levy, G.S. (1967). Radio occultation measurement of the Martian atmosphere over two regions by the Mariner IV space probe. In *Moon and Planets* (ed A. Dollfus), 226, North-Holland, Amsterdam.

Kopal, Z. (1966). *An Introduction to the Study of the Moon*, Reidel, Netherlands.

Kothari, D.S. (1938). The theory of pressure ionization and its application. *Proc. Roy. Soc. (Lond.)*, A, **165**, 486–500.

Kovach, R.L. and Anderson, D.L. (1965). The interiors of the terrestrial planets. *J. Geophys. Res.*, **70**, 2873–2882.

Kovalevsky, J. (1972). A system of planetary masses and related quantities. *Phys. Earth Planet. Interiors*, **6**, 29–35.

Kozlovskaya, S.K. (1969). On the internal constitution and chemical composition of Mercury. *Astrophys. Letters*, **4**, 1–3.

Kronig, R., de Boer, J. and Korringa, J. (1946). On the internal constitution of the Earth. *Physica*, **12**, 245–256.

Kuhn, W. and Rittmann, A. (1941). Über den Zustand des Erdinnern und seine Entstehung aus einem homogenen Urzustand. *Geol. Rundschau*, **32**, 215–256.

Kuiper, G.P. (1956). The formation of the planets. *Roy. Astr. Soc. Canada J.*, **50**, 57–68, 105–121, 158–176.

Kuiper, G.P. and Middlehurst, B.M. (eds.) (1961). *Planets and Satellites*, Chicago University Press.

Levin, B. (1971). Internal constitution and thermal histories of the terrestrial planets. In *Highlights of Astronomy* (ed. C. de Jager), 204–227.

Lewis, J.S. (1972). Metal/silicate fractionation in the Solar System. *Earth Planet. Sci. Letters.*, **15**, 286–290.

Lighthill, M.J. (1950). On the instability of small planetary cores. *Mon. Not. R. Astr. Soc.*, **110**, 339–342.

Luoff, N. (1932). *Russ. Astr. J.*, **9**, 68.

Lyttleton, R.A. (1936). On the possible results of an encounter of Pluto with the Neptunian system. *Mon. Not. R. Astr. Soc.*, **97**, 108–115.

Lyttleton, R.A. (1965). On the phase-change hypothesis of the structure of the Earth. *Proc. Roy. Soc.* (Lond.), A, **287**, 471–493.

Lyttleton, R.A. (1967). Dynamical capture of the Moon by the Earth. *Proc. Roy. Soc.* (Lond.), A, **296**, 285–292.

Lyttleton, R.A. (1969). On the internal structures of Mercury and Venus. *Astrophys. and Space Sci.*, **5**, 18–35.

McCrea, W.H. (1960). The origin of the solar system. *Proc. Roy. Soc.* (Lond.), A, **256**, 245–266.

MacDonald, G.J.F. (1962). On the internal constitution of the inner planets. *J. Geophys. Res.*, **67**, 2945–2974.

MacDonald, G.J.F. (1963). The internal constitutions of the inner planets and the Moon. *Space Sci. Rev.*, **2**, 473–577.

Michael, W.H., Blackshear, T. and Gapcynski, J.P. (1969). Paper C.2.1, XII COSPAR (Prague).

Miles, B. and Ramsey, W.H. (1952). On the internal structures of Jupiter and Saturn. *Mon. Not. R. Astr. Soc.*, **112**, 234–243.

Mizutani, H., Matsui, T. and Takeuchi, H. (1972). Accretion process of the Moon. *The Moon*, **4**, 476–489.

Nakamura, Y. and Latham, G.V. (1969). Internal constitution of the Moon. *J. Geophys. Res.*, **74**, 3771–3780.

Öpik, E.J. (1955). The origin of meteorites and the constitution of the terrestrial planets. *Irish Astron. J.*, **3**, 206–225.

Öpik, E.J. (1962). Jupiter: chemical composition, structure and origin of a giant planet. *Icarus*, **1**, 200–257.

Pecker, J.-C. (ed.) (1966). *Astronomers' Handbook, Trans I.A.U.*, XII C, 21–23.

Peebles, P.J.E. (1964). The structure and composition of Jupiter and Saturn. *Astrophys. J.*, **140**, 328–347.

Press, F. (1971). The Earth and the Moon. *Q. J. Roy. Astr. Soc.*, **12**, 232–243.

Ramsey, W.H. (1948). On the constitution of the terrestrial planets. *Mon. Not. R. Astr. Soc.*, **108**, 406–413.

Ramsey, W.H. (1950a). The planets and the white dwarfs. *Mon. Not. R. Astr. Soc.*, **110**, 444–454.

Ramsey, W.H. (1950b). On the instability of small planetary cores. *Mon. Not. R. Astr. Soc.*, **110**, 326–338.

Ramsey, W.H. (1951). On the constitution of the major planets. *Mon. Not. R. Astr. Soc.*, **111**, 427–447.

Ramsey, W.H. (1963). On the densities of methane, metallic ammonium, water and neon at planetary pressures. *Mon. Not. R. Astr. Soc.*, **125**, 469–485.

Ramsey, W.H. (1967). On the constitutions of Uranus and Neptune. *Planet. Space Sci.*, **15**, 1609–1623.

Reynolds, R.T. and Summers, A.L. (1965). Models of Uranus and Neptune. *J. Geophys. Res.*, **70**, 199–208.

REFERENCES

Reynolds, R.T. and Summers, A.L. (1969). Calculations on the composition of the terrestrial planets. *J. Geophys. Res.*, 74, 2494–2511.

Ringwood, A.E. (1966). Chemical evolution of the terrestrial planets. *Geochim. et Cosmochim. Acta*, 30, 41–104.

Runcorn, S.K. (ed.) (1967). *Mantles of the Earth and Terrestrial Planets*, Interscience, London.

Runcorn, S.K. (ed.) (1969). *The Application of Modern Physics to the Earth and Planetary Interiors*, Wiley-Interscience, London.

Runcorn, S.K. (1972). Evidence on the deeper planetary interiors. *Phys. Earth Planet. Interiors*, 6, 100–102.

Russell, H.N. (1935). *Observatory*, 58, 259.

Schatzman, E. (1951). Sur la stabilité de certains modèles de planètes. *Bull. Acad. Roy. Belg.*, Sér. 5, 37, 599–609.

Shimazu, Y. (1967). Thermodynamical aspects of formation processes of the terrestrial planets and meteorites. *Icarus*, 6, 143–174.

Smoluchowski, R. (1967). Internal structure and energy emission of Jupiter. *Nature* (Lond.), 215, 691.

Solomon, S.C. and Toksöz, M.N. (1968). On the density distribution in the Moon. *Phys. Earth Planet. Interiors*, 1, 475–484.

Soroktin, O.G. (1971). Physical and chemical processes of Earth's core formation and chemistry of gravitational differentiation of Earth's material. *Dokl. Akad. Nauk SSSR*, 198, No. 6, 1327–1330.

Toksöz, M.N., Latham, G. and others (1972). Velocity structure and properties of the lunar crust. *The Moon*, 4, 490–504.

Urey, H.C. (1952). *The Planets, Their Origin and Development*, Yale University Press.

Urey, H.C. (1960). On the chemical evolution and densities of the planets. *Geochim. et Cosmochim. Acta*, 18, 151–153.

Urey, H.C. (1966). Chemical evidence relative to the origin of the Solar System. *Mon. Not. R. Astr. Soc.*, 131, 199–223.

Urey, H.C. (1967). Origin of the Moon. In *Mantles of the Earth and Terrestrial Planets* (ed S.K. Runcorn), 251–260, Interscience, London.

Urey, H.C. (1968). Mascons and the history of the Moon. *Science*, 162, 1408–1410.

Urey, H.C. (1969). Early temperature history of the Moon. *Science*, 165, 1275.

Wigner, E.W. and Huntington, H.B. (1938). On the possibility of a metallic modification of hydrogen. *J. Chem. Phys.*, 3, 764–770.

Wildt, R. (1938). On the state of matter in the interior planets. *Astrophys. J.*, 87, 508–516.

Wildt, R. (1947). The constitution of the planets. *Mon. Not. R. Astr. Soc.*, 107, 84–102.

Wildt, R. (1962). Planetary interiors. In *The Solar System* (eds. G.P. Kuiper and B.M. Middlehurst), 159–212, Chicago: University Press.

Wood, J.A. (1971). Planetary interiors. *Trans. Amer. Geophys. Un.*, **52**, 468–476.

Woolard, E.W. (1944). The secular perturbations of the satellites of Mars. *Astron. J.*, **51**, 33–36.

Zharkov, V.N. and Trubitsyn, V.P. (1974). Internal constitution and figures of the giant planets. *Phys. Earth Planet. Interiors*, **8**, 105–107.

Index

Italic numbers relate to entries in the lists of references

Abdallah al Mamum, 3
Abel, N.H., 146
Abnormal density gradients, 188, 282, 313, 317, 333–4
— temperature gradients, 160, 173–7
— velocity variation, 136–40, 146, 149, 162, 163–4, 178, 188–9, 212–7, 229, 241
Adams, L.H., 84–5, 108, 154, 167, 170, 376; *86, 398*
Adams, R.D., 241–2, 342, 365; *255, 345, 346, 365*
Adiabatic changes, 103–4, 185, 190–2
— incompressibility, 103–5
Age of Universe, 345
Airy, Sir, G.B., 16, 80; *19*
Aki, K., 277, 280, 343; *283, 346*
Akimoto, S., 213; *223*
Alaskan earthquake, 301
Alexandria, 2
Alfvén, H., 398, *398*
Alpine regions, 229
Alsop, L.E., 121, 244, 271, 280; *127, 257, 284, 319*
Alterman, Z., 269, 272–3, 280, 289, 291, 292, 294–300, 304; *283, 319, 321*
Alt'schuler, L.V., 201, 218, 247; *223, 255*
Amplitudes, 138, 219, 234–5, 238–9, 243, 341

— and density jumps, 337–8
Anaximander, 1
Anderson, D.L., 95, 204, 230–1, 238, 267, 273, 278, 280, 282–3, 317–8, 324, 327, 334, 357–60, 383, 395, 398; *107, 223, 225, 259, 283, 285, 286, 320, 347, 349, 366, 398–9, 401*
Anderson, O.L., 236, 323–4, 325–6; *256, 346, 350*
Anelasticity, 92–3
Angle of emergence, 125
— of incidence, 125
Angular velocity of Earth, 8, 48
— — of planets, 370–2
Ansell, J.H., 342; *346*
Anticentre, 271
Antisymmetrical tensor, 91
Arabia, 3
Archambeau, C.B., 95; *107, 256*
Aristarchus, 7
Aristotle, 1–2
Arkani-Hamed, J., 344; *346*
Arnold, E.P., 232, 234; *256*
Array stations, 146, 219, 228, 235, 250, 342
Arthur's Seat, 16
Artificial satellites, 56–7, 77, 312, 344, 377, 385, 387, 388
Asbel, I.J., *346*
Associated Legendre function, 26, 290

405

INDEX

Asteroids, 396
Astrophysics, 68, 367
Aswan, 2
Asymptotic approximations, 23, 121, 267
Atmospheric oscillations, 301
Atomic forces, 326
— number, 196, 200−2, 205−11, 216
— spectra, 291
— weight, 166, 214
Attenuation, 95, 250
— factor, 95
Autocorrelation functions, 340
Average earthquake, 144
— travel times, 145
Averaging processes, 353−4, 358
Axial rotation, 291−2

Backus, G., 289, 292, 311, 317, 335, 337, 357; *319, 346, 365*
Baille, J., 18; *19*
Baily, F., 18; *19*
Bardeen, J., 326; *346*
Basin-range, 228−9
Bateman, H., 146; *150*
Båth, M., 235, 267, 274, 337−8; *28, 256, 283, 284, 346*
Benioff, H., 289, 299; *319*
Ben-menaham, A., 95, 271, 280; *107, 28 283, 286*
Berlage, H.P., 398; *399*
Bernal, J.D., 213; *223*
Bertrand, A.E.S., 244; *256*
Bessel, F.W., 10
Biehler, S., 334; *350*
Bikini explosion, 230
Binary system, 213
Binder, A.B., 372; *399*
Biot, J.B., 11
Birch, F., 106−7, 155, 162, 165, 167, 174−5, 178, 184−6, 191, 205, 213−5, 218, 236−8, 247−9, 253, 312, 322−5, 327−30, 334−5; *107, 180−1, 223, 256, 346*
Blackman, R., 301; *319*
Blackshear, T., 373; *402*
Bodily seismic waves, 112−5, 123−7

Bogert, B.P., 299; *319*
Bolt, B.A., 219, 232, 241−3, 248−9, 251−3, 273, 283, 312, 343; *223, 256, 284, 319, 346*
Born, M., 326; *347*
Boschi, E., 201, 326−7; *223, 347*
Boscovich, R.J., 10
Bouguer correction, 14, 32
Bouguer, P., 9, 10, 13−16, 32; *19*
Boys, C.V., 18−19; *19*
Braun, 18
Bridgman, P.W., 106, 198; *107, 223*
Brown, R.J., 344; *347*
Brune, J.N., 121, 228−9, 232, 244, 270−1, 276, 278, 280−1, 334, 337; *127, 256, 258, 284, 348*
Buchbinder, G.G.R., 235, 238−9, 243, 244, 338, 344; *256, 347*
Bulk modulus, 95
— sound velocity, 247, 327
Bullard, Sir E.C., 171, 312; *181*
Bullen, K.E., 58, 114, 143−7, 149, 152, 158, 160, 165−8, 169−72, 185, 188, 194−211, 213−22, 229−30, 236−8, 240−1, 243, 245, 247, 249−55, 270, 274, 302, 307, 310, 313−8, 332−3, 342, 354, 358−62, 365, 369, 372, 376−80, 382, 384−5, 389, 396, 398; *58, 127, 150, 181−2, 223−5, 256−8, 319−20, 347−8, 365, 399*
Burke-Gaffney, T.N., 158; *181*
Byerly, P., 160; *182*

Cain, D.L., 277, 398; *400, 401*
Callandreau, O., 75−6; *86*
Caloi, P., 219, 283; *225, 284*
Caputo, M., 201, 292, 304, 326−7; *223, 319, 321, 347*
Carder, D.S., 230−1, 244; *257*
Carlini, F., 16
Carr, R.E., 294, 298; *319*
Cartesian coordinates, 22, 88, 290, 294
Cassini, J., 9
Cassini, J.D., 7, 9
Cauchy, A.-L., 11
Cavendish, H., 10, 13, 15, 16−18; *19*

INDEX

Celestial equator, 53
Central core of Earth, 55, 141, 156—7, 197, 273
— density of Earth, 74, 79—80, 156, 168—9, 204, 216, 220—2, 246—9, 253, 300, 378
— pressure of planet, 385
Centrifugal force, 8, 44, 46
Chandrasekhar, S., 218; *225*
Chemical composition, 84, 153; 155—6, 186, 201—2
— explosions, 109, 228
— phase, 153, 155—6, 186—7
Chernov, L.A., 339; *347*
Chilean earthquake, 287, 299, 301, 306, 312
Chimborazo, 14—15
China, 2—3
Chonne material, 393
Chrzanowski, P., 18; *19*
Chung, D.H., 328; *347*
Circumference of Earth, 1—3
Clairault, A.C., 9, 10, 47—50, 79; *11, 58*
Clairault's equation on ellipticities, 48—50, 64—6, 75—7
— theorem on gravity, 47
Clark, S.P., Jr., 178, 214, 330, 333; *182, 225, 347*
Clarke, R.E., 16; *19*
Cleary, J.R., 189, 231—2, 245—6, 339, 342—3, 365; *225, 257—9, 347—9, 365*
Clowes, R.M., 244; *256*
Coefficient(s) J_2, 47, 51, 53—7, 371—2
— J_m, 47, 57
— p_2, 38
— η (inhomogeneity), 207—11, 216—7, 237, 282, 315, 364, 374
Colatitude, 21
Cold bodies, 367, 391
Columbus, C., 2
Communication theory, 235
Compensation, 14, 16, 159
Composition of Earth, 373—8
— — —'s core, 215, 235—9, 375—80
— — —'s mantle, 167, 213—4

Compressibility, definition of, 95
— in Earth, 184—223
Compressibility-pressure hypothesis, 197—205, 217, 219—22, 239
— relations, 184—223, 325—6, 375
— —, empirical, 222—3
Compression, 66, 67, 90, 101
— in Earth, 185—7, 194—6, 202—3
Conduction of heat, 103
Conductivity, thermal, 174
Confidence levels, 336
Constant h, 8, 46, 48, 53—4, 64, 77, 368
— H, 53—6, 64, 77
— m, 48
—, precession, 53—8
Continental crust, 157, 159, 165, 229, 274—5
— shield, 229, 262
Convection currents in core, 163, 218, 382
Cook, A.H., 41, 56, 197, 201—2, 204, 223, 324, 369, 371—2; *42, 58, 225, 347, 400*
Cook, G.E., 56; *59*
Coordinates, cartesian, 22, 88, 290, 294
— polar, 21—2, 290, 294
Copernicus, N., 7
Core oscillations, 300
Cornu, A., 18; *19*
Correlation distance, 340
Coulomb, C.A., 11
Coulomb, J., 294; *319*
Crampin, S., 267; *284*
Credibility criterion, 357
Critical pressure, 376, 378, 391, 396
Crowding of oscillation periods, 308
Crust of Earth, 141, 157, 159—60
Crustal layers, 157, 165
— structure, 273—6
— thickness, 229, 315, 358
Crystals, 91
Cubical expansion, 104, 154—5
Curvature of Earth, 271
— of seismic ray, 134—5, 138—9
Cusps, 137—40, 160, 161, 235

Dahlen, F.A., 292; *320*

407

INDEX

Dahm, C.G., 149; *150*
D'Alembert, J.L., 11
Daly, R.A., 300; *347*
Damping of Earth oscillations, 245–6
Darwin density model, 70–1, 73, 76–7
Darwin, Sir George, 11, 41, 50, 51, 52, 70, 73, 76–7; *11, 42, 58, 86*
Datta, A.N., 396; *400*
Davies, D., 250; *259*
Davies, G.F., 204, 324, 326; *225, 347*
Davis, D.R., 372; *399*
Decompression, 186
de Boer, J., 391; *401*
Deep-ocean basin, 229, 274
Deep-focus earthquakes, 344
Degenerate oscillating systems, 28, 291
Deimos, 371
De la Condamine, 9
de Lisle, J.F., 274; *284*
De Maupertius, P.L.M., 9
DeMarcus, W.C., 391–2; *400*
Density at top of mantle, 84–5, 166
– distribution in Earth, 73–4, 85, 173, 254, 316, 329–30, 359, 361
– – in Moon, 388
– – in planets, 381–2, 383–4, 386, 392–4
– gradient in inhomogeneous regions, 206–11
– –, theory of, 67–9, 153–6
– inversion, 282, 317, 333–4
–, mean, of Earth, 5–7, 11, 13–19, 48
– – of Moon, 167, 369
– – of planets, 369
– near Earth's surface, 84–5, 165
– -velocity relations, 165–6, 327, 329–30
Derr, J.S., 301, 306, 307, 317, 318; *320*
Des Hayes, 8
Deviatoric strain, 90
– stress, 89, 97–8, 354, 372, 388, 390
Differentiation inside planets, 368, 374, 385, 394
Diffracted waves, 232
Diffraction, 130, 158
Dilatation, definition of, 90

Dilatational waves, 113–4
Dirac, P.A.M., 345; *347*
Discontinuity, 20°, 160, 268, 278
–, Mohorovičić, 157, 159, 275, 364
Dispersion, 118, 121–3
– curves, 123, 263–6
Dollfus, A., 377; *400*
Doornbos, D.J., 342; *347*
Dorman, J., 244, 269, 270, 273, 276–7, 278; *257, 284*
Doyle, H.A., 232; *257*
Duncombe, R.L., 369; *400*
Dunite, 166, 167, 169
Dynamical ellipticity, 53
Dynamic flattening, 372
Dynamo theory, 218
Dziewonski, A.M., 232, 238, 244, 279–80, 301–4, 306–8, 317–8, 337, 354, 357–60, 365; *258, 284, 320, 348, 365*

Earth model, Darwin, 70–1, 73, 76–7
– –, Legendre-Laplace, 64–6, 69, 72, 75, 193
– –, Lipschitz-Lévy, 70
– –, meteorite, 330
– –, optimum, 352–62
– –, Roche, 70, 73, 75, 80–1, 170
– –, rotating, 44–5
– –, spherical (SS), 32–4, 135, 143, 353
– –, spheroidal, 39–41, 135, 143
– –, standard, 57–8, 352, 363–5
– –, Wiechert type, 72, 73–4, 77–8, 80
– –, Williamson-Adams, 85
– – with constant $dk/d\rho$, 193
– – A, 153, 171, 186, 197, 220–1, 246, 312, 376
– – A', 153, 171
– – A'', 153, 172–3, 198, 255, 287, 312, 314–6, 329, 388
– –, A type, 152–80, 375
– – B, 219–23, 246, 312, 376
– – B_2, 253–5, 388
– –, B type, 219–23, 251–5
– – B1, 317, 357–62

408

– – B497, 317, 357–62
– – CIT-11, 278, 282
– – BI-11, 317
– – HB_1, 288, 305–11, 314–8, 358–62
– – HB'_1, 358–62
– – M1, 313
– – M3, 313, 317
– – M3 (G-LSN), 317
– – Q1, 317
– – UTD124(A', B'), 317
– – I, II (Birch), 329–30
– – I, II (Clark-Ringwood), 330
– – 5·08, 282
– – 8099, 277
Earthquakes, 108–10
Ebn-Junis, 3
Eckert, W.J., 398; *400*
Ecliptic, 53
Eclogite, 166, 167
Egyed, L., 345; *348*
Elastic afterworking, 93
Elasticity of Earth, 87–107, 173, 184–223, 254, 360, 362
– parameters, 91–3, 95–7
–, perfect, 91–2
Elastic strain energy, 105–6, 109
Electronic computers, 69, 171, 193, 269, 331
Ellipticity, 8
– corrections, 135–6, 145
–, dynamical, 53
– of Earth, 7–9, 11, 38–41, 43–59, 64–5, 71, 77, 135–6, 152, 291–2, 343, 354
– of planets, 370–2
Elsasser, W.M., 200; *225*
Emden equation, 68–9, 193, 208, 386
Empirical relations, incompressibility-pressure, 222–3
– –, density-velocity, 165–6, 327, 329–30
Energy function, atomic, 326
–, Helmholtz, 105–6, 324
– in earthquakes, 109
– in nuclear explosions, 109
–, internal, 102, 105

Engdahl, E.R., 232, 241, 244, 245, 248, 343, 344, 365; *258, 260, 348, 365*
Entropy, 103, 153
Eötvös, R. von, 18; *19*
Epicentre, 109
Equations of motion, 110, 294–5
– – state, 67–8, 106–7, 322–8, 367–8, 374, 376, 391
– – –, Hugoniot, 247
Equilibrium configurations of planets, 395–6
– tide, 82
Equinoxial points, 53
Equivoluminal waves, 113
Eratosthenes, 2
Euler, L., 9, 10
Everndon, J.F., 214; *225*
Ewing, J., 244; *257*
Ewing, W.M., 267–9, 274, 276, 278, 289, 299; *284, 285, 319*
Exact sciences, 62
Expansions in spherical harmonics, 27
Explosions, chemical, 109
–, nuclear, 109–10
External gravity field, 46–7, 56–7

Families of rays, 133–4
Fayalite, 213
Fedotov, S.A., 95; *107*
Fermat's principle, 132–3
Fernel, J., 3
Feynman, R.P., 198, 200; *225*
Fe_2O theory, 200, 215, 378–80, 382–4, 386–7, 389
Figure of Earth, 43–59
Finite-strain theory, 90, 99–101, 106–7, 185, 191–7, 323–6, 353
First-order discontinuity, 160
Flamsteed, J., 7
Flat Earth, 115, 117, 131, 271–3, 277
Flattening, 7–9, 370–2, 377; *see also* Ellipticity
– approximation, 273
–, dynamic, 372
Flinn, E.A., 244, 248, 343; *256, 258, 348, 349*

409

INDEX

Flow, 97–8
Fluid, definition of, 96, 99
Fluidity and S waves, 113
— of outer core, 83, 157–8, 174, 279, 338
Focal depth, 109, 133
— region, 109
Focus, 109
Foot, Chinese, 3
—, French, 4
—, Greek and Roman, 1–2
Foucault, J.L., 7
Fowler, R.H., 69; *86*
Fracture, 98
Franklin, J.N., 337; *349, 350*
Free-air correction, 14
Free Earth oscillations, 27–8, 82, 172, 213, 232, 244–5, 251, 271–2, 275, 279, 282–3, 287–319, 331, 344, 353–4
— energy function, 105–6, 324
Freeman, K.C., 398; *400*
Fu, C.Y., 3
Fujiyama, 16
Fundamental oscillations (modes, 28, 288, 304–7

Galileo, 4
Gapcynski, J.P., 373; *402*
Garnet, 167
Gas, definition of, 95
Gauss, C.F., 10, 30, 32, 34
Gavrilov, I.V., 398; *400*
Geocentric latitude, 21, 44, 57, 135
Geochemistry, 155, 213–4, 221, 330, 353, 377, 394–5
Geodetic reference system, 57
Geodesy, 11, 48, 353, 363
Geographic latitude, 57
Geological time-scale, 94, 96, 98
Geopotential function, 44–9
Gilbert, F., 232, 238, 244, 289, 292, 301–4, 307–8, 311, 317–8, 335, 337, 339, 354, 357–60; *258, 319, 320, 346, 348, 349, 365*
Gilvarry, J.J., 174, 326, 398; *182, 348, 400*

Gnome explosion, 281
Goldreich, P., 56; *59*
Goody, R.M., 398; *400*
Granite, 166
Granitic layer, 148, 165
Gravest mode, 288
Gravitational intensity, 4–5, 13, 32–4, 45, 47–8, 204
— —, distribution in Earth, 173, 254, 360, 362
— — on planets, 397
— —, Saigey's theorem, 80
— potential, 31–41, 44–5, 110–14
Gravitation constant, 4–6, 13–19, 48, 57, 345
— theory, 4, 9, 11, 29–42
Gravity, 5, 29–41, 45, 47–8, 271
— meter, 300, 302
Green's lemma, 31
Group velocity, 120–3, 261, 263–8, 279, 306
Grüneisen's equation of state, 327
— ratio, 155, 175, 190–1, 327
Guilio, C.I., 16
Gutenberg, B., 84, 108, 143, 145, 147–9, 157–8, 161–4, 188–9, 204, 216–7, 229–30, 235, 236–8, 240–2, 244, 274, 276–7, 312, 330, 363; *86, 127, 150, 258, 285*

Haalck, H., 84; *86*
Haddon, R.A.W., 58, 147, 165, 172, 199, 219, 222, 238, 241–3, 245, 251, 253–5, 302, 307, 310, 313, 314–7, 318, 332–3, 339–43, 358–62, 365; *58, 150, 181–2, 225, 257–8, 319–20, 347–9, 365*
Hager, C.L., 304; *321*
Hales, A.L., 231–4, 245, 365; *257–8, 365*
Halley, E., 157; *182*
Harrison, J.C., 289; *321*
Harun al-Rashid, 3
Haskell, N.A., 269; *285*
Hayford, J.F., 57; *59*
Heat conduction, 103
— outflow, 173–4

410

INDEX

Heiskanen, W.A., 18, 275; *19, 285*
Helium in planets, 391, 393
Hell, 61
Helmholtz free energy, 105–6, 324
Herglotz, G., 146; *150*
Herrin, E., 231, 233; *258*
Heyl, P.R., 18–19; *19*
Hide, R., 398; *400*
Higgins, G., 174–5, 191, 218; *182, 225*
Higher-mode surface waves, 279–81
Hipparchus, 2
Homer, 1
Homogeneity, seismic test for, 211–2
Homogeneous, definition of, 156
— regions of Earth, 177–8, 187–93
Honda, H., 338; *348*
Hooke, R., 94; *107*
Hooke's law, 60, 91
Hoskins, L.M., 293; *300*
Hubbard, W.B., 398; *400*
Hudson, J.A., 339; *348, 349*
Huestis, S., 228, 344; *258, 348*
Hugoniot conditions, 204, 247
Huntington, H.B., 391; *403*
Husebye, E.S., 342; *347*
Hutton, C., 15–16; *19*
Huygens, C., 8–10, 52
Hydrodynamic velocity, 247, 327
Hydrogen in planets, 391–4
—, metallic, 376, 391
— theory of Earth's core, 376
Hydrostatic stress, 43, 89, 99–100, 185, 295, 368, 372, 376
— theory, 9, 43–56, 58, 63, 67, 153, 185, 206, 295

Ibrahim, A.K., 232, 338; *258, 348*
Ida, Y., 213; *223*
Identification of normal modes, 280, 302
Igneous rocks, 166
Inch (French), 8
Incident waves, 124–7
Incompressibility, adiabatic, 104–5
—, definition of, 95–6
— in Earth, 81, 184–223
— — —, distribution of, 173, 254, 360, 362
—, isothermal, 104–5
— -pressure hypothesis, 197–205, 217, 219–22, 239
— ratios, 194–7
— —, trend toward unity, 196
— relations, 325–6, 375
— —, empirical, 222–3
Index η, 207–11, 216–7, 237, 282, 315, 364, 374
Inductive inference, 62
Infinitesimal strain theory, 89
Inhomogeneity index; *see* Index η
— in Earth's core, 215
— — — mantle, 213
— — — outer core, 178, 189, 235–9
Inhomogeneous, definition of, 156
— regions of Earth, 162, 167–8, 177–8, 188–9, 193, 205–11, 212–7
Inner core boundary, 201, 217–9, 250, 368
— — of Earth, 141, 157–8, 164–5, 197
— — of planets, 379
— — proper, 163, 216
— —, solidity of, 205, 217–9, 220, 248–51, 303, 317–8, 365
Intermediate layer, 147, 165
Internal gravity field, 48
— energy, 102, 105
— regions of Earth, 156–65, 173
International Association of Geodesy, 57, 363
— Astronomical Union, 57
— ellipsoid, 57
— gravity formula, 57
— reference systems, 57–8
— Seismological Summary, 145, 232, 363
— Union of Geodesy and Geophysics, 57
Interplay, 337, 364
Inverse-square law, 4, 30
Inversion, density, 282, 317, 333–4
— of seismic data, 146, 334–7
Ionic theory, 205, 326–7
Ion-core theory, 375–6, 381, 383

411

INDEX

Irrotational waves, 113
Isentropic, 103, 154
Isothermal changes, 103–5, 185, 190–190–2
— incompressibility, 104–5
Isotropic, 91, 110, 294
— random function, 339–40

Jackson, D.D., 95, 337; *107, 349*
Jacobs, J.A., 174, 218; *182, 225*
James, R.E., 16; *19*
Jardetzky, W.S., 268; *284*
Jarosch, H., 269, 272–3, 289, 291–2, 294–300, 304; *283, 319, 321*
J.B. tables, 143–5, 147–9, 159, 232, 239–40, 356, 363
Jeans, Sir James, 272, 289; *285, 320*
Jeffreys, B.S., 23; *28*
Jeffreys, Sir H., 19, 23, 41, 46–7, 50–1, 54, 56, 74, 78, 94–5, 112, 122, 143, 146–9, 156–64, 167–9, 188–9, 204, 212–3, 216–7, 221, 228, 231–45, 253, 267, 269, 277, 290, 294, 303, 339, 345, 354–6, 390–3, 386–90, 398; *19, 28, 42, 59, 86, 107, 127, 150, 182, 225, 256–9, 285, 320, 348, 366, 400–1*
Jeffreys velocity model, 147–9, 156, 159, 161, 163, 188–9, 204, 212, 221, 228–42, 253, 277
Jobert, G., 294; *319*
Jobert, N., 273, 289; *285, 320*
Jolly, P. von, 18; *19*
Johnson, L.E., 245, 337; *259, 349*
Johnson, L.R., 245; *259*
Jordan, T.H., 238, 317–8, 334, 337, 357–60; *259, 320, 349, 366*
Julian, B.R., 246, 250; *259*
Jupiter, 7, 367–72, 390–3, 397

Kamitsuki, A., 232; *259*
Kanamori, H., 201, 278, 280–1; *226, 285*
Kaula, W.M., 398; *401*
Kausel, E., 267; *285*
Keane, A., 325–6; *349*

Keilis-Borok, V.I., 331, 337, 355; *346, 349*
Kellogg, O.D., *42*
Kelvin, Lord (William Thomson), 11, 72, 73, 82–3, 93, 94, 158, 269; *85, 86, 286*
Kennedy, G.C., 174–5, 191, 218; *182, 225*
Kieffer, H.H., 398, *401*
Kiloton, 109
King, D.W., 342–3; *347, 349*
King-Hele, D.G., 56–7; *59*
Kishimoto, Y., 232; *259*
Klepczynski, W.J., 369; *400*
Klussmann, W., 84; *85*
Kliore, A., 377; *401*
Knopoff, L., 95, 196, 200, 215, 218, 267, 270, 280–1, 323, 326–8, 337, 339, 343; *107, 225, 285, 348, 349*
Knott, C.G., 150
Kobayashi, N., 278; *286*
Kogan, S.D., 244; *259*
Kopal, Z., 398; *401*
Kormer, S.V., 218, 247; *223, 255*
Korringa, J., 391; *401*
Kothari, D.S., 391; *401*
Kovach, R.L., 232, 267, 273–4, 283, 383; *259, 285, 339, 401*
Kovalevsky, K.J., 398; *59, 401*
Kozlovskaya, S.K., 386; *401*
Kraut, E., 218
Kronig, R., 391; *401*
Kuhn, W., 376; *401*
Kuiper, G.P., 377, 384, 394, 398; *401*
Kuo, J., 278; *284*
Kurile Islands earthquake, 306

Lagrange, J.-L., 11
Lagrangian equations, 324
— integral, 314
Lamb, Sir Horace, 289, 293, 323; *320, 349*
Lambert, D.G., *256*
Lambert, W.D., 82
Lamé parameters, 92, 170, 180
Landisman, M., 214, 278–80, 304,

306–7, 313–4, 317; *225, 284, 286, 320*
Lane-Ritter theory, 68
Laplace, P.-S., 11, 22, 31, 33, 64–7, 70; *86*
Laplace's equation, 22, 24, 31, 33, 290
Lapland, 9, 10
Lapwood, A.R., 365; *365*
Lateral variations, 160, 228–30, 280, 292–3, 343–4, 353–4
Latham, G.V., 390, 398; *402, 403*
Latitude, geocentric, 21, 44, 57, 135
–, geographic, 57
–, length of degree of, 2–4, 9
Lattice theory, 93, 324
Laws of nature, 60–1
Lee, A.W., 122, *127*
Legendre, A.-M, 11, 23–4, 64–6; *86*
Legendre associated function, 26
– -Laplace density model, 64–6, 69, 72, 75, 193
– polynomials, 23–4
– 's equation, 24, 26
Lehmann, I., 157–8, 164, 217, 229–30, 232, 243, 276–7; *183, 259*
Leiden, 3
Levin, B., 377; *401*
Levy, G.S., 377; *401*
Lévy, M., 70
Lewis, J.S., 398; *401*
Librations, 377
Lighthill, Sir James, 398; *401*
Linear strain theory, 89, 90
Line (French), 8
Lipschitz, R.O., 70
Lipschitz-Lévy density model, 70
Liquid, definition of, 95
Longitude, 21, 44
Longitudinal waves, 114
Long-period surface waves, 279
Love, A.E.H., 118–23, 289, 298; *127, 320*
Love waves, 118–23, 263–7
– 's numbers, 82–3
Low, A.H., 384; *400*
Lower core, 163, 216–7, 239–44, 251–3

– mantle, 84, 162, 199
Low velocity layer, 161, 229, 235, 276–7, 333
Lubimova, H.A., 174, 218; *183, 225*
Luoff, N., 371; *401*
Lyngå, G., 398; *400*
Lyttleton, R.A., 69, 378, 386, 394, 398; *86, 401–2*
Lyubimov, V.M., *321*

MacCullagh, J., 35–6, 51
McCrea, W.H., 398; *402*
MacDonald, G.J.F., 82, 196, 200, 215, 218, 291, 304, 372, 398; *86, 225, 320, 321, 402*
Macdonald, J.R., 326–7, 328; *349, 350*
Maclaurin, C., 9, 10
McMechan, G.A., 337; *350*
McQueen, R.G., 201, 247; *226, 259*
Macroscopic, microscopic theories, 87, 93, 326–7
Magnetic field of Earth, 163, 218, 291
– – of Mars, 385
– – of Venus, 382
Major planets, 368, 390–4
Mantle of Earth, 141, 157, 160–3
– -core boundary, 192–3, 196–7, 215, 220, 222, 237–9, 279, 323–4, 337–8, 345, 359–60, 376–80, 382
– velocities, 228–32
– waves, 262
Mars, 202, 367–72, 377, 379, 380–1, 383–5, 397
Marsh, S.P., 201, 247; *226. 259*
Marussi, A., 58, 365; *59, 366*
Maskelyne, N., 15; *19*
Mass of Earth, 5–7, 13–19, 48, 57
– of Moon, 369
– of planets, 369
Massé, R.P., 244
Mathematical model, 60–3
Matsui, T., 398; *402*
Matumoto, T., 289; *321*
Maxwell, J.C., 93
Mean density of Earth, 5–7, 11, 13–19, 48

413

INDEX

– – of Moon, 167, 369
– – of planets, 369
– rigidity of Earth, 11, 258
Melik-Gajkazan, I.A., 219; *226*
Melting-point curve, 218
Mendenhall, T.C., 16
Mercury, 367–72, 385–7, 397
–, primitive, 387
Meteorites, 156, 330, 375
Metre, 10
Metropolis, N., 198; *225*
Michael, W.H., 373; *402*
Michell, J., 16–18
Mid-continental regions, 229
Mid-oceanic ridges, 229
Middlehurst, B.M., 398; *401*
Miles, B., 372, 391–3; *402*
Milne seismograph, 108
Mises function, 98
Mixed paths, 262, 280–1, 306
Mizutani, H., 398; *402*
Mohorovičić, A., 157; *183*
Mohorovičić discontinuity, 157, 159, 275, 364
Molnar, P., 228, 344; *258, 259, 348*
Molodenski, M.S., 158; *183*
Moment of inertia coefficient, 51, 55, 56 56
– – – of Earth, 51–6, 64, 71, 77, 84–5, 168–70, 172, 199, 219, 221, 253, 255, 312, 313
– – – of planets, 370–2, 377
Monochromatic waves, 116
Monte Carlo techniques, 278, 331–4, 356–7
Moon, 53, 82, 167, 202, 367–9, 373, 385, 388–91, 396
Moritz, H., 58, 365; *59, 366*
Mountain building, 378
Mountainous regions, 157, 274
Mueller, S., 270; *285*
Muirhead, K.J., 245; *259*
Müller, G., 342; *350*
Munk, W.H., 82; *86*
Murnaghan, F.D., 106, 185, 191, 322–3, 325–6; *107, 350*

Nafe, J., 121, 214, 270–1, 280, 304, 313–4; *127, 225, 284, 320*
Nakamura, K., 338; *348*
Nakamura, Y., 398; *402*
Negative density gradient, 282, 317, 333–4
– velocity gradient in Earth's core, 163–4, 216, 241
Neptune, 367–72, 393–4, 397
Ness, N.F., 289, 291; *320, 321*
Newton, Sir Isaac, 4–9, 16, 46, 52, 60, 61; *11*
Nguyen Hai, 240–2, 343; *259, 350*
Niazi, M., 231, 232; *259*
Nishimura, E., 232; *259*
Nodal surfaces, 28, 267
Nomenclature of Earth's regions, 158–65
– of seismic phases, 141–3, 149
Non-linearity, 171
Non-uniqueness, 283, 334–7, 354–5
Normal modes, 266–7, 279–80, 290–1
– velocity variation, 136, 178, 211, 212–7
Norwood, R., 3
Nuclear explosions, 109–10, 158, 228, 230, 244, 343

Oblateness, 4, 7–9; see also Ellipticity
Oceanic crust, 159, 229, 274–5
– regions, 157, 159, 262
Oceanus Procellarium, 390
Oldham, R.D., 94, 108, 143, 156–7; *107, 127*
Oliver, J., 228, 230, 267, 269, 270, 276, 344; *258, 259, 284, 285, 348*
Olivine, 85, 167, 214, 330[
O'Neill, M.E., 219, 343; *223, 346*
Ono, 376
Öpik, E.J., 398; *402*
Optical flattening, 372
Optimum Earth models, 352–62
Origin of Moon, 389, 396
Oscillations, free Earth; see Free Earth oscillations
– of dynamical system, 288–9
– parameters, 288–9, 294

414

INDEX

Outer core of Earth, 83, 141, 157–8, 163, 174, 215, 279, 338, 375–80
– – of planets, 379, 382
– – proper, 163, 216
Overtones, 28, 288, 307–11

Parameter(s), elasticity, 91–3, 95–7
–, Lamé, 92, 170, 180
–, in scientific laws, 61–3
–, of seismic ray, 131–5
–, of oscillation, 288–9, 294
–, of splitting, 292
– η (inhomogeneity), 207–11, 216–7,
– η (Radau), 49–50, 52–8, 74–8, 371
Parametric profligacy, 161
Pear-shaped Earth, 57
Pecker, J.-C., 369; *402*
Peebles, P.J.E., 393; *402*
Pekeris, C.L., 269, 272–3, 282, 289, 291–2, 294–301, 304, 317; *283, 285, 319, 321*
Pencil (ray), 129
Perfect elasticity, 91–2, 110, 294
– fluid, 92, 97, 104, 113
Peridotite, 166, 167
Peru, 9, 10, 13–15
Phase changes, chemical, 186–7, 200, 213, 374
–, chemical, 153, 155–6
–, seismic, 128, 141–3
– shift, 271
– transition (hydrogen), 376, 391
– – theory, 200, 215, 376–8, 381–7, 383–4, 386–7, 389, 395–6
– velocity, 261, 264, 268–72, 279–81, 306, 309–11
– P, S, etc., 141–3
– PdP, 343
– $PKiKP$, 142, 219, 244, 248
– $PKJKP$, 142, 218, 250–1
– $PmKP$, 344
– P_m, S_n, 228–9
Phobos, 371
Physical justification, 251
Picard, J., 3–4
Pilant, W.L., 270; *285*
Planets, 6–8, 51, 155, 202, 223, 353, 367–98
–, angular velocities, 370
–, density distributions, 381–2, 383–4, 386, 392–4
–, flattening, 370–2
–, gravitational intensity, 397
–, major, 368, 390–4
–, masses, 369
–, mean densities, 369
–, moments of inertia, 370–2
–, pressures in, 381–3, 386, 392–3
–, radii, 369
–, rotation periods, 370–2
–, seismic velocities in, 382, 385
–, surface temperatures, 382, 390, 397
–, terrestrial, 368, 380–7
Plane waves, 114
Playfair, A., 15
Plumb-line, 45
– deflection, 6, 14–16
Pluto, 367, 369, 372, 394, 397
P_n velocities, 228–9
Poincaré, H., 75; *86*
Poisson, S.D., 11, 31–2, 33, 35
Poisson's equation, 31–2, 33
– ratio, 97, 171, 173, 180, 254, 360, 362
– relation, 118
Polar coordinates, 21–2, 290, 294
Polarized waves, 114
Polytropic index, 68
Porra, K., 232; *257*
Posidonius, 2
Powell, D.R., 326, 328; *349*
Power-spectrum techniques, 301
Poynting, J.H., 18; *19*
Pratt, J.H., 16; *12, 19*
Precession constant, 53–8
– of equinoxes, 53
Precision problems, 302
Precursors to $PKIKP$, 239–44, 255, 341–2
Press, F., 167, 245, 268–9, 270–1, 274, 276–8, 282–3, 299, 331–4, 338, 398; *183, 260, 283, 284, 285–6, 319, 350, 402*
Pressure, critical, 376, 378, 381, 391, 396

415

INDEX

—, definition of, 89, 99—100
—, distribution in Earth, 81, 173, 254, 316, 360—1
— in Moon, 388
— in planets, 381—3, 386, 392—3
Primary waves, 112
Primitive compositions of planets, 396—8
— Earth-Moon planet, 389—96
— Mercury, 387
Principal strain, 90
— stress, 88
Probability, 61
P-SV waves, 118, 122, 273—83, 294, 306
Ptolemy, 2
Pure paths, 262, 280—1
P velocity, 111—3
— — distribution in Earth, 128—150, 173, 227—32, 254, 316, 359, 361
P waves, 112—5, 123—7
Pyrolite, 330
Pyroxene, 330
Pyroxenite, 167

Qamar, A., 248, 343; *256, 346*
Quasi-uniqueness, 335—6, 357

Radau, R.R., 49—52, 64, 72, 74, 79, 80; *59, 86*
Radau approximation, 50—2, 64, 74—8, 369, 370, 372
— parameter, 49—50, 52—8, 74—8, 371
— transformation, 50
— -Darwin approximation, 52
Radius of Earth, 57—8
— — Earth's core, 157, 244—6, 313—5, 317, 358
— — — inner core, 158, 239—44, 317
— — planets, 369
Ramsey, W.H., 372, 376—7, 391—3, 395—6, 397; *402*
Randall, M.J., 232—4, 238—9, 241—2; *255, 260*
Random function, 339—40
— walk, 332
Range of (oscillation) activity, 304—6, 314
Rapp, R.H., 58, 365; *59, 366*
Ray parameter, 131—5
—, seismic, 128—35
— theory, 128—40, 341—2
Rayleigh, Lord (Strutt, J.W.), 117—8, 122; *127*
— 's principle, 269
Read, L., 232; *257*
Reality, 63
Rees, J.M., 56; *59*
Reflected waves, 123—7
Refracted waves, 123—7
Regions A-G, 158—65, 173, 212—7
— B', B'', 161, 315
— C', C'', 162, 231, 315, 343
— D', 163, 177, 188, 190, 192, 199, 214, 231, 373
— D'', 149, 163, 178, 188—90, 192, 215, 221, 231, 232, 245, 340—2, 364, 373
Reich, F., 18
Relative displacement, 89
Reliability of Models A, B, 179—80, 312
Representation of Earth's crust, 275
— — — surface, 36—8, 45—6
Representative atomic number, 196, 200—2, 205—11, 216, 375
Retrograde motion, 118
Reynolds, R.T., 398; *402*
Richarz, F., 18; *20*
Richer, J., 8—9
Richter, C.F., 145, 149, 158, 230; *150, 182*
Rigid body, 96
Rigidity, definition of, 96
—, distribution in Earth, 173, 254, 360, 362
—, mean, of Earth, 11, 83, 158
— of Earth's core, 81—3, 158, 163
— — — inner core, 205, 217—9, 220, 248—51, 303, 317—8, 360, 362, 365
Ringwood, A.E., 178, 204, 213, 215, 330, 333, 395; *182, 226, 347, 402*
Ritter, 68

416

INDEX

Rittmann, A., 376; *401*
Roberts, J.L., 231–4, 245; *258*
Robinson, E.A., 235; *260*
Robinson, R., 232; *259*
Roche, E., 70, 73; *86*
Roche density model, 70, 73, 75, 80–1, 170
Rodrigues' formula, 23, 26
Romney, C., 244, 248; *258*
Rotation, 90–1
— tensor, 91, 111
Rotational periods of planets, 370
— waves, 113–4
Runcorn, S.K., 205, 398; *226, 348, 402–3*
Russell, H.N., 391; *403*

Saa, G., 243; *260*
Sabine, 16
Sacks, I.S., 232, 235, 243, 244, 250; *260*
Saigey, 80
Sammis, C.G., 324, 326, 327; *350*
Santô, T., 274, 281; *286*
Satellites, 371
—, artificial, 56–7, 77, 312, 344, 377, 385, 387, 388
Satô, Y., 214, 271, 272, 278, 289, 292, 304, 306, 313–4; *225, 286, 320, 321*
Saturn, 367–72, 390–3, 397
Scalar potential, 114–5
Scattering of seismic waves, 130, 243, 255, 339–43
Schatzmann, E., 398; *403*
Schiehallion, 15–16
Schwab, F., 267; *285*
Scientific inference, 60–3, 251, 328, 345
Scott, D.W., *59*
Second approximations, 227–55
— -order discontinuity, 160
— — ellipticity theory, 47–8, 54–5
Secondary waves, 112
Sectorial harmonic, 27
Seidelmann, P.K., 369; *400*
Seismic bodily waves, 112–5, 123–7

— phase(s), 128
— — $P, S, pP, PcP, PKP, SKS, PKIKP$, etc., 141–3
— — PdP, 343
— — $PKiKP$, 142, 219, 244, 248
— — $PKJKP$, 142, 218, 250–1
— — $PmKP$, 344
— — P_m, S_n, 228–9
— rays, 128–35
— spectrum, 122, 289
— surface waves, 115–23, 261–83
— test for homogeneity, 211–2
— velocity distributions, 173, 254, 359, 361
— — — in Moon, 390
— — — in planets, 382, 385
— wave transmission, 108–27, 289
Self-energy in planets, 395–6
Sengupta, M.K., 246; *259*
Sezawa, K., 122; *127*
Shadow zone, 140, 157, 339
Shapiro, J.N., 327; *349*
Shell of Earth, 156–7
Sheppard, R.M., 250; *259*
Shida, T., 82
Shimazu, Y., 327, 398; *350, 403*
Shimshoni, M., 232; *256*
Shock waves, 204, 215, 218, 247–8
Shurbet, G.L., 165; *183*
SH waves, 114, 117–20, 263–83, 294, 306, 324, 353, 375, 378
Signature, 141
Sima, H., 338; *338; 348*
Simon, Lord, (F.E.), 218; *226*
Simplicity requirements, 61–3, 353, 356, 358, 363
Simpson, T., 10
Sinoidal waves, 116
S.I. units, 10
Slichter, L.B., 218, 289–90, 292, 294, 299–300, 304, 337; *226, 321, 350*
Slowness, 146, 228, 341–2, 343
Smith, M.L., 337; *350*
Smoluchowski, R., 398; *403*
Snell, W., 3, 126–7
Snell's law, 126–7, 131–3

417

INDEX

S_n velocities, 228–9
Solid, definition of, 96, 99
— harmonic, 22
— -state theory, 93, 353
Solidity and S waves, 113
— of inner core, 205, 217–9, 220, 248–51, 303, 317–8, 360, 362, 365
— of mantle, 157
Solomon, S.C., 398; *403*
Somerville, M.R., 232; *256*
Soroktin, O.G., 378–9; *403*
Sound velocity, 247, 327
Spectral curves, 301–2
Spectrum, seismic, 122, 289
Specific heat, 104, 154–5, 175
Spherical harmonics, 21–8
— shell, 33
— symmetry (SS), 353
Sphericity, 271
Spheriodal shell, 40
— oscillations, 293–4, 296–8, 304–9
Spheroid, equation of, 38
Spinal, 213
Split modes, 292
Splitting parameter, 292
Stacey, F.D., 95; *107*
Stadion, 1–2
Standard Earth model, 57–8, 352, 363–5
Sterneck, R. von, 16, 80
Stewart, R.M., 205; *226*
Stieltjes, R.J., 79–80
Stirling, J., 9, 10
Stishovite, 202
Stoneley, R., 122, 267, 274, 293; *127, 286, 321*
Strabo, 7
Strain, 89–90, 111
—, deviatoric, 90, 111
— energy, 102, 109
— tensor, 88–90
Strength, 97–8
Stress, 88–9, 100–1
—, deviatoric, 56, 89, 111, 372, 383, 388, 390
— -difference, 56, 98

—, hydrostatic, 43, 89, 100, 185, 206, 368, 372, 376
— -strain relations, 91–4, 295
— tensor, 88–9
Strips, 336, 355
Sucessive approximation, 171, 355–7
Sudo, K., 337; *350*
Summation convention, 88
Summers, A.L., 398; *402*
Super-adiabatic temperature gradient, 154–5, 173–8, 199, 209, 213, 374
Surface harmonic, 22
— temperatures of planets, 382, 390, 397
— waves, 115–23, 261–83
—s of equal density, 41, 45, 58
Sutton, G.H., *319*
S velocity, 111–3
— — distribution in Earth, 128–50, 173, 227–31, 232–4, 254, 316, 359, 361
SV waves, 114
S waves, 113–5, 123–7
Syene, 2
Sykes, L., 278; *286*
Symmetrical tensor, 88

Table-land, 13–14, 34–5
Taggart, J.N., 232, 245; *260*
Tait, P.G., 72; *86*
Takeuchi, H., 158, 163, 201, 278, 289, 337, 398; *183, 226, 286, 321, 350, 402*
Tams, E., 274
Tang-li, 3
Teller, E., 198; *225*
Temperature conditions, 101–7, 153–5, 166, 172–8, 189–91, 208–9, 218, 230, 282, 368–9, 374
Ternary system, 214
Terrestrial planets, 368, 380–7
Tesseral harmonic, 25
Thatcher, W., 280; *286*
Thermal conductivity, 174
Thermodynamical conditions, 91, 101–7

INDEX

Thomsen, L., 323–4, 325, 326; *346, 350*
Thomson, Sir W. (Lord Kelvin), 11, 72, 73, 82–3, 93, 94, 158, 269; *85, 86, 286*
Tidal gravimeter, 300, 302
Tides, 11, 82–3, 94, 158, 353
Tisserand, F., 50, 75, 79, 80; *59, 86*
Todhunter, I., 10; *12*
Toise, 4
Töksöz, M.N., 228, 271, 278, 280, 337, 390, 398; *260, 283, 286, 350, 403*
Toomre, A., 56; *59*
Toroidal oscillations, 293
Torsional oscillations, 293–4, 298–300, 304–9
Total reflexion, 125
Trade-off, 337, 355, 364
Transition zones in Earth's core, 148–9, 163–4, 216–7, 239–44, 247, 364, 373
—s, explosive, 395–6
Transverse waves, 114
Travel-time, distance relations, 135
— tables, 143–5; *see also* J.B. tables
Triplication of travel times, 137–8, 146, 160
Trubitsyn, V.P., 398; *403*
Tukey, J., 301; *319*
Turner, H.H., 143

Uffen, R.J., 174, 200; *183, 225*
Ultrabasic rock, 85, 197, 199
Uncompressed densities, 186
Uniqueness, 62, 283, 334–7
Upper layers of Earth, 157
— mantle, 84, 162, 167–8, 253
Uranus, 367–72, 393–4, 397
Urey, H.C., 377, 200; *183, 225*
Usami, T., 292; *321*

Valle, P.E., 270; *286*
Varin, 8
Vector potential, 114–5
Velocity gradient condition, 138
— -depth relations, 138, 145–50, 268

Velocities of seismic waves, 111–15, 118, 120–6, 128–50, 173, 227–32, 254, 316, 359, 316, 390
Vening Meinesz, F.A., 18, 275; *19, 285*
Verhoogen, J., 174; *183*
Venus, 202, 367–72, 377, 379, 380–2, 296, 397
Vicente, R.O., 58, 294, 365; *59, 320, 366*
Viscosity, 93
Vlaar, N.J., 342; *347*
Voigt, W., 93
Volatilization losses from planets, 387, 395

Wadati, K., 147; *150*
Wang, C.-Y., 328, 330; *350*
Wave guide, 115
— length, 115–6, 261
— number, 116
— period, 116
— velocity, 116
Wesson, R.A., 345; *350*
Whitcomb, J.H., 343; *350*
White dwarfs, 391
Wiechert, D.H., 344; *350*
Wiechert, E., 72, 78, 83, 146, 157; *86*
Wiechert-type density models, 72, 73–4, 77–8, 80
Wiggins, R.A., 333, 337; *350*
Wigner, E.W., 391; *403*
Wildt, R., 391; *403*
Williamson, E.D., 84–5, 108, 154, 167, 170; *86*
Williamson-Adams density model, 85
— equation, 67, 166, 206–9, 216, 330
Willmore, P.L., 15
Wilson, James T., 274; *286*
Witte, H., 147; *150*
Wollaston, W.H., 18
Wood, J.A., 398; *403*
Woolard, E.W., 371; *403*
Worldwide Network of Standardized Seismographs, 342
Worzel, L.J., 165; *183*
Wright, C., 338, 343; *351*

Yanovitskaya, G.T., 398; *400*

419

INDEX

Yanovskaya, T.B., 242–3, 355; *260, 346, 349, 366*
Yi-Hsing, 3
Young's modulus, 97, 171, 173, 180, 254

Zahradnicek, 18
Zeeman effect, 291
Zharkov, V.N., 298; *321, 403*
Zonal harmonic, 24, 27
Zöppritz, K., 143, 157; *150, 183*
Zöppritz-Turner tables, 157, 355–6